# FORENSIC ENGINEERING INVESTIGATION

# FORENSIC ENGINEERING INVESTIGATION

## Randall K. Noon

CRC Press

Boca Raton   London   New York   Washington, D.C.

**Library of Congress Cataloging-in-Publication Data**

Noon, Randall
        Forensic engineering investigation / Randall Noon.
            p.   cm.
        Includes bibliographical references and index.
        ISBN 0-8493-0911-5 (alk. paper)
        1. Forensic engineering. I. Title.

    TA219 .N64 2000
    620—dc21
                                                                    00-044457
                                                                         CIP

© 2001 by CRC Press LLC

No claim to original U.S. Government works
International Standard Book Number 0-8493-0911-5
Library of Congress Card Number 00-044457
Printed in the United States of America  1  2  3  4  5  6  7  8  9  0
Printed on acid-free paper

# Preface

Forensic engineering is the application of engineering principles, knowledge, skills, and methodologies to answer questions of fact that may have legal ramifications. Forensic engineers typically are called upon to analyze car accidents, building collapses, fires, explosions, industrial accidents, and various calamities involving injuries or significant property losses. Fundamentally, the job of a forensic engineer is to answer the question, what caused this to happen?

A forensic engineer is not a specialist in any one science or engineering discipline. The solution of "real-world" forensic engineering problems often requires the simultaneous or sequential application of several scientific disciplines. Information gleaned from the application of one discipline may provide the basis for another to be applied, which in turn may provide the basis for still another to be applied. The logical relationships developed among these various lines of investigation usually form the basis for the solution of what caused the event to occur. Because of this, skilled forensic engineers are usually excellent engineering generalists.

A forensic engineering assignment is perhaps akin to solving a picture puzzle. Initially, there are dozens, or perhaps even hundreds, of seemingly disjointed pieces piled in a heap. When examined individually, each piece may not provide much information. Methodically, the various pieces are sorted and patiently fitted together in a logical context. Slowly, an overall picture emerges. When a significant portion of the puzzle has been solved, it then becomes easier to see where the remaining pieces fit.

As the title indicates, the following text is about the analyses and methods used in the practice of forensic engineering. It is intended for practicing forensic engineers, loss prevention professionals, and interested students who are familiar with basic undergraduate science, mathematics, and engineering. The emphasis is how to apply subject matter with which the reader already has some familiarity. As noted by Samuel Johnson, "We need more to be reminded than instructed!"

As would be expected in a compendium, the intention is to provide a succinct, instructional text rather than a strictly academic one. For this reason, there are only a handful of footnotes. While a number of useful references are provided at the end of each chapter, they are not intended to represent an exhaustive, scholarly bibliography. They are, however, a good starting point for the interested reader. Usually, I have listed references commonly used in "the business" that are available in most libraries or through inter-library loans. In a few cases I have listed some hard-to-get items that are noteworthy because they contain some informational gems relevant to the business or represent fundamental references for the subject.

The subjects selected for inclusion in this text were chosen on the basis of frequency. They are some of the more common types of failures, cata-strophic events, and losses a general practicing forensic engineer may be called upon to assess. However, they are not necessarily, the most common types of failures or property losses that occur. Forensic engineers are not usually called upon to figure out the "easy ones." If it was an easy problem to figure out, the services of a forensic engineer would not be needed.

In general, the topics include fires, explosions, vehicular accidents, industrial accidents, wind and hail damage to structures, lightning damage, and construction blasting effects on structures. While the analysis in each chapter is directed toward the usual questions posed in such cases, the principles and methodologies employed usually have broader applications than the topic at hand.

It is the intention that each chapter can be read individually as the need for that type of information arises. Because of that, some topics or principles may be repeated in slightly different versions here and there in the text, and the same references are sometimes repeated in several chapters. Of course, some of the subjects in the various chapters naturally go together or lead into one another. In that regard, I have tried to arrange related chapters so that they may be read as a group, if so desired.

I have many people to thank for directly or indirectly helping me with this project. I am in debted to my wife Leslie, who encouraged me to under-take the writing of this book despite my initial reluctance. I also thank the people at CRC Press, both present and past, who have been especially sup-portive in developing the professional literature associated with forensic sci-ence and engineering. And of course, here's to the engineers, techs, investigators, and support staff who have worked with me over the years and have been so helpful. I'll see you all on St. Paddy's at the usual place.

R. N.

# About the Author

Mr. Noon has written three previous texts in the area of forensic engineering: *Introduction to Forensic Engineering, Engineering Analysis of Fires and Explosions*, and *Engineering Analysis of Vehicular Accidents*. All three are available through CRC Press, Boca Raton, FL.

*Remember me when I am gone away,*
*Gone far away into the silent land;*
*When you can no more hold me by the hand,*
*Nor I half turn to go, yet turning stay.*
*Remember me when no more, day by day,*
*You tell me of our future that you planned;*
*Only remember me; you understand*
*It will be late to counsel then or pray.*

*Yet, if you should forget me for a while*
*And afterwards remember, do not grieve;*
*For if the darkness and corruption leave*
*A vestige of the thought that once I had,*
*Better by far that you should forget and smile*
*Than that you should remember and be sad.*

—*Christina Rossetti 1830–1894*

# Table of Contents

**1 Introduction**     **1**

1.1    Definition of Forensic Engineering     1
1.2    Investigation Pyramid     3
1.3    Eyewitness Information     6
1.4    Role in the Legal System     8
1.5    The Scientific Method     9
1.6    Applying the Scientific Method to Forensic Engineering     10
1.7    The Scientific Method and the Legal System     12
1.8    *A Priori* Biases     13
1.9    The Engineer as Expert Witness     14
1.10   Reporting the Results of a Forensic Engineering Investigation     16
Further Information and References     21

**2 Wind Damage to Residential Structures**     **23**

2.1    Code Requirements for Wind Resistance     23
2.2    Some Basics about Wind     26
2.3    Variation of Wind Speed with Height     32
2.4    Estimating Wind Speed from Localized Damages     33
2.5    Additional Remarks     34
Further Information and References     35

**3 Lightning Damage to Well Pumps**     **37**

3.1    Correlation is Not Causation     37
3.2    Converse of Coincidence Argument     38
3.3    Underlying Reasons for Presuming Cause and Effect     39
3.4    A Little about Well Pumps     40
3.5    Lightning Access to a Well Pump     40
3.6    Well Pump Failures     43
3.7    Failure Due to Lightning     44
Further Information and References     46

**4   Evaluating Blasting Damage**                                      **47**

    4.1   Pre-Blast and Post-Blast Surveys                          47
    4.2   Effective Surveys                                         49
    4.3   Types of Damages Caused by Blasting                       50
    4.4   Flyrock Damage                                            51
    4.5   Surface Blast Craters                                     53
    4.6   Air Concussion Damage                                     54
    4.7   Air Shock Wave Damage                                     57
    4.8   Ground Vibrations                                         58
    4.9   Blast Monitoring with Seismographs                        59
    4.10  Blasting Study by U.S. Bureau of Mines, Bulletin 442      60
    4.11  Blasting Study by U.S. Bureau of Mines, Bulletin 656      61
    4.12  Safe Blasting Formula from Bulletin 656                   62
    4.13  OSM Modifications of the Safe Blasting Formula in
           Bulletin 656                                          63
    4.14  Human Perception of Blasting Noise and Vibrations         64
    4.15  Damages Typical of Blasting                               66
    4.16  Types of Damage Often Mistakenly Attributed to
           Blasting                                              69
    4.17  Continuity                                                72
    Further Information and References                             74

**5   Building Collapse Due to Roof Leakage**                           **75**

    5.1   Typical Commercial Buildings 1877–1917                    75
    5.2   Lime Mortar                                               77
    5.3   Roof Leaks                                                80
    5.4   Deferred Maintenance Business Strategy                    80
    5.5   Structural Damage Due to Roof Leaks                       82
    5.6   Structural Considerations                                 84
    5.7   Restoration Efforts                                       87
    Further Information and References                             87

**6   Putting Machines and People Together**                            **89**

    6.1   Some Background                                            89
    6.2   Vision                                                    92
    6.3   Sound                                                     93
    6.4   Sequencing                                                95
    6.5   The Audi 5000 Example                                     95
    6.6   Guarding                                                  97
    6.7   Employer's Responsibilities                               99

6.8     Manufacturer's Responsibilities                          100
6.9     New Ergonomic Challenges                                 101
Further Information and References                               101

**7     Determining the Point of Origin of a Fire        103**

7.1     General                                                 103
7.2     Burning Velocities and "V" Patterns                     104
7.3     Burning Velocities and Flame Velocities                 107
7.4     Flame Spread Ratings of Materials                       110
7.5     A Little Heat Transfer Theory: Conduction and
        Convection                                              114
7.6     Radiation                                               118
7.7     Initial Reconnoiter of the Fire Scene                   122
7.8     Centroid Method                                         124
7.9     Ignition Sources                                        125
7.10    The Warehouse or Box Method                             127
7.11    Weighted Centroid Method                                128
7.12    Fire Spread Indicators — Sequential Analysis            130
7.13    Combination of Methods                                  133
Further Information and References                               133

**8     Electrical Shorting                            135**

8.1     General                                                 135
8.2     Thermodynamics of a "Simple Resistive" Circuit          138
8.3     Parallel Short Circuits                                 146
8.4     Series Short Circuits                                   149
8.5     Beading                                                 152
8.6     Fuses, Breakers, and Overcurrent Protection             156
8.7     Example Situation Involving Overcurrent Protection      161
8.8     Ground Fault Circuit Interrupters                       162
8.9     "Grandfathering" of GFCIs                               163
8.10    Other Devices                                           163
8.11    Lightning Type Surges                                   165
8.12    Common Places Where Shorting Occurs                     165
Further Information and References                               172

**9     Explosions                                     175**

9.1     General                                                 175
9.2     High Pressure Gas Expansion Explosions                  177
9.3     Deflagrations and Detonations                           178

9.4    Some Basic Parameters                                    182
9.5    Overpressure Front                                       185
Further Information and References                               188


# 10    Determining the Point of Ignition of an Explosion                                           191

10.1   General                                                  191
10.2   Diffusion and Fick's Law                                 192
10.3   Flame Fronts and Fire Vectors                            194
10.4   Pressure Vectors                                         195
10.5   The Epicenter                                            196
10.6   Energy Considerations                                    197
Further Information and References                               199


# 11    Arson and Incendiary Fires                              201

11.1   General                                                  201
11.2   Arsonist Profile                                         203
11.3   Basic Problems of Committing an Arson for Profit         204
11.4   The Prisoner's Dilemma                                   206
11.5   Typical Characteristics of an Arson or Incendiary Fire   207
11.6   Daisy Chains and Other Arson Precursors                  209
11.7   Arson Reporting Immunity Laws                            211
11.8   Liquid Accelerant Pour Patterns                          212
11.9   Spalling                                                 214
11.10  Detecting Accelerants after a Fire                       218
Further Information and References                               222


# 12    Simple Skids                                            223

12.1   General                                                  223
12.2   Basic Equations                                          223
12.3   Simple Skids                                             224
12.4   Tire Friction                                            226
12.5   Multiple Surfaces                                        227
12.6   Calculation of Skid Deceleration                         229
12.7   Speed Reduction by Skidding                              229
12.8   Some Considerations of Data Error                        229
12.9   Curved Skids                                             230
12.10  Brake Failures                                           231
12.11  Changes in Elevation                                     233
12.12  Load Shift                                               235

12.13  Antilock Brake Systems (ABS)                          236
Further Information and References                            237

## 13  Simple Vehicular Falls                                239

13.1  General                                                239
13.2  Basic Equations                                        239
13.3  Ramp Effects                                           241
13.4  Air Resistance                                         244
Further Information and References                           246

## 14  Vehicle Performance                                   247

14.1  General                                                247
14.2  Engine Limitations                                     247
14.3  Deviations from Theoretical Model                      251
14.4  Example Vehicle Analysis                               252
14.5  Braking                                                253
14.6  Stuck Accelerators                                     254
14.7  Brakes vs. the Engine                                  255
14.8  Power Brakes                                           257
14.9  Linkage Problems                                       258
14.10 Cruise Control                                         258
14.11 Transmission Problems                                  259
14.12 Miscellaneous Problems                                 260
14.13 NHTSA Study                                            260
14.14 Maximum Climb                                          261
14.15 Estimating Transmission Efficiency                     263
14.16 Estimating Engine Thermal Efficiency                   265
14.17 Peel-Out                                               265
14.18 Lateral Tire Friction                                  266
14.19 Bootlegger's Turn                                      266
Further Information and References                           268

## 15  Momentum Methods                                      271

15.1  General                                                271
15.2  Basic Momentum Equations                               272
15.3  Properties of an Elastic Collision                     273
15.4  Coefficient of Restitution                             275
15.5  Properties of a Plastic Collision                      276
15.6  Analysis of Forces during a Fixed Barrier Impact       278
15.7  Energy Losses and "$\varepsilon$"                      279

15.8   Center of Gravity                                      281
15.9   Moment of Inertia                                      283
15.10  Torque                                                 285
15.11  Angular Momentum Equations                             287
15.12  Solution of Velocities Using the Coefficient
       of Restitution                                         288
15.13  Estimation of a Collision Coefficient of Restitution
       from Fixed Barrier Data                                291
15.14  Discussion of Coefficient of Restitution Methods       293
Further Information and References                            294

## 16   Energy Methods                                         295

16.1   General                                                295
16.2   Some Theoretical Underpinnings                         297
16.3   General Types of Irreversible Work                     303
16.4   Rollovers                                              304
16.5   Flips                                                  310
16.6   Modeling Vehicular Crush                               316
16.7   Post-Buckling Behavior of Columns                      318
16.8   Going from Soda Cans to the Old 'Can You Drive?'       320
16.9   Evaluation of Actual Crash Data                        322
16.10  Low Velocity Impacts — Accounting for the Elastic
       Component                                              323
16.11  Representative Stiffness Coefficients                  324
16.12  Some Additional Comments                               326
Further Information and References                            327

## 17   Curves and Turns                                       329

17.1   Transverse Sliding on a Curve                          329
17.2   Turnovers                                              333
17.3   Load Shifting                                          334
17.4   Side vs. Longitudinal Friction                         335
17.5   Cornering and Side Slip                                336
17.6   Turning Resistance                                     337
17.7   Turning Radius                                         338
17.8   Measuring Roadway Curvature                            339
17.9   Motorcycle Turns                                       340
Further Information and References                            340

## 18   Visual Perception and Motorcycle Accidents             343

18.1   General                                                343

18.2    Background Information                                      344
18.3    Headlight Perception                                       345
18.4    Daylight Perception                                        347
18.5    Review of the Factors in Common                            348
18.6    Difficulty Finding a Solution                              349
Further Information and References                                 350

## 19    Interpreting Lamp Filament Damages                        351

19.1    General                                                    351
19.2    Filaments                                                  351
19.3    Oxidation of Tungsten                                      353
19.4    Brittleness in Tungsten                                    355
19.5    Ductility in Tungsten                                      355
19.6    Turn Signals                                               357
19.7    Other Applications                                         357
19.8    Melted Glass                                               357
19.9    Sources of Error                                           358
Further Information and References                                 359

## 20    Automotive Fires                                          361

20.1    General                                                    361
20.2    Vehicle Arson and Incendiary Fires                         362
20.3    Fuel-Related Fires                                         364
20.4    Other Fire Loads under the Hood                            368
20.5    Electrical Fires                                           368
20.6    Mechanical and Other Causes                                370
Further Information and References                                 371

## 21    Hail Damage                                               373

21.1    General                                                    373
21.2    Hail Size                                                  375
21.3    Hail Frequency                                             378
21.4    Hail Damage Fundamentals                                   380
21.5    Size Threshold for Hail Damage to Roofs                    384
21.6    Assessing Hail Damage                                      387
21.7    Cosmetic Hail Damage — Burnish Marks                       395
21.8    The Haig Report                                            398
21.9    Damage to the Sheet Metal of Automobiles and
        Buildings                                                  401
21.10   Foam Roofing Systems                                       404
Further Information and References                                 405

**22   Blaming Brick Freeze-Thaw Deterioration
      on Hail**                                                      **407**

      22.1   Some General Information about Bricks              407
      22.2   Brick Grades                                       408
      22.3   Basic Problem                                      409
      22.4   Experiment                                         410
      Further Information and References                        410

**23   Management's Role in Accidents and
      Catastrophic Events**                                         **413**

      23.1   General                                            413
      23.2   Human Error vs. Working Conditions                 417
      23.3   Job Abilities vs. Job Demands                      417
      23.4   Management's Role in the Causation of Accidents
             and Catastrophic Events                            419
      23.5   Example to Consider                                420
      Further Information and References                        423

      **Further Information and References**                        **425**

      **Index**                                                     **447**

# Introduction

1

Every man has a right to his opinion, but no man has a right to be wrong in his facts.
— **Bernard Baruch,** *1870–1965*

A great many people think they are thinking when they are merely rearranging their prejudices.
— **William James,** *1842–1910*

## 1.1  Definition of Forensic Engineering

Forensic engineering is the application of engineering principles and methodologies to answer questions of fact. These questions of fact are usually associated with accidents, crimes, catastrophic events, degradation of property, and various types of failures.

Initially, only the end result is known. This might be a burned-out house, damaged machinery, collapsed structure, or wrecked vehicle. From this starting point, the forensic engineer gathers evidence to "reverse engineer" how the failure occurred. Like a good journalist, a forensic engineer endeavors to determine who, what, where, when, why, and how. When a particular failure has been explained, it is said that the failure has been "reconstructed." Because of this, forensic engineers are also sometimes called reconstruction experts.

Forensic engineering is similar to failure analysis and root cause analysis with respect to the science and engineering methodologies employed. Often the terms are used interchangeably. However, there are sometimes implied differences in emphasis among the three descriptors.

"Failure analysis" usually connotes the determination of how a specific part or component has failed. It is usually concerned with material selection, design, product usage, methods of production, and the mechanics of the failure within the part itself.

"Root cause analysis" on the other hand, places more emphasis on the managerial aspects of failures. The term is often associated with the analysis of system failures rather than the failure of a specific part, and how procedures and managerial techniques can be improved to prevent the problem from reoccurring. Root cause analysis is often used in association with large systems, like power plants, construction projects, and manufacturing facilities, where there is a heavy emphasis on safety and quality assurance through formalized procedures.

The modifier "forensic" in forensic engineering typically connotes that something about the investigation of how the event came about will relate to the law, courts, adversarial debate or public debate, and disclosure. Forensic engineering can be either specific in scope, like failure analysis, or general in scope, like root cause analysis. It all depends upon the nature of the dispute.

To establish a sound basis for analysis, a forensic engineer relies mostly upon the actual physical evidence found at the scene, verifiable facts related to the matter, and well-proven scientific principles. The forensic engineer then applies accepted scientific methodologies and principles to interpret the physical evidence and facts. Often, the analysis requires the simultaneous application of several scientific disciplines. In this respect, the practice of forensic engineering is highly interdisciplinary.

A familiarity with codes, standards, and usual work practices is also required. This includes building codes, mechanical equipment codes, fire safety codes, electrical codes, material storage specifications, product codes and specifications, installation methodologies, and various safety rules, work rules, laws, regulations, and company policies. There are even guidelines promulgated by various organizations that recommend how some types of forensic investigations are to be conducted. Sometimes the various codes have conflicting requirements.

In essence, a forensic engineer:

- assesses what was there before the event, and the condition it was in prior to the event.
- assesses what is present after the event, and in what condition it is in.
- hypothesizes plausible ways in which the pre-event conditions can become the post-event conditions.
- searches for evidence that either denies or supports the various hypotheses.
- applies engineering knowledge and skill to relate the various facts and evidence into a cohesive scenario of how the event may have occurred.

Implicit in the above list of what a forensic engineer does is the application of logic. Logic provides order and coherence to all the facts, principles, and methodologies affecting a particular case.

In the beginning of a case, the available facts and information are like pieces of a puzzle found scattered about the floor: a piece here, a piece there, and perhaps one that has mysteriously slid under the refrigerator. At first, the pieces are simply collected, gathered up, and placed in a heap on the table. Then, each piece is fitted to all the other pieces until a few pieces match up with one another. When several pieces match up, a part of the picture begins to emerge. Eventually, when all the pieces are fitted together, the puzzle is solved and the picture is plain to see.

## 1.2 Investigation Pyramid

It is for this reason that the scientific investigation and analysis of an accident, crime, catastrophic event, or failure is structured like a pyramid (Figure 1.1). There should be a large foundation of verifiable facts and evidence at the bottom. These facts then form the basis for analysis according to proven scientific principles. The facts and analysis, taken together, support a small number of conclusions that form the apex of the pyramid.

Conclusions should be directly based on the facts and analysis, and not on other conclusions or hypotheses. If the facts are arranged logically and systematically, the conclusions should be almost self-evident. Conclusions based on other conclusions or hypotheses, that in turn are only based upon a few selected facts and very generalized principles, are a house of cards. When one point is proven wrong, the logical construct collapses.

Consider the following example. It is true that propane gas systems are involved in some explosions and fires. A particular house that was equipped with a propane system sustained an explosion and subsequent fire. The focus of the explosion, the point of greatest explosive pressure, was located in a basement room that contained a propane furnace. From this information, the investigator concludes that the explosion and fire were caused by the propane system, and in particular, the furnace.

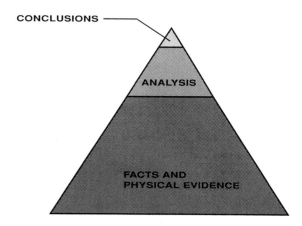

**Figure 1.1** Investigation pyramid.

The investigator's conclusion, however, is based upon faulty logic. There is not sufficient information to firmly conclude that the propane system was the cause of the explosion, despite the fact that the basic facts and the generalized principle upon which the conclusion is based are all true.

Consider again the given facts and principles in the example, rearranged in the following way.

Principle:   Some propane systems cause explosions and fires.
Fact:        This house had a propane system.
Fact:        This house sustained a fire and explosion.
Fact:        The explosion originated in the same room as a piece of equipment that used propane, the furnace.
Conclusion: The explosion and fire were caused by the propane system.

The principle upon which the whole conclusion depends asserts only that *some* propane systems cause explosions, *not all* of them. In point of fact, the majority of propane systems are reliable and work fine without causing an explosion or fire for the lifetime of the house. Arguing from a statistical standpoint, it is more likely that a given propane system will *not* cause an explosion and fire.

In our example, the investigator has not yet actually checked to see if this propane system was one of the "some" that work fine or one of the "some" that cause explosions and fires. Thus, a direct connection between the general premise and the specific case at hand has not been made. It has only been assumed. A verification step in the logic has been deleted.

Of course, not all explosions and fires are caused by propane systems. Propane systems have not cornered the market in this category. There is a distinct possibility that the explosion may have been caused by some factor not related to the propane system, which is unknown to the investigator at this point. The fact that the explosion originated in the same room as the furnace may simply be a coincidence.

Using the same generalized principle and available facts, it can equally be concluded by the investigator (albeit also incorrectly) that the propane system did not cause the explosion. Why? Because, it is equally true that some propane systems never cause explosions and fires. Since this house has a propane system, it could be concluded in the same manner that this propane system could not have been the cause of the explosion and fire.

As is plain, our impasse in the example is due to the application of a generalized principle for which there is insufficient information to properly deduce a unique, logical conclusion. The conclusion that the propane system caused the explosion and fire is based implicitly on the conclusion that the location of the explosion focus and propane furnace is no coincidence. It is further based upon another conclusion, that the propane system is one of the "some" that cause explosions and fires, and not one of the "some" that never cause explosions and fires. In short, in our example we have a conclusion, based on a conclusion, based on another conclusion.

The remedy for this dilemma is simple: get more facts. Additional information must be gathered to either uniquely confirm that it *was* the propane system, or uniquely *eliminate* it as the cause of the explosion and fire.

Returning to the example, compressed air tests at the scene find that the propane piping found after the fire and explosion does not leak despite all it has been through. Since propane piping that leaks before an explosion will not heal itself so that it does not leak after the explosion, this test eliminates the piping as a potential cause.

Testing of the furnace and other applicances find that they all work in good order also. This now puts the propane equipment in the category of the "some" that do not cause explosions and fires. We have now confirmed that the conclusion that assumed a cause-and-effect relationship between the location of the epicenter and the location of the propane furnace was wrong. It was simply a coincidence that the explosion occurred in the same room as the furnace.

Further checks by the investigator even show that there was no propane missing from the tank, which one would expect to occur had the propane been leaking for some time. Thus, now there is an accumulation of facts developing that show the propane system was not involved in the explosion and fire.

Finally, a thorough check of the debris in the focus area finds that within the furnace room there were several open, five-gallon containers of paint thinner, which the owner had presumed to be empty when he finished doing some painting work. Closer inspection of one of the containers finds that it is distended as if it had experienced a rapid expansion of vapors within its enclosed volume.

During follow-up questioning, the owner recalls that the various containers were placed only a few feet from a high wattage light bulb, which was turned on just prior to the time of the explosion. A review of the safety labels finds that the containers held solvents that would form dangerous, explosive vapors at room temperature, even when the container appeared empty. The vapors evolve from a residual coating on the interior walls of the container.

A "back of the envelope" calculation finds that the amount of residual solvent in just one container would be more than enough to provide a cloud of vapor exceeding the lower threshold of the solvent's explosion limits. A check of the surface temperature of the light bulb finds that when turned on, it quickly rises to the temperature needed to ignite such fumes. A subsequent laboratory test confirms that fumes from an erstwhile empty container set the same distance away can be ignited by the same type of light bulb and cause a flash fire.

The above example demonstrates the value of the "pyramid" method of investigation. When a large base of facts and information is gathered, the conclusion almost suggests itself. When only a few facts are gathered to back up a very generalized premise, the investigator can steer the conclusion to nearly anything he wants. Unfortunately, there are some forensic engineers who do the latter very adroitly.

As a general rule, an accident or failure is not the result of a single cause or event. It is usually the combination of several causes or events acting in concert, that is, at the same time or in sequence, one after another in a chain. An example of causes acting in sequence might be a gas explosion.

- Accumulated gas is ignited by a spark from a pilot light.
- The gas originated from a leak in a corroded pipe.
- The pipe corroded because it was poorly maintained.
- The poor maintenance resulted from an inadequate maintenance budget that gave other items a higher priority.

An example of causes acting in concert might be an automobile accident.

- Both drivers simultaneously take dangerous actions. Driver A has to yield to approaching traffic making a left turn and has waited for the light to turn yellow to do so. He also doesn't signal his turn. At the end of the green light, he suddenly turns left assuming there will be a gap during the light change. Coming from the opposite direction, driver B enters the intersection at the tail end of the yellow light. They collide in the middle of the intersection.
- Driver A is drunk.
- Driver B's car has bad brakes, which do not operate well during hard braking. Driver B is also driving without his glasses, which he needs to see objects well at a distance.

Often, failures and accidents involve both sequential events and events acting in concert in various combinations.

## 1.3   Eyewitness Information

Eyewitness accounts are important sources of information, but they must be carefully scrutinized and evaluated. Sometimes eyewitnesses form their own opinions and conclusions about what occurred. They may then intertwine these conclusions and opinions into their account of what they say they observed. Skillful questioning of the eyewitness can sometimes separate the factual observations from the personal asumptions.

Consider the following example. An eyewitness initially reports seeing Bill leave the building just before the fire broke out. However, careful questioning reveals that the eyewitness did not actually see Bill leave the building at all. The witness simply saw someone drive away from the building in a car similar to Bill's. The witness presumed it must have been Bill. Of course, the person driving the car could have been Bill, but it also could have been someone with a car like Bill's, or someone who had borrowed Bill's car.

Of course, some eyewitnesses are not impartial. They may be relatives, friends, or even enemies of persons involved in the event. They may have a personal stake in the outcome of the investigation. For example, it is not unusual for the arsonist who set the fire to be interviewed as an eyewitness to the fire. Let us also not forget the eyewitnesses who may swear to anything to pursue their own agendas or get attention.

What an honest and otherwise impartial eyewitness reports observing may also be a function of his location with respect to the event. His perceptions of the event may also be colored by his education and training, his life experiences, his physical condition with respect to eyesight or hearing, and any social or cultural biases. For example, the sound of a gas explosion might variously be reported as a sonic boom, cannon fire, blasting work, or an exploding sky rocket. Because of this, eyewitnesses to the same event may sometimes disagree on the most fundamental facts.

Further, the suggestibility of the eyewitness in response to questions is also an important factor. Consider the following two exchanges during statementizing. "Statementizing" is a term that refers to interviewing a witness to find out what the witness knows about the incident. The interview is often recorded on tape, which is later transcribed to a written statement. Usually, it is not done under oath, but it is often done in the presence of witnesses. It is important to "freeze" a witness's account of the incident as soon as possible after the event. Time and subsequent conversations with others will often cause the witness's account of the incident to change.

## Exchange I

Interviewer: Did you hear a gas explosion last night at about 3:00 A.M.?
Witness: Yeah, that's what I heard. I heard a gas explosion. It did occur at 3:00 A.M.

**Exchange II**

Interviewer: What happened last night?
Witness: Something loud woke me up.
Interviewer: What was it?
Witness: I don't know. I was asleep at the time.
Interviewer: What time did you hear it?
Witness: I don't know exactly. It was sometime in the middle of the night.
    I went right back to sleep afterwards.

In the first exchange, the interviewer suggested the answers to his question. Since the implied answers seem logical, and since the witness may presume that the interviewer knows more about the event than himself, the witness agrees to the suggested answers. In the second exchange, the interviewer did not provide any clues to what he was looking for. He allowed the witness to draw upon his own memories and did not suggest any.

## 1.4  Role in the Legal System

From time to time, a person who does this type of engineering analysis is called upon to testify in deposition or court about the specifics of his or her findings. Normally the testimony consists of answers to questions posed by an attorney for an involved party. The attorney will often be interested in the following:

- the engineer's qualifications to do this type of analysis.
- the basic facts and assumptions relied upon by the engineer.
- the reasonableness of the engineer's conclusions.
- plausible alternative explanations for the accident or failure not considered by the engineer, which often will be his client's version of the event.

By virtue of the appropriate education and experience, a person may be qualified as an "expert witness" by the court. In some states, such an expert witness is the only person allowed to render an opinion to the court during proceedings. Because the U.S. legal system is adversarial, each attorney will attempt to elicit from the expert witness testimony to either benefit his client or disparage his adversary's client.

In such a role, despite the fact that one of the attorneys may be paying the expert's fee, the expert witness has an obligation to the court to be as objective as possible, and to refrain from being an advocate. The best rule to follow is to be honest and professional both in preparing the original analysis and in testifying. Prior to giving testimony, however, the expert witness has an obligation to fully discuss with his or her client both the favorable and the unfavorable aspects of the analysis.

Sometimes the forensic engineer involved in preparing an accident or failure analysis is requested to review the report of analysis of the same event by the expert witness for the other side. This should also be done honestly and professionally. Petty one-upmanship concerning academic qualifications, personal attacks, and unfounded criticisms are unproductive and can be embarrassing to the person who engages in them. When preparing a criticism of someone else's work, consider what it would sound like when read to a jury in open court.

Honest disagreements between two qualified experts can and do occur. When such disagreements occur, the focus of the criticism should be the theoretical or factual basis for the differences.

## 1.5   The Scientific Method

The roots of the scientific method go back to ancient Greece, in Aristotle's elucidation of the inductive method. In this method, a general rule or conclusion is established based on an accumulation of evidence obtained by making many observations and gathering many corroborative facts. In assessing all these observations and facts, an underlying commonality is shown to exist that demonstrates a principle or proposition.

However, a possible pitfall of the inductive method is that the number of observations may be too small or too selective for a true generalization or conclusion to be made. A false conclusion may be reached if the observations or facts are representative of a special subset rather than the general set.

Roger Bacon, a 13th century English Franciscan monk, is often credited with defining the modern scientific method. He believed that scientific knowledge should be obtained by close observation and experimentation. He experimented with gunpowder, lodestones, and optics to mention a few items, and was dubbed "Doctor Admirabilis" because of his extensive knowledge.

For his efforts to put knowledge on a verifiable basis, the curious friar was accused of necromancy, heresy, and black magic by the chiefs of his order. He was confined to a monastery for ten years in Paris so that he could be watched. He attempted to persuade Pope Clement IV to allow experimental science to be taught at the university. However, his efforts failed.

After Pope Clement IV died, he was imprisoned for another ten years by the next pope, Nicholas III. Nicholas III also specifically forbade the reading of his papers and books. This was somewhat moot, however, since his work had generally been banned from publication anyway. Some scholars believe that Roger Bacon was the original inspiration for the Doctor Faustus legend.

Why all the fuss over a friar who wants to play in the laboratory? Because it threatened the "correctness" of theories promulgated by the church. At the time of Bacon, the doctrine of *a priorism* was the accepted basis for inquiry. *A priorism* is the belief that the underlying causes for observed effects are already known, or at least can be deduced from some first principles. Under *a priorism*, if a person's observations conflict with the accepted theory, then somehow the person's observations must be imperfect. They may even be the product of the Devil tempting a foolish mortal.

For example, in Bacon's time it was assumed *a priori* that the sun revolved about the earth and that all celestial motions followed perfect circles. Thus, all other theories concerning the universe and the planets had to encompass these *a priori* assumptions. This led, of course, to many inaccuracies and difficulties in accounting for the motions of the planets. Why then were these things assumed to be correct in the first place? It was because they were deemed reasonable and compatible to accepted religious dogmas. In fact, various Bible verses were cited to "prove" these assumptions.

Because holy scriptures had been invoked to prove that the earth was the center of the universe, it was then reasoned that to cast doubt upon the assumption of the earth being at the center of the universe was to cast doubt by inference upon the church itself. Because the holy scriptures and the church were deemed above reproach, the problem was considered to lie within the person who put forward such heretical ideas.

Fortunately, we no longer burn people at the stake for suggesting that the sun is the center of our solar system, or that planets have elliptical orbits. The modern scientific method does not accept the *a priori* method of inquiry. The modern scientific method works as follows. First, careful and detailed observations are made. Then, based upon the observations, a working hypothesis is formulated to explain the observations. Experiments or additional observations are then made to test the predictive ability of the working hypothesis.

As more observations are collected and studied, it may be necessary to modify, amplify, or even discard the original hypothesis in favor of a new one that can account for all the observations and data. Unless the data or observations are proven to be inaccurate, a hypothesis is not considered valid unless it accounts for all the relevant observations and data.

## 1.6  Applying the Scientific Method to Forensic Engineering

In a laboratory setting, it is usual to design experiments where the variable being studied is not obscured or complicated by other effects acting simultaneously. The variable is singled out to be free from other influences. Various experiments are then conducted to determine what occurs when the variable is changed. Numerous tests of the effects of changing the variable provide a statistical basis for concluding how the variable works, and predicting what will occur under other circumstances.

In this way, theoretically, any accident, failure, crime, or catastrophic event could be experimentally duplicated or reconstructed. The variables would simply be changed and combined until the "right" combination is found that faithfully reconstructs the event. When the actual event is experimentally duplicated, it might be said that the reconstruction of the failure event has been solved.

There are problems with this approach, however. Foremost is the fact that many accidents and failures are singular events. From considerations of cost and safety, the event can not be repeated over and over in different ways just so an engineer can play with the variables and make measurements.

It can be argued, however, that if there is a large body of observational evidence and facts about a particular accident or failure, this is a suitable substitute for direct experimental data. The premise is that only the correct reconstruction hypothesis will account for all of the observations, and also be consistent with accepted scientific laws and knowledge.

An analogous example is the determination of an algebraic equation from a plot of points on a Cartesian plane. The more data points there are on the graph, the better the curve fit will be. Inductively, a large number of data points with excellent correspondence to a certain curve or equation would be proof that the fitted curve was equal to the original function that generated the data.

Thus, the scientific method, as it applied to the reconstruction of accidents and failures, is as follows:

- a general working hypothesis is proposed based on "first cut" verified information.
- as more information is gathered, the original working hypothesis is modified to encompass the growing body of observations.
- after a certain time, the working hypothesis could be tested by using it to predict the presence of evidence that may not have been obvious or was overlooked during the initial information gathering effort.

A hypothesis is considered a complete reconstruction when the following are satisfied:

- the hypothesis accounts for all the verified observations.
- when possible, the hypothesis accurately predicts the existence of additional evidence not previously known.

- the hypothesis is consistent with accepted scientific principles, knowledge, and methodologies.

The scientific method as noted above is not without some shortcomings. The reality of some types of failures and accidents is that the event itself may destroy evidence about itself. The fatigue fracture that may have caused a drive shaft to fail, for example, may be rubbed away by friction with another part after the failure has occurred. Because of this, the fatigue fracture itself may not be directly observable. Or, perhaps the defective part responsible for the failure is lost or obscured from discovery in the accident debris. Clean up or emergency repair activities may also inadvertently destroy or obscure important evidence. In short, there can be observational gaps.

Using the previous graph analogy, this is like having areas of the graph with no data points, or few data points. Of course, if the data points are too few, perhaps several curves might be fitted to the available data. For example, two points not only determine a simple line, but also an infinite number of polynomial curves and transcendental functions.

Thus, it is possible that the available observations can be explained by several hypotheses. Gaps or paucity in the observational data may not allow a unique solution. For this reason, two qualified and otherwise forthright experts can sometimes proffer two conflicting reconstruction hypotheses, both equally consistent with the available data.

## 1.7   The Scientific Method and the Legal System

Having several plausible explanations for an accident or failure may not necessarily be a disadvantage in our legal system. It can sometimes happen that an investigator does not have to know exactly what happened, but rather what did not happen.

In our adversarial legal system, one person or party is the accuser, prosecutor, or plaintiff, and the other is the accused, or defendant. The plaintiff or prosecutor is required to prove that the defendant has done some wrong to him, his client, or the state. However, the defendant has merely to prove that he, himself, did not do the wrong or have a part in it; he does not have to prove who else or what else did the wrong, although it often is advantageous for him to do so.

For example, suppose that a gas range company is being sued for a design defect that allegedly caused a house to burn down. As far as the gas range company is concerned, as long as it can prove that the range did not cause the fire, the gas range company likely has no further concern as to what did cause the fire. ("We don't know what caused the fire, but it wasn't our gas range that did it.") Likewise, if the plaintiffs cannot prove that the particular gas range caused the fire, they also may quickly lose interest in litigation. This is because there may be no one else for the plaintiffs to sue; there is insufficient evidence to identify any other specific causation.

Thus, even if the observational data is not sufficient to provide a unique reconstruction of the failure, it may be sufficient to deny a particular one. That may be all that is needed.

## 1.8 *A Priori* Biases

One of the thornier problems in the reconstruction of a failure or catastrophic event is the insidious application of *a priori* methodology. This occurs when legal counsel hires a forensic engineer to find only information beneficial to his client's position. The counselor will not specifically state what findings are to be made, but may suggest that since the other side will be giving information detrimental to his client, there is no pressing need to repeat that work.

While the argument may seem innocent enough, it serves to bias the original data. This is because only beneficial data will be considered and detrimental data will be ignored. If enough bad data is ignored, the remaining observations will eventually force a beneficial fire or explosion reconstruction. Like the previous graph analogy, if enough data points are erased or ignored, almost any curve can be fitted to the remaining data.

A second version of this *a priori* problem occurs when a client does not provide all the basic observational data to the forensic engineer for evaluation. Important facts are withheld. This similarly reduces the observational database, and enlarges the number of plausible hypotheses that might explain the facts.

A third variant of *a priori* reasoning is when the forensic engineer becomes an advocate for his client. In such cases, the forensic engineer assumes that his client's legal posture is true even before he has evaluated the data. This occurs because of friendship, sympathy, or a desire to please his client in hopes of future assignments.

To guard against this, most states require that licensed professional engineers accept payment only on a time and materials basis. Unlike attorneys, licensed engineers may not work on a contingency basis. This, at least, removes the temptation of a reward or a share of the winnings.

Further, it is common for both the adversarial parties to question and carefully examine technical experts. During court examination by the attorneys for each party, the judge or jury can decide for themselves whether the expert is biased. During such court examinations, the terms of hire of the expert are questioned, his qualifications are examined, any unusual relationships with the client are discussed, all observations and facts he considered in reaching his conclusions are questioned, etc.

While an expert is considered a special type of witness due to his training and experience, he is not held exempt from adversarial challenges. While not perfect, this system does provide a way to check such biases and *a priori* assumptions.

## 1.9   The Engineer as Expert Witness

To the lay public and even to many engineers, engineering is often considered an objective science. This is perhaps fostered by the quantitative problem-solving methods used by engineers in design work. People hire engineers to tell them exact answers to their questions. For example, how many cubic meters of earth must be dug out, how much steel is required, or what size bearing is needed?

The undergraduate training of engineers also emphasizes exact problem-solving techniques. Students spend many hours calculating the correct answer to a specific problem, learning the correct fact, or applying the correct theorem to answer their homework problems. Where there is doubt, often the correct answer is found at the back of the textbook.

Engineers often refer to their discipline as a "hard" science: one that provides a "hard" or exact solution to a problem. This is in contrast to disciplines like psychology, sociology, or economics, which are sometimes considered "soft" sciences because of their inability to supply exact or specific answers. While an engineer can calculate when a beam will break because of excessive load stress, a sociologist is unable to similarly calculate the date when a community will riot because of analogous social stress.

Because of this traditional bias, some engineers are wholly taken aback when they present their case findings and conclusions in a courtroom for the first time. Their well-reasoned, scientifically sound investigation of an accident, failure, or catastrophic event may be pronounced unsound or fallacious, and may even be dismissed out of hand.

In fact, their qualifications, which may impress colleagues, will be belittled. Their experience, which may be considerable, will be minimized or characterized as inappropriate. Their character and professionalism, no matter how impeccable, will be questioned and doubted. It is not unknown for the experience to be so unpleasant to some that they never again undertake a forensic assignment.

In a courtroom, extremely well-qualified and distinguished professional engineers may testify on behalf of one side of an issue. They may radically disagree with another set of equally well-qualified and distinguished engineers who may testify on behalf of the other side of the issue. Bystanders might presume that the spectacle of strong disagreement among practitioners of such a hard science indicates that one side or the other has been bought off, is incompetent, or is just outright lying.

While the engineering profession is certainly not immune from the same dishonesty that plagues other professions and mankind in general, the basis for disagreement is often not due to corruption or malfeasance. Rather, it is a highly visible demonstration of the subjective aspects of engineering. Nowhere else is the subjectivity in engineering so naked as in a courtroom. To some engineers and lay persons, it is embarrassing to discover, perhaps for the first time, that engineering does indeed share some of the same attributes and uncertainties as the soft sciences.

Because of the adversarial role, no attorney will allow another party to present evidence hurtful to his client's interests without challenging and probing its validity. If the conclusions of a forensic engineer witness cause his client to lose $10 million, it is a sure bet that the attorney will not let those conclusions stand unchallenged! This point should be well considered by the forensic engineer in all aspects of an investigation. It is unreasonable to expect otherwise.

It is not the duty of the attorney to judge his client; that is the prerogative of the judge and jury. However, it is the attorney's duty to be his client's advocate. In one sense, the attorney is his client: the attorney is supposed to do for his client what the client would do for himself had he had the same training and expertise. When all attorneys in a dispute present their cases as well as possible, the judge and jury can make the most informed decision possible.

An engineer cannot accept a cut of the winnings or a bonus for a favorable outcome. He can only be paid for his time and expenses. If it is found that he has accepted remuneration on some kind of contingency basis, it is grounds for having his professional engineer's license suspended or revoked. The premise of this policy is that if a forensic engineer has a stake in the outcome of a trial, he cannot be relied upon to give honest answers in court.

Attorneys, on the other hand, can and do accept cases on a contingency basis. It is not uncommon for an attorney to accept an assignment on the promise of 30–40% of the take plus expenses if the suit is successful. This is allowed so that poor people who have meritorious cases can still obtain legal representation.

However, this situation can create friction between the attorney and the forensic engineer. First, the attorney may try to delay paying the engineer's bill until after the case. This is a version of "when I get paid, you get paid" and may be a *de facto* type of contingency fee arrangement. For this reason, it is best to agree beforehand on a schedule of payments from the attorney for service rendered. Follow the rule: "would it sound bad in court if the other side brought it up?"

Secondly, since the lawyer is the advocate for the case and may have a financial stake in the outcome, he may pressure the engineer to manufacture some theory to better position his client. If the engineer caves in to this temptation, he is actually doing the attorney a disservice.

A forensic engineer does his job best when he informs the attorney of all aspects of the case he has uncovered. The "other side" may also have the benefit of an excellent engineer who will certainly point out the "bad stuff" in court. Thus, if the attorney is not properly informed of the "bad stuff," he cannot properly prepare the case for presentation in court.

## 1.10   Reporting the Results of a Forensic Engineering Investigation

There are several formats used to report the results of a forensic engineering investigation. The easiest is a simple narrative, where the engineer simply describes all his investigative endeavors in chronological order. He starts from when he received the telephone call from the client, and continues until the last item in the investigation is complete. The report can be composed daily or piecewise when something important occurs as the investigation progresses, like a diary or journal. Insurance adjusters, fire investigators, and detectives often keep such chronological journals in their case files.

A narrative report works well when the investigation involves only a few matters and the evidence is straightforward. However, it becomes difficult for the reader to imagine the reconstruction when a lot of evidence and facts must be considered, along with test results, eyewitness accounts, and the application of scientific principles. Often the connections among the various items are not readily apparent, and the chronology of the investigation often does not logically develop the chronology of the accident itself.

Alternately, the report could be prepared like an academic paper, replete with technical jargon, equations, graphs, and reference footnotes. While this type of report might impress colleagues or the editors of technical journals, it is usually unsatisfactory for this application. It does not readily convey the findings and assessment of the investigation to the people who need to read it to make decisions. They are usually not professional scholars.

To determine what kind of format to use, it is often best to first consider who will be reading the forensic investigation report. In general, the audience includes the following.

1.  **Claims adjuster:** The adjuster will use the report to determine whether a claim should be paid under the terms and conditions of the insurance policy. If he suspects there is subrogation potential, he will forward

the report to the company's attorney for evaluation. In some insurance companies, such reports are automatically evaluated for subrogation potential. Subrogation is a type of lawsuit filed by an insurance company to get back the money they paid out for a claim by suing a third party that might have something to do with causing the loss. For example, if a wind storm blows the roof off a house, the insurance company will pay the claim to the homeowner, but may then sue the original contractor because the roof was supposed to withstand such storms without being damaged.

2. **Attorneys:** This includes attorneys for both the plaintiff and the defendant. The attorneys will scrutinize every line and every word used in the report. Often, they will inculcate meaning into a word or phrase that the engineer-author never intended. Sometimes the engineer-author will unadvisedly use a word in an engineering context that also has a specific legal meaning. The legal meaning may be different from the engineering meaning. Lawyers are wordsmiths by trade. Engineers as a group are renown for being poor writers. This disparity in language skill often provides the attorneys for either side plenty of sport in reinterpreting the engineer's report to mean what they need it to mean.

3. **Technical experts:** The report will also be read by the various technical experts working for the attorneys. They will want to know on what facts and observations the engineer relied, which regulations and standards he consulted and applied, and what scientific principles or methodologies were used to reach the conclusions about the cause of the loss or failure. The experts for the other side, of course, will challenge each and every facet of the report that is detrimental to their client and will attempt to prove that the report is a worthless sham. Whatever standard the engineer used in his report will, of course, be shown to be incorrect, incorrectly applied, or not as good as the one used by the other side's technical expert. One common technique that is used to discredit a report is to segment the report into minute component parts, none of which, when examined individually, are detrimental to their side. This technique is designed to disconnect the interrelationships of the various components and destroy the overall meaning and context. It is akin to examining individual heart cells in a person's body to determine if the person is in love.

4. **The author:** Several years after the report has been turned in to the client and the matter has been completely forgotten about, the forensic engineer who originally authored the report may have to deal with it again. Court cases can routinely take several years for the investigating engineer to be involved. Thus, several years after the original investigation, the engineer may be called upon to testify in deposition or

court about his findings, methodologies, and analytical processes. Since so much has happened in the meantime, the engineer may have to rely on his own report to recall the particulars of the case and what he did.

5.  **Judge and jury:** If the matter does end up in trial, the judge will decide if the report can be admitted into evidence, which means that the jury will be allowed to read it. Since this is done in a closed jury room, the report must be understandable and convey the author's reasoning and conclusions solely within the four corners of each page. Bear in mind that the members of an average jury have less than a 12th grade educational level. Most jurors are uncomfortable with equations and statistical data. Some jurors may believe there is something valid in astrology and alien visitations, will be distrustful of intellectual authorities from out of town, and since high school, their main source of new scientific knowledge has consisted of television shows and tabloids.

In order to satisfy the various audiences, the following report format is often used, which is consistent with the pyramid method of investigation noted previously. The format is based on the classical style of argument used in the Roman Senate almost 2000 years ago to present bills. As it did then, the format successfully conveys information about the case to a varied audience, who can chose the level of detail they wish to obtain from the report by reading the appropriate sections.

1.  **Report identifiers:** This includes the title and date of the report, the names and addresses of the author and client, and any identifying information such as case number, file number, date of loss, etc. The identifying information can be easily incorporated into the inside address section if the report is written as a business letter. Alternately, the identifying information is sometimes listed on a separate page preceding the main body of the report. This allows the report to be separate from other correspondence. A cover letter is then usually attached.

2.  **Purpose:** This is a succinct statement of what the investigator seeks to accomplish. It is usually a single statement or a very short paragraph. For example, "to determine the nature and cause of the fire that damaged the Smith home, 1313 Bluebird Lane, on January 22, 1999." From this point on, all the parts of the report should directly relate to this "mission statement." If any sentence, paragraph, or section of the report does not advance the report toward satisfying the stated purpose, those parts should be edited out. The conclusions at the end of the report should explicitly answer the question inferred

in the purpose statement. For example, "the fire at the Smith house was caused by an electrical short in the kitchen ventilation fan."

3. **Background Information:** This part of the report sets the stage for the rest of the report. It contains general information as to what happened so that the reader understands what is being discussed. A thumbnail outline of the basic events and the various parties involved in the matter are included. It may also contain a brief chronological outline of the work done by the investigator. It differs from an abstract or summary in that it contains no analysis, conclusions, or anything persuasive.

4. **Findings and Observations:** This is a list of all the factual findings and observations made related to the investigation. No opinions or analysis is included: "just the facts, ma'm." However, the arrangement of the facts is important. A useful technique is to list the more general observations and findings first, and the more detailed items later on. As a rule, going from the "big picture" to the details is easier for the reader to follow than randomly jumping from minute detail to big picture item and then back to a detail item again. It is sometimes useful to organize the data into related sections, again, listing generalized data first, and then more detailed items. Movie directors often use the same technique to quickly convey detailed information to the viewer. An overview scene of where the action takes place is first shown, and then the camera begins to move closer to where things are going on.

5. **Analysis:** This is the section wherein the investigating engineer gets to explain how the various facts relate to one another. The facts are analyzed and their significance is explained to the reader. Highly technical calculations or extensive data are normally listed in an appendix, but the salient points are summarized and explained here for the reader's consideration.

6. **Conclusions:** In a few sentences, perhaps even one, the findings are summarized and the conclusion stated. The conclusion should be stated clearly, with no equivocation, using the indicative mode. For example, a conclusion stated like, "the fire could have been caused by the hot water tank," is simply a guess, not a conclusion. It suggests that it also could have been caused by something other than the hot water tank. Anyone can make a guess. Professional forensic engineers offer conclusions. As noted before, the conclusions should answer the inferred question posed in the purpose section of the report. If the report has been written cohesively up to this point, the conclusion should be already obvious to the reader because it should rest securely on the pyramid of facts, observations, and analysis already firmly established.

7. **Remarks:** This is a cleanup, administrative section that sometimes is required to take care of case details, e.g., "the evidence has been moved

and is now being stored at the Acme garage," or, "it is advisable to put guards on that machine before any more poodles are sucked in." Sometimes during the course of the investigation, insight is developed into related matters that may affect safety and general welfare. In the nuclear industry, the term used to describe this is "extent of condition." Most states require a licensed engineer to promptly warn the appropriate officials and persons of conditions adverse to safety and general welfare to prevent loss of life, loss of property, or environmental damage. This is usually required even if the discovery is detrimental to his own client.

8. **Appendix:** If there are detailed calculations or extensive data relevant to the report, they go here. The results of the calculations or analysis of data is described and summarized in the analysis section of the report. By putting the calculations and data here, the general reading flow of the report is not disrupted for those readers who cannot follow the detailed calculations, or are simply not interested in them. And, for those who wish to plunge into the details, they are readily available for examination.

9. **Attachments:** This is the place to put photographs and photograph descriptions, excerpts of regulations and codes, lab reports, and other related items that are too big or inconvenient to directly insert into the body of the report, but are nonetheless relevant. Often, in the findings and observations portion of the report, reference is made to "photograph 1" or "diagram 2B, which is included in the attachments."

In many states, a report detailing the findings and conclusions of a forensic engineering investigation are required to be signed and sealed by a licensed professional engineer. This is because by state law, engineering investigations are the sole prerogative of licensed, professional engineers. Thus, on the last page in the main body of the report, usually just after the conclusions section, the report is often signed, dated, and sealed by the responsible licensed professional engineer(s) who performed the investigation. Often, the other technical professionals who worked under the direction of the responsible professional engineer(s) are also listed, if they have not been noted previously in the report.

Some consulting companies purport to provide investigative technical services, investigative consulting services, or scientific consulting services. Their reports may be signed by persons with various initials or titles after their names. These designations have varying degrees of legal status or legitimacy vis-à-vis engineering investigations depending upon the particular state or jurisdiction. Thus, it is important to know the professional status of the person who signs the report. A forensic engineering report signed by a person without the requisite professional or legally required credentials in the particular jurisdiction may lack credibility and perhaps even legal legitimacy.

In cases where the report is long and complex, an executive summary may be added to the front of the report as well as perhaps a table of contents. The executive summary, which is generally a few paragraphs and no more than a page, notes the highlights of the investigation, including the conclusions. A table of contents indicates the organization of the report and allows the reader to rapidly find sections and items he wishes to review.

## Further Information and References

"Chemist in the Courtroom," by Robert Athey, Jr., *American Scientist*, 87(5), September-October 1999, pp. 390–391, Sigma Xi. For more detailed information please see Further Information and References in the back of the book.

*The Columbia History of the World*, Garraty and Gay, Eds., Harper and Row, New York, 1981. For more detailed information please see Further Information and References in the back of the book.

"Daubert and Kumho," by Henry Petroski, *American Scientist*, 87(5), September-October 1999, pp. 402–406, Sigma Xi. For more detailed information please see Further Information and References in the back of the book.

*The Engineering Handbook*, Richard Dorf, Ed., CRC Press, Boca Raton, FL, 1995. For more detailed information please see Further Information and References in the back of the book.

*Forensic Engineering*, Kenneth Carper, Ed., Elsevier, New York, 1989. For more detailed information please see Further Information and References in the back of the book.

*Galileo's Revenge*, by Peter Huber, Basic Books, New York, 1991. For more detailed information please see Further Information and References in the back of the book.

*General Chemistry*, by Linus Pauling, Dover Publications, New York, 1970. For more detailed information please see Further Information and References in the back of the book.

*Introduction to Mathematical Statistics*, by Paul Hoel, John Wiley & Sons, New York, 1971. For more detailed information please see Further Information and References in the back of the book.

*On Man in the Universe*, Introduction by Louside Loomis, Walter Black, Inc., Roslyn, NY, 1943. For more detailed information please see Further Information and References in the back of the book.

*Procedures for Performing a Failure Mode, Effects and Criticality Analysis (FMECA)*, MIL-STD-1629A, November 24, 1980. For more detailed information please see Further Information and References in the back of the book.

*Reporting Technical Information*, by Houp and Pearsall, Glencoe Press, Beverly Hills, California, 1968. For more detailed information please see Further Information and References in the back of the book.

*Reason and Responsibility*, Joel Feinburg, Ed., Dickenson Publishing, Encino, CA, 1971. For more detailed information please see Further Information and References in the back of the book.

*To Engineer is Human*, by Henry Petroski, Vintage Books, 1992. For more detailed information please see Further Information and References in the back of the book.

"Trial and Error," by Saunders and Genser, *The Sciences*, September/October 1999, 39(5), 18–23, the New York Academy of Sciences. For more detailed information please see Further Information and References in the back of the book.

"When is Seeing Believing?" by William Mitchell, *Scientific American*, Feb. 1994, 270(2), pp. 68–75. For more detailed information please see Further Information and References in the back of the book.

# Wind Damage to Residential Structures

2

You know how to whistle don't you? Just put your lips together and blow.
— **Lauren Bacall to Humphrey Bogart, in** *To Have and Have Not*
*Warner Bros. Pictures, 1945*

## 2.1   Code Requirements for Wind Resistance

Most nationally recognized U.S. building codes, such as the Unified Building Code (UBC) and the Building Officials and Code Administrators (BOCA) code require that buildings be able to withstand certain minimum wind speeds without damage occurring to the roof or structure. In the Midwest, around Kansas City for example, the minimum wind speed threshold required by most codes is 80 mph. For comparison, hurricane level winds are considered to begin at 75 mph.

According to the National Oceanic and Atmospheric Administration (NOAA) weather records, the record wind speed to date measured at the weather recording station at Kansas City International Airport is 75 mph. This occurred in July 1992. Considering together the Kansas City building code requirements and the Kansas City weather records, it would appear that if a building is properly "built to code" in the Kansas City area, it should endure all winds except record-breaking winds, or winds associated with a direct hit by a tornado.

Unfortunately, many buildings do not comply with building code standards for wind resistance. Some communities have not legally adopted formal building codes, and therefore have no minimum wind resistance standard. This allows contractors, more or less, to do as they please with respect to wind resistance design. This is especially true in single-family residential structures because most states do not require that they be designed by licensed architects or engineers. Essentially, anyone can design and build a house. Further, in some states, anyone can be a contractor.

It is also likely that many older buildings in a community were constructed well before the current building code was adopted. The fact that they have survived this long suggests that they have withstood at least some

23

**Plate 2.1** Severe wind damage to structure.

severe wind conditions in the past. Their weaker contemporaries have perhaps already been thinned out by previous storms. Most codes allow buildings that were constructed before the current code was adopted and that
appear to be safe to be "grandfathered." In essence, if the building adheres
to construction practices that were in good standing at the time it was built,
the code does not require it to be rebuilt to meet the new code's requirements.

Of course, while some buildings are in areas where there is indeed a
legally adopted code, the code may not be enforced due to a number of
reasons, including graft, inspector malfeasance, poorly trained inspectors, or
a lack of enforcement resources. Due to poor training, not all contractors
know how to properly comply with a building code. Sometimes, contractors
who know how to comply, simply ignore the code requirements to save
money. In the latter case, Hurricane Andrew is a prime example of what
occurs when some contractors ignore or subvert the wind standards contained in the code.

Hurricane Andrew struck the Florida coast in August 1992. Damages in
south Florida alone were estimated at $20.6 billion in 1992 dollars, with an
estimated $7.3 billion in private insurance claims. This made it the most
costly U.S. hurricane to date. Several insurance companies in Florida went
bankrupt because of this, and several simply pulled out of the state altogether.
Notably, this record level of insurance damage claims occurred despite the
fact that Andrew was a less powerful storm than Hugo, which struck the
Carolinas in September 1989.

**Plate 2.2** Relatively moderate wind caused collapse of tank during construction due to insufficient bracing.

Andrew caused widespread damage to residential and light commercial structures in Florida, even in areas that had experienced measurable wind speeds less than the minimum threshold required by local codes. This is notable because Florida building codes are some of the strictest in the U.S. concerning wind resistance. Additionally, Florida is one of the few states that also requires contractors to pass an examination to certify the fact that they are familiar with the building code. Despite all these paper qualifications, however, in examining the debris of buildings that were damaged, it was found that noncompliance with the code contributed greatly to the severity and extent of wind damage insurance claims.

The plains and prairie regions west of Kansas City are famous for wind, even to the point of having a "tall tale" written about it, the *Legend of Windwagon Smith*. According to the story, Windwagon Smith was a sailor turned pioneer who attached a ship's sail to a Conestoga wagon. Instead of oxen, he harnessed the wind to roam the Great Plains, navigating his wind-driven wagon like a sloop.

An old squatters' yarn about how windy it is in Western Kansas says that wind speed is measured by tying a log chain to a fence post. If the log chain is blowing straight out, it's just an average day. If the links snap off, its a windy day. In fact, even the state's name, "Kansas," is a Sioux word that means people of the south wind.

According to a publication from Sandia Laboratories (see references), Kansas ranks third in windy states for overall wind power, 176.6 watts per square meter. The other most windy states with respect to overall wind power are North Dakota (1), Nebraska (2), South Dakota (4), Oklahoma (5), and Iowa (6). Because of Kansas' windy reputation, it is hard to imagine any contractor based in Kansas, or any of the other windy Midwestern or sea-board states for that matter, who is not aware of the wind and its effects on structures, windows, roofs, or unbraced works in progress.

## 2.2  Some Basics about Wind

Air has two types of energy, potential and kinetic. The potential energy associated with air comes from its pressure, which at sea level is about 14.7 pounds per square inch or 1013.3 millibars. At sea level, the air is squashed down by all the weight of the air that lies above it, sort of like the guy at the bottom of a football pile-up. Like a compressed spring, compressed air stores energy that can be released later.

The kinetic energy associated with air comes from its motion. When air is still, it has no kinetic energy. When it is in motion, it has kinetic energy that is proportional to its mass and the square of its velocity. When the velocity of air is doubled, the kinetic energy is quadrupled. This is why an 80-mph wind packs *four times* the punch of a 40-mph wind.

The relationship between the potential and kinetic energies of air was first formalized by Daniel Bernoulli, in what is now called Bernoulli's equation. In essence, Bernoulli's equation states that because the total amount of energy remains the same, when air speeds up and increases its kinetic energy, it does so at the expense of its potential energy. Thus, when air moves, its pressure decreases. The faster it moves, the lower its pressure becomes. Like-wise, when air slows down, its pressure increases. When it is dead still, its pressure is greatest.

The equation developed by Daniel Bernoulli that describes this "sloshing" of energy between kinetic and potential when air is flowing more or less horizontally is given in Equation (i), which follows.

$$\text{total energy} = \text{potential} + \text{kinetic}$$

$$[P_{atmos}/\rho] = [P/\rho] + v^2/2g_c \tag{i}$$

where $P_{atmos}$ = local pressure of air when still, $\rho$ = density of air, about 0.076 lbf/ft³, P = pressure of air in motion, v = velocity of air in motion, and $g_c$ = gravitational constant for units conversion, 32.17 ft/(lbf-sec²).

It should be noted that Equation (i), assumes that gas compressibility effects are negligible, which considerably simplifies the mathematics. For wind speeds associated with storms near the surface of the earth and where

**wind streamlines**

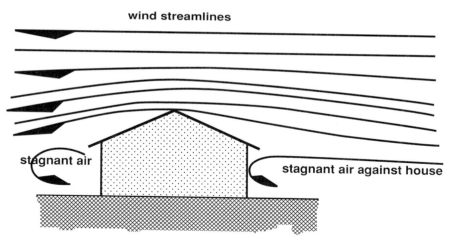

**Figure 2.1** Side view of wind going over house.

air pressure changes are relatively small, the incompressibility assumption implicit in Equation (i) is reasonable and introduces no significant error.

Wading through the algebra and the English engineering units conversions, it is seen that a 30-mph wind has a kinetic energy of 30 lbf-ft. Since the total potential energy of still air at 14.7 lbf/in$^2$ is 27,852 lbf-ft, then the reduction in air pressure when air has a velocity of 30 mph is 0.0158 lbf/in$^2$ or 2.27 lbf/ft$^2$. Similarly at 60 mph, the reduction in air pressure is 0.0635 lbf/in$^2$ or 9.15 lbf/ft$^2$.

What these figures mean becomes more clear when a simplified situation is considered. Figure 2.1, shows the side view of a house with wind blowing over it. As the wind approaches the house, several things occur.

First, some of the wind impinges directly against the vertical side wall of the house and comes more or less to a stop. The change in momentum associated with air coming to a complete stop against a vertical wall results in a pressure being exerted on the wall. The basic flow momentum equation that describes this situation is given below.

$$P = k\rho(v^2) \tag{ii}$$

where P = average pressure on vertical wall, k = units conversion factor, $\rho$ = mass density of air, about 0.0023 slugs/ft$^3$, and v = velocity of air in motion.

Working through the English engineering units, Equation (ii) reduces to the following.

$$P = (0.00233)v^2 \tag{iii}$$

where P = pressure in lbf/square feet, v = wind velocity in ft/sec.

**Table 2.1  Perpendicular Wind Speed
Versus Average Pressure on Surface**

| Wind Speed ft/sec | Resulting Pressure lbf/sq ft |
|---|---|
| 10 | 0.23 |
| 20 | 0.93 |
| 30 | 2.10 |
| 40 | 3.73 |
| 50 | 5.83 |
| 60 | 8.39 |
| 70 | 11.4 |
| 80 | 14.9 |
| 90 | 18.9 |
| 100 | 23.3 |
| 120 | 33.6 |
| 150 | 52.4 |

By solving Equation (iii) for a number of wind speeds, Table 2.1 is generated. The table shows the relationship between a wind impinging perpendicularly on a flat surface and coming to a complete stop, and the resulting average pressure on that surface.

In practice, the pressure numbers generated by Equation (iii) and listed in Table 2.1 are higher than that actually encountered. This is because the wind does not fully impact the wall and then bounce off at a negligible speed, as was assumed. What actually occurs is that a portion of the wind "parts" or diverts from the flow and smoothly flows over and away from the wall without actually slamming into it, as is depicted in Figure 2.1. Therefore, to be more accurate, Equation (ii) can be modified as follows.

$$P = k\rho(v_1^2 - v_2^2) \text{ or } = C\rho(v_1^2) \qquad \text{(iv)}$$

where P = average pressure on vertical wall, k = units conversion factor, $\rho$ = density of air, about 0.0023 slugs/ft$^3$, $v_1$ = average velocity of air flow as it approaches wall, $v_2$ = average velocity of air flow as it departs wall, and C = overall factor which accounts for the velocity of the departing flow and the fraction of the flow that diverts.

In general, the actual average pressure on a vertical wall when the wind is steady is about 60–70% of that generated by Equation (iii) or listed in Table 2.1. However, in consideration of the momentary pressure increases caused by gusting and other factors, using the figures generated by Equation (iii) is conservative and similar to those used in actual design.

This is because most codes introduce a multiplier factor in the wall pressure calculations to account for pressure increases due to gusting, building geometry, and aerodynamic drag. Often, the end result of using this multiplier is a vertical wall design pressure criteria similar, if not the same, as that generated by Equation (iii). In a sense, the very simplified model equation ends up producing nearly the same results as that of the complicated model equation, with all the individual components factored in. This is, perhaps, an example of the *fuzzy central limit theorem* of statistics at work.

Getting back to the second thing that wind does when it approaches a house, some of the wind flows up and over the house and gains speed as it becomes constricted between the rising roof and the air flowing straight over the house along an undiverted streamline. Again, assuming that the air is relatively incompressible in this range, as the cross-sectional area through which the air flows decrease, the air speed must increase proportionally in order to keep the mass flow rate the same, as per Equation (v).

$$\Delta m / \Delta t = \rho A v \qquad (v)$$

where $\Delta m / \Delta t$ = mass flow rate per unit time, A = area perpendicular to flow through which the air is moving (an imaginary "window," if you please), $\rho$ = average density of air, and v = velocity of air.

Constriction of air flow over the house is often greatest at the roof ridge. Because of the increase in flow speed as the wind goes over the top of the roof, the air pressure drops in accordance with Bernoulli's equation, Equation (i). Where the air speed is greatest, the pressure drop is greatest.

Thirdly, air also flows around the house, in a fashion similar to the way the air flows over the house.

Lastly, on the leeward side of the house, there is a stagnant air pocket next to the house where there is no significant air flow at all. Sometimes this is called the wind shadow. A low pressure zone occurs next to this leeward air pocket because of the Bernoulli effect of the moving air going over and around the house.

A similar effect occurs when a person is smoking in a closed car, and then opens the window just a crack. The air inside the car is not moving much, so it is at high pressure. However, the fast moving air flowing across the slightly opened window is at a lower pressure. This difference in relative pressures causes air to flow from the higher pressure area inside the car to the low pressure area outside the car. The result is that smoke from the cigarette flows toward and exits the slightly opened window.

If a wind is blowing at 30 mph and impinges against the vertical side wall of a house like that shown in Figure 2.1, from the simplified momentum flow considerations noted in Equation (iii), an average pressure of 4.5 lbf/ft² will be exerted on the windward side vertical wall.

If the same 30-mph wind increases in speed to 40 mph as it goes over the roof, which is typical, the air pressure is reduced by 4.0 lbf/ft². Because the air under the roof deck and even under the shingles is not moving, the air pressure under those items is the same as that of still air, 14.7 lbf/in² or 2116.8 lbf/ft². The air pressure under the roof and under the shingles then pushes upward against the slightly lower air pressure of the moving air going over the roof. This pressure difference causes the same kind of lift that occurs in an airplane wing. This lifting force tries to lift up the roof itself, and also the individual shingles.

While 4.0 lbf/ft² of lift may not seem like much, averaged over a roof area of perhaps 25 × 50 ft, this amounts to a total force of 5000 lbf trying to lift the roof. At a wind speed of 80 mph, the usual threshold for code compliance in the Midwest, the pressure difference is 16 lbf/ft² and the total lifting force for the same roof is 20,000 lbf.

If the roof in question does not weigh at least 20,000 lbf, or is not held down such that the combined total weight and holding force exceed 20,000 lbf in upward resistance, the roof will lift. This is why in Florida, where the code threshold is 90 mph, extra hurricane brackets are required to hold down the roof. The usual weight of the roof along with typical nailed connections is not usually enough to withstand the lift generated by 90 mph winds.

It is notable that the total force trying to push the side wall inward, as in our example, is usually less than the total lift force on the roof and the shingles. This is a consequence of the fact that the area of the roof is usually significantly larger than the area of the windward side wall (total force = ave. pressure × area). Additionally, a side wall will usually offer more structural resistance to inward pressure than a roof will provide against lift. For these reasons, it is typical that in high winds a roof will lift off a house before a side wall will cave in.

Lift is also the reason why shingles on a house usually come off before any structural wind damages occur. Individual asphalt shingles, for example, are much easier to pull up than roof decking nailed to trusses. Shingles tend to lift first at roof corners, ridges, valleys, and edges. This is because wind speeds are higher in locations where there is a sharp change in slope. Even if the workmanship related to shingle installation is consistent, shingles will lift in some places but not in others due to the variations in wind speed over them.

Most good quality windows will not break until a pressure difference of about 0.5 lbf/sq in, or 72 lbf/sq ft occurs. However, poorly fitted, single pane glass may break at pressures as low as 0.1 lbf/sq in, or about 14 lbf/sq ft. This means that loosely fitted single pane glass will not normally break out until wind gusts are at least over 53 mph, and most glass windows will not break out until the minimum wind design speed is exceeded.

Assuming the wind approaches the house from the side, as depicted in Figure 2.1, as the wind goes around the house, the wind will speed up at the corners. Because of the sharpness of the corners with respect to the wind flow, the prevalent 30-mph wind may speed up to 40 mph or perhaps even 50 mph at the corners, and then slow down as it flows away from the corners and toward the middle of the wall. It may then speed up again in the same manner as it approaches the rear corner of the house.

Because of this effect, where wind blows parallel across the vertical side walls of a house, the pressure just behind the lead corner will decrease. As the wind flows from this corner across the wall, the pressure will increase again as the distance from the corner increases. However, the pressure will then drop again as the wind approaches the next corner and speeds up. This speed-up–slow-down–speed-up effect due to house geometry causes a variation in pressure, both on the roof and on the side walls.

These effects can actually be seen when there is small-sized snow in the air when a strong wind is blowing. The snow will be driven more or less horizontal in the areas where wind speed is high, but will roil, swirl, and appear cloud-like in the areas where the wind speed significantly slows down. Snow will generally drift and pile up in the zones around the house where the air speed significantly slows down, that is, the stagnation areas. The air speed in those areas is not sufficient to keep the snow flakes suspended. During a blizzard when there is not much else to do anyway, a person can at least entertain himself by watching snow blow around a neighbor's house and mapping out the high and low air flow speed areas.

**Plate 2.3** Roof over boat docks loaded with ice and snow, collapsed in moderate wind.

Because a blowing wind is not steady, the distribution of low pressure and high pressure areas on the roof and side walls can shift position and vary from moment to moment. As a consequence of this, a house will typically shake and vibrate in a high wind. The effect is similar to that observed when a flag flaps in the wind, or the flutter that occurs in airplane wings. It is the flutter or vibration caused by unsteady wind that usually causes poorly fitted windows to break out.

Because of all the foregoing reasons, when wind damages a residential or light commercial structure, the order of damage is usually as follows:

1. lifting of shingles.
2. damage to single pane, loose-fitting glass windows.
3. lifting of awnings and roof deck.
4. damage to side walls.

Depending upon the installation quality of the contractor, of course, sometimes items 1 and 2 will reverse.

Unless there are special circumstances, the wind does not cause structural damage to a house without first having caused extensive damage to the shingles, windows, or roof. In other words, the small stuff gets damaged before the stronger stuff gets damaged. There is an order in the way wind causes damage to a structure. When damages are claimed that appear to not follow such a logical order, it is well worth investigating why.

## 2.3   Variation of Wind Speed with Height

Wind blows slower near the surface of the ground than it does higher up. This is because the wind is slowed down by friction with the ground and other features attached to the ground, like trees, bushes, dunes, tall grass, and buildings. Because of this, wind speeds measured at, say 50 feet from the ground, are usually higher than wind speeds measured at only 20 feet from the ground. In fact, the wind speed measured at 50 feet will usually be 14% higher than the speed at 20 feet, assuming clear, level ground, and even wind flow. As a general rule, the wind speed over clear ground will vary with 1/7th the power of the height from the ground. This is called the "1/7th power rule."

$$v = k[h]^{1/7} \qquad\qquad (vi)$$

where v = wind speed measurement, h = height from ground, and k = units conversion and proportionality constant.

For this reason, when wind data from a local weather station is being compared to a specific site, it is well to note that most standard wind measurements are made at a height of 10 meters or 32.8 feet. If, for example, it is necessary to know what the wind speed was at a height of 15 feet, then by applying the 1/7th power rule, it is found that the wind speed at 15 feet would have been about 11% less than that measured at 32.8 feet, all other things being the same.

If a wind speed is measured to be 81 mph at a standard weather reporting station, that does not automatically mean that a nearby building was also subject to winds that exceeded the code threshold. If the building was only 10 feet high, then the wind at that height would likely have been about 16% less or 68 mph, which is well below the code threshold. If there were also nearby windbreaks or other wind-obstructing barriers, it could have been even less.

Local geography can significantly influence wind speed. Some geographic features, such as long gradual inclines, can speed up the wind. This is why wind turbines are usually sited at the crests of hills that have long inclines on the windward side. The arrangement of buildings in a downtown area can also increase or decrease wind speed at various locations by either blocking the wind or funneling it. Thus, the wind speed recorded at a weather station does not automatically mean that it was the same at another location, even if the two sites are relatively close. The relative elevations, the placement of wind obstructing or funneling structures, and the local geography have to be considered.

## 2.4  Estimating Wind Speed from Localized Damages

One of the problems in dealing with wind damages is the estimation of wind speed when the subject building is located far from a weather reporting station, or is in an area that obviously experienced wind conditions different from that of the nearest weather station. In such cases, wind speed can actually be estimated from nearby collateral damage by the application of the Beaufort wind scale.

The Beaufort wind scale is a recognized system introduced in 1806 by Admiral Beaufort to estimate wind speed from its effects. Originally it was used to estimate wind speeds at sea. The methodology, however, has been extended to estimating wind speeds over land as well. The Beaufort wind scale is divided into 12 levels, where each level corresponds to a range of wind speeds and their observable effects. A brief version of the currently accepted Beaufort wind scale is provided below.

In reviewing the Beaufort scale, it is notable that tree damage begins to occur at level 8 and uprooting begins at level 10. However, most building

**Table 2.2   Beaufort Wind Scale**

| Scale Value | Wind Range | Effects Noted |
|---|---|---|
| 0, calm | 0–1 mph | Smoke rises vertically, smooth water, no perceptible movement |
| 1, light air | 1–3 mph | Smoke shows the direction of the wind, barely moves leaves |
| 2, light breeze | 4–7 mph | Wind is felt on the face, rustles trees, small twigs move |
| 3, gentle breeze | 8–12 mph | Wind extends a light flag, leaves, and small twigs in motion |
| 4, moderate breeze | 13–18 mph | Loose paper blows around, whitecaps appear, moves small branches |
| 5, fresh breeze | 19–24 mph | Small trees sway, whitecaps form on inland water |
| 6, strong breeze | 25–31 mph | Telephone wires whistle, large branches in motion |
| 7, moderate gale | 32–38 mph | Large trees sway |
| 8, fresh gale | 39–46 mph | Twigs break from trees, difficult to walk |
| 9, strong gale | 47–54 mph | Branches break from trees, litters ground with broken branches |
| 10, whole gale | 55–63 mph | Trees are uprooted |
| 11, storm | 64–75 mph | Widespread damage |
| 12, hurricane | 75 mph + | Structural damage occurs |

codes require a residential structure and roof to withstand wind levels up to 12. This means that the mere presence of wind damage in nearby trees does not automatically indicate that there should be structural or roof wind damage to a building located near the trees. Because kinetic energy increases with the square of velocity, a level 9 wind has only about half the "punch" of a level 12 wind.

## 2.5   Additional Remarks

Most of the major building codes do not simply use a single wind speed of 80 mph for design purposes. Within the codes there are usually multipliers that account for many factors, including the height and shape of the building, gusting, and the building class. For example, in the UBC a factor of 1.15 is to be applied to the pressure exerted by the wind when "important" buildings are being designed, such as schools, hospitals, and government buildings.

Generally, most codes require that public buildings, such as schools and hospitals, be built stronger than other buildings, in the hope that they will survive storms and calamities when others will not. Thus, when this factor is figured in and the calculations are backtracked, it is found that the actual wind speed being presumed is much greater than the design base speed of 80 mph, or whatever the speed.

## Further Information and References

ANSI A58.1, *American National Standard Minimum Design Loads for Buildings and Other Structures*, American National Standards Institute. For more detailed information please see Further Information and References in the back of the book.

*BOCA National Building Code*, 1993 edition. For more detailed information please see Further Information and References in the back of the book.

*Kansas Wind Energy Handbook*, Kansas Energy Office, 1981. For more detailed information please see Further Information and References in the back of the book.

*Peterson Field Guide to the Atmosphere*, by Schaefer and Day, Houghton-Mifflin, 1981. For more detailed information please see Further Information and References in the back of the book.

*UBC*, International Conference of Building Officials, 1991 ed., Appendix D, Figure 23-1. For more detailed information please see Further Information and References in the back of the book.

*Wind Power Climatology of the United States — Supplement*, by Jack Reed, Sandia Laboratories, April 1979. For more detailed information please see Further Information and References in the back of the book.

Winds and Air Movement, in *Van Nostrand's Scientific Encyclopedia*, Fifth Edition, Van Nostrand Reinhold Company. For more detailed information please see Further Information and References in the back of the book.

# Lightning Damage to Well Pumps

# 3

There was a young man from Peru
Who was zapped by a bolt from the Blue.
All the MOSFETs were blown
In his IBM clone,
And the bolt grounded out through his shoe.

**— Anonymous**

## 3.1 Correlation is not Causation

In the Midwest, especially in the region around Kansas City, there are 50 to 60 thunderstorms in the course of an average year. Since a typical year has 365 days, then simply by random chance it is expected that 60/365 or about 1/6 of all well pumps that fail from age and normal wear will fail on a day in which there has been a thunderstorm.

This same correlation, however, also applies to other singular events such as stopped clocks, blown tires, and weddings. In other words, random chance also dictates that there is a similar 1/6 probability in an average year that a clock will stop working, a tire will blow out, or a wedding will be held on a day in which a thunderstorm occurs. Of course when these other events occur on the same day as a thunderstorm, no one seriously believes they have actually been caused by the thunderstorm. It is generally acknowledged that they are just coincidences.

However, when an electrical appliance, such as a well pump, fails on a day in which a thunderstorm occurs, there is a tendency to presume that lightning must have had something to do with the failure, that is, that there is a cause-and-effect relationship between the two events. The rationale often applied in such cases is, "it worked fine until that storm occurred." Of course, this argument ignores the role that coincidence plays in the two events, or the obvious fact that the appliance had endured many such thunderstorms before and had not failed.

The probability that a well pump will fail for any reason after a thunderstorm increases as the length of time between the thunderstorm and the date the well pump failed increases. For example, by applying some basic proba-

**Plate 3.1** ARC-over from circulating panel lid, which was grounded.

bility rules, specifically Bayes' theorem, to the weather statistics already cited, it is found that when a well pump has failed due to age and wear, there is a 30% chance that there was a thunderstorm in the immediate area two days or less prior to the discovery of the failure.

Similarly, by again applying Bayes' theorem, it is found that there is a 50% chance that there will be thunderstorm in the area four days or less immediately prior to a well pump failure. In other words, when a well pump fails, there is a significant probability that a thunderstorm has occurred a few days prior to the failure simply by random chance.

Thus, the mere fact that a well pump has failed a day or so after a thunderstorm has occurred is not evidence that a thunderstorm caused the failure. In fact, it is arguable whether it has any significance at all. Random chance alone indicates that it is likely that well pumps will fail a few days directly after a thunderstorm, no matter what has caused the failure.

## 3.2   Converse of Coincidence Argument

With respect to whether thunderstorm/well pump failure coincidences have any significance at all, it is interesting to consider the converse, that is, to turn the basic proposition around by considering how often thunderstorms occur directly after a well pump has failed. The probability figures concerning this "reverse" coincidence are, in fact, exactly the same. In other words, when a well pump fails, there is a 1/6 chance that a thunderstorm will occur on the same day. Or, that when a well pump fails, there is a 50% chance that a thunderstorm will follow in four days or less.

Despite this impressive rate of coincidence, however, no one seriously believes that well pump failures cause thunderstorms. No one claims in a similar fashion that, "the weather was fine until that well pump failed." In a nutshell, this is an example of what is meant by "correlation is not causation." While some things might *appear* to occur in an ordered, cause-and-effect sequence, closer scrutiny sometimes reveals that there is no actual cause-and-effect relationship present.

In the case of well pumps, the only proper way to verify if a well pump has actually been damaged by lightning is to check for internal damage that is specific or unique to lightning. In the same way that a stabbing creates a wound very different from a gunshot, lightning damage is very different from other modes of failure. A well pump that has been damaged by lightning, will have physically verifiable damage consistent with lightning. Likewise, a well pump that has failed due to wear will exhibit the usual signs and symptoms of wear.

## 3.3   Underlying Reasons for Presuming Cause and Effect

At this point it is worth examining the underlying reasons why people readily presume a cause-and-effect relationship between thunderstorms and the finding of a failed well pump afterward. First is the fact that lightning and electrical appliances do have electricity in common. Lightning can cause electrical surges to occur when it strikes a power line, and electrical surges, when strong enough, can cause damage to electrical appliances and motors. Thus, when there has been a thunderstorm in the area and a well pump has failed afterwards, it is simply presumed that this coincidence signifies that lightning struck a power line, created a surge, and that this surge traveled along the power line until it reached the well pump and caused it to fail.

Secondly, and likely the most powerful reason, however, is the fact that insurance companies usually pay for lightning damaged well pumps, but usually do not pay for other types of well pump failures. Since replacement of a well pump is not a trivial expense, this is a powerful incentive for a homeowner or service person to overlook evidence that might otherwise indicate a normal failure from wear or age. Why would any person go out of his way to deliberately look for evidence that would cause him to lose money from the insurance company? Thus, this is why most claims of lightning damage to well pumps do not cite any specific, verifiable physical evidence to confirm the cause of damage, but do prominently note the coincidence of a thunderstorm shortly before the pump failed.

## 3.4   A Little about Well Pumps

Well pumps are basically long, skinny electric motors with an impeller-type water pump mounted on the outboard motor shaft. The diameter of a well pump ranges from 4 inches for the smaller sizes to 8 inches for the largest ones. Usually they are encased in an electrically conducting stainless steel shell or housing.

Small well pumps, in the range of 1/3 hp to 1-1/2 hp, can be two-wire, single phase motors. However, most well pumps are of the three wire type, which have separate "start" and "run" operational modes. Three-wire, single-phase well pumps range from 1/3 hp to 15 hp. However, three-phase well pumps are preferred in sizes greater than 2 hp because they are more economical in terms of electricity consumption.

Well pumps are lowered many feet below the surface of the ground into a drilled well hole and pump water from the underground aquifer to the surface. In this underground location, totally submerged in water, it is nearly impossible for lightning to strike a well pump directly. The intervening layers of rock, dirt, and water shield it. When lightning strikes the surface of the earth, the electrical charge is quickly dissipated at the surface. A lightning strike cannot penetrate to the depths where well pumps are usually placed.

## 3.5   Lightning Access to a Well Pump

By being submerged in water many feet below the surface, well pumps are in intimate electrical contact with the mother of all electrical grounds — the earth. When lightning strikes the surface of the earth near a well pump, there is no incentive for the lightning to flow from its grounding point at the surface to the well pump situated many feet below the surface. Once a lightning strike has grounded out and dissipated at the surface, it does not regroup itself and flow through grounded earth just to get to other locations to ground out again.

Some people posit that a lightning strike could enter a well pump by way of the water pipe that runs down to the pump, especially if the water pipe is metal. However, it should be remembered that the water pipe itself is usually grounded at the surface. As is shown in Figure 3.1, a water pipe is normally in direct contact with the earth at various points at the surface. It is also in direct contact with the well water, which, in turn, is in direct contact with

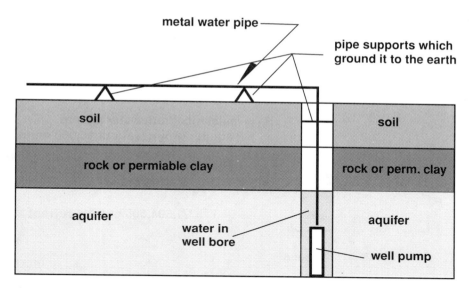

**Figure 3.1** Schematic of metal water pipe to well pump.

the earth. Thus, any electrical current flow into the metal water pipe from a lightning strike at the surface would follow the path of least resistance, which is to one of these grounding points.

Furthermore, electrical flow from the water pipe into the well pump motor windings is actually the path of highest resistance, as an examination of the relative resistances shows. The resistance between the water pipe to the pump and the motor windings is typically tens of millions of ohms, with 30,000,000 ohms being somewhat typical. The resistance through the well water is only several thousand ohms depending upon mineral content of the water, with 10,000 ohms being typical. The resistance through pipe connections to the ground is typically in tens of ohms, usually in the range of 20 ohms.

Figure 3.2 is a schematic showing the essential circuit arrangement of a lightning strike following a metal water pipe to ground. The strike has three choices. It can follow the pipe and ground out in the motor, water, or where it is in contact with the earth. This type of circuit arrangement, where the resistances are in parallel, is often called a circuit divider. When the parallel resistance circuit equations for Figure 3.2 are algebraically solved, it is found that the amount of current that would enter the motor windings in this way is trivial, well less than one-millionth of the total current. This is not enough to cause damage.

Thus, there is no physical reason why current from a lightning strike would flow in any significant amount through the water pipe, through the housing, into the rotor shaft, and ground out in the windings. Most of the

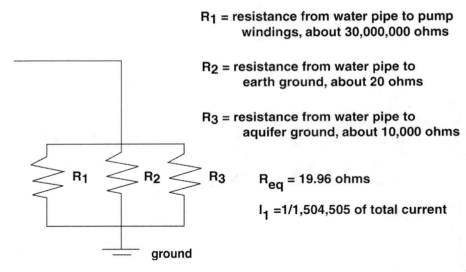

$R_1$ = resistance from water pipe to pump windings, about 30,000,000 ohms

$R_2$ = resistance from water pipe to earth ground, about 20 ohms

$R_3$ = resistance from water pipe to aquifer ground, about 10,000 ohms

$R_{eq}$ = 19.96 ohms

$I_1$ = 1/1,504,505 of total current

**Figure 3.2** Electrical schematic of water pipe circuit to pump and ground.

current will simply ground out through the pipe supports that provide earth grounding, or ground out through the water in the well.

Additionally, the electrically conductive stainless steel shell around the well pump acts as a grounded "Faraday box" or coaxial cable shield. In other words, the grounded outer shell of the pump serves to prevent externally applied stray currents and fields from entering the well pump motor, either directly or inductively. Because of this, electric currents flowing through the well water cannot enter the pump through the outer shell.

Thus, the only way a lightning strike can cause damage to a well pump located many feet below the surface of the earth is to enter by way of the electrical supply circuit. In other words, a lightning surge has to go through the electrical supply wires to get to the pump motor.

In this regard, it is interesting to note that since 1972, Franklin Electric, one of the world's largest manufacturers of well pumps, has equipped their well pumps with internal lightning arrestors to prevent such lightning surges from damaging their motors. Several other well pump companies do the same. If the well pump service person is at all slightly familiar with the company literature, he will know this fact. Additionally, since 1975, Franklin Electric has equipped their motors with self-healing, antitrack resin, to prevent the kind of arcing within the windings commonly associated with very high voltage surges.

In short, the manufacturers of some, if not most, well pumps have equipped their well pump motors with devices to prevent damage by lightning surges that enter by means of the electrical service. Thus, the only way these well pumps can be damaged by a lightning surge is for the internal lightning arrestor to be defective or deficient. While possible, the manufacturers' quality control measures make this highly unlikely.

## 3.6  Well Pump Failures

Lightning is not the only reason well pumps fail. Like all appliances, well pumps eventually succumb to age and wear. One of the more common ways for a well pump to fail is for the windings in the electric motor to simply wear out.

When an electric motor begins to fail, it consumes more electricity than normal and operates at a higher temperature. As the motor further degenerates, the operating temperature of the motor rises until the dielectric within the motor windings breaks down and a short within the windings occurs. This causes the electric motor to "ground out." It is notable that some service persons will cite that a "grounded out" motor is evidence of lightning damage. This is not exactly true. What is true, is that "grounded out windings" are a symptom common to both lightning and normal motor failure.

When aboveground electric motors, such as window fans, begin to fail, usually the motor casing feels hot, perhaps a slight electrical odor is noted, and when started, the fan hesitates, taking longer to reach operating speed. Unfortunately, because well pumps are located underground, these same symptoms will not be noted by the owner. This is why it is often a total surprise to the owner.

Some well pump motors have a temperature sensor within the windings. When the temperature of the windings becomes too high, a switch opens the circuit and turns off the motor. This is an indication that the motor is near the end of its life and will soon fail. However, when the motor windings cool back down, the switch will often be of the type that resets, and the motor will operate for a time again. Unfortunately, this type of behavior in the motor is also rarely noted by the owner due to the location of the equipment.

However, some of the symptoms that clearly indicate routine failure by wear and age in a well pump include the following:

- worn motor bearings (rotor shaft is hard to turn, or does not freely spin).
- water leakage into the motor housing and windings.
- worn pump impeller bearings.
- worn packing around bearings and shaft allowing silt to get into bearings and internals.
- discoloration of the internal lubrication oil (if any) due to operation at elevated temperatures.

**Plate 3.2** Surge damage due to loose utility cable that fell onto lower line and caused high voltage surge. Initially diagnosed as lightning damage.

- deteriorated relays due to higher than normal operating currents, or higher than normal duty cycling.
- widespread deterioration of dielectric in windings by overheating (instead of a single, breakthrough point of damage like lightning makes).
- higher than normal electric bills (when accurate comparisons can be made), unusual tripping of breakers, or other items that indicate higher than normal current usage.

## 3.7   Failure Due to Lightning

When lightning has actually damaged a well pump, it is often verified by noting the following items:

- verifiable reports of damaging electrical surges at both ends of the electrical line that services the well pump (note: surges flow both ways along a line).
- evidence of high voltage surges in the line, e.g., flashovers at terminals and relay gaps (the "jump" distance can be used to estimate the peak voltage of the surge).
- high voltage damage to the internal lightning arrestor.
- high voltage damage to a capacitor, usually the "run" capacitor.

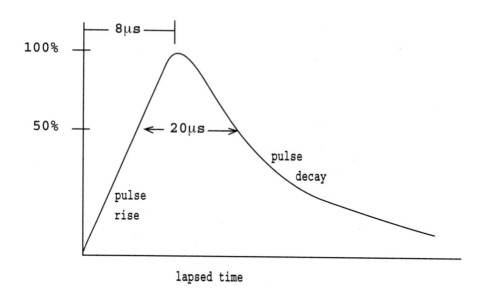

**Figure 3.3** Typical lightning strike pulse.

- verification of surge on the power line by the power company (many power companies monitor their lines for surges and electrical anomalies).
- high voltage damage to delicate circuit boards in the control equipment.
- evidence that one winding, usually the "start" winding, has normal dielectric resistance readings while the other winding, usually the "run" winding, is grounded out at a single breakdown point.

A *bona fide* lightning surge typically has a wave form like that shown in Figure 3.3. Figure 3.3 is the typical lightning strike pulse wave form as proposed by the Institute of Electrical and Electronic Engineers, IEEE. As is noted in Figure 3.3, the peak power in a typical lightning surge occurs perhaps in just 8 microseconds. Lightning-induced power line spikes of up to 6,000 volts have been measured with rise times as short as 500 nanoseconds.

It is notable that blown fuses or breakers *do not* indicate the presence of a lightning type surge. These components are not designed to be lightning protection devices. They simply cannot respond fast enough to react to the rapid rise of a lightning pulse. In fact, for technical reasons that involve motor reactances and start current on-rush, these devices are specifically designed not to react quickly to a power surge on the line.

The performance of surge protection equipment is usually gaged to the IEEE lightning pulse curve shown in Figure 3.3. Generally, for a device to afford some protection against lightning surges, the device must be able to shunt a power surge to ground in less time than 4μs, the time it typically takes for the surge to reach 50% of its full power.

Fuses or breakers, however, do respond and open the circuit when abnormal levels of current are used over a relatively long period of time, which occurs when a well pump motor fails due to normal age and wear. This is what they are supposed to do.

It is unusual for an electric motor in even fair condition to be damaged by a lightning surge, while other, more delicate electronic components connected to the same feeder line are undamaged. This is because the damage threshold for electric motors is between 1,000,000 and 10,000,000 watts per microsecond, while the damage threshold for relays is between 100 and 1,000 watts per microsecond. Even more delicate are computerized control electronic components, whose damage threshold is between 0.001 and 0.1 watts per microsecond.

## Further Information and References

"The Essentials of Lightning Protection," *Record*, May–June 1981, pp. 12–18. For more detailed information please see Further Information and References in the back of the book.

"In Search of Electrical Surges," by Ivars Peterson, *Science News*, December 12, 1987, 132, pp. 378–379. For more detailed information please see Further Information and References in the back of the book.

*Lightning*, by Martin Uman, Dover Publications, New York, 1984. For more detailed information please see Further Information and References in the back of the book.

*Mechanics of Structures with High Electric Currents*, by Francis Moon, National Science Foundation, Princeton University, Princeton, NJ, July 1973. For more detailed information please see Further Information and References in the back of the book.

*Note on Vulnerability of Computers to Lightning*, by R. D. Hill, General Research Corporation, Santa Babara, CA, April 1971. For more detailed information please see Further Information and References in the back of the book.

# Evaluating Blasting Damage

# 4

There are as many opinions as there are experts.
— **Franklin Roosevelt,** *June 12, 1942*

Quot homines tot sententiae.
(There are as many opinions as there are men.)
— **Terence,** *a Roman playwright, ca. 2160 years ago*

## 4.1   Pre-Blast and Post-Blast Surveys

The simplest method to assess and document blasting damages in a structure is the "before and after" approach. The structure in question is thoroughly inspected immediately prior to blasting work being done. Then, the same structure is reinspected in the same way immediately after blasting work is completed. The two surveys are then compared to determine if there are cracks and fissures present after blasting that were not present before blasting. It is then presumed that any new cracks or fissures discovered in the second inspection were caused by the blasting work.

The first inspection is generally called the pre-blast survey. Naturally enough, the follow-up inspection that is done when blasting work is completed is called the post-blast survey. In some communities, local law prescribes that blasting contractors conduct pre-blast surveys of all buildings and structures in the general area in which blasting work will be done. Post-blast surveys may then be done according to the following schedule:

- on a complaint basis, that is, specifically when people complain that they think their home or structure has been damaged in some way by nearby blasting work.
- on a small number of structures nearest the blasting work, that is, on the structures that would have been exposed to the highest level of ground vibrations resulting from the blasting work.
- repeated within the whole area adjacent to the blasting work to match all the pre-blast surveys that were done.

Conducting post-blast surveys on a complaint-by-complaint basis is more commonly preferred by blasting contractors, ostensibly to reduce costs. The contractor usually hopes that if he conducts the blasting work within accepted standards and practices only a few people will complain, and those complaints will be quickly handled. Unfortunately, even when a contractor adheres strictly and conservatively to all the accepted standards and guidelines, this is often not the case.

Repeating the whole pre-blast survey after the blasting work is completed is sometimes preferred by the blasting contractor's insurer to minimize overall potential liability. This is because property owners' complaints may be filed months or even years after the blasting work has been completed. Unless the entire pre-blast survey was repeated immediately after the blasting work was completed, a timely post-blast survey completed immediately after blasting work was finished will not be available for comparison to the pre-blast survey.

A post-blast survey done in response to a complaint filed months or even years after the blasting work has been completed may turn up cracks and fissures that are not the result of the blasting work. Of course, the longer the time delay in doing a post-blast survey, the higher this risk will be. Any newly discovered cracks and fissures in a delayed post-blast survey may be the result of normal differential settlement or other natural effects at work during the interim and not at all related to blasting. In such cases, additional evidence will have to be gathered and additional engineering work will have to be done to discriminate the cracks caused by blasting from cracks caused by settlement, or whatever cause.

When this occurs, the simplicity and intuitiveness of the "before and after" method is lost, and the concomitant investigation costs of the complaint increase greatly. Further, if the case goes to court, it is much easier for the plaintiff (the property owner) to argue, "I didn't have cracks before, and now I do," than it is for an expert witness engineer to discuss soil expansion and compaction effects, active lateral pressure effects, differential settlement effects, heaving, original deficiencies in the foundation design, or the differences between wind wracking a building and surface blast vibrations wracking a building.

In general, simple "horse sense" arguments are generally more persuasive to juries than complicated, jargon filled, technical arguments replete with complicated algebraic formulas and number-filled tables. Lawyers who are on the short end of a technical argument love to embroil expert witnesses in lengthy, arcane arguments about technical minutiae. This is because many jurors can't follow the arguments, lose interest, and eventually consider the whole argument a wash. Also, because most jurors will be unfamiliar with the technical information being presented and bandied about, there is a latent fear of being bamboozled by fast talking, pointy-headed professors full of book learnin' but no common sense. For these reasons, it is best to avoid having to use such evidence whenever possible, no matter how scientifically rigorous, in the legal system.

The other alternative, doing post-blast surveys on structures nearest the blasting work, can be a useful intermediate step. Obviously, structures nearest the blasting work would be affected more than structures located farther away. Thus, if no damages are noted in the closest structures, it is logical to conclude that no damages occurred in structures located farther away where blasting effects are even less intense.

However, some property owners may contend that their structures are intrinsically weaker than the ones closer to the blasting work, which were checked, or that some other "special effect" is at work that would allow damage to occur to their structure, while not causing damage to the ones checked.

If such a complaint is filed, it is likely that a post-blast survey of the structure in question will have to be done anyway. Likewise, if the complaint is filed a long time after the blasting work has been completed, there is a risk that other types of settlement effects have been at work in the meantime, and, as noted before, additional evidence gathering and engineering work will have to be done to discriminate cracks caused by blasting versus cracks caused by other effects.

## 4.2   Effective Surveys

Companies that do pre-blast and post-blast survey work are often hired by the blasting contractor, the local unit of government, a home owners' association, or an insurance company, usually the one representing the blasting contractor. The reports and documents generated by blast survey companies are usually open to all parties involved, not just the client. This is usually stipulated in the local ordinance or the contractual agreement.

The field people that do pre-blast and post-blast surveys are usually engineers, engineering technologists, or engineering technicians with at least some drafting or construction background. Sometimes they simply prepare drawings, sketches, and written descriptions of cracks and fissures found in a structure. It is arguable whether or not this is the least expensive method for such surveys. It does seem, however, from the experiences of this author, that this is the least effective and least convincing method used for such surveys.

More effective documentation methods for conducting such surveys include the use of traditional emulsion photographs, slide photographs, digital photographs, and videotape coordinated with written or verbal descriptions recorded on tape. In this way, before and after images of the disputed problem areas can be shown side by side for easy comparison.

As alluded to previously, for the "before and after" comparison method to be an effective assessment of blasting damage, the following conditions should be met.

1. The initial pre-blast survey inspection must be thorough. If significant cracks or fissures are missed in the pre-blast survey and are then discovered in the post-blast survey, the property owner will generally argue that the blasting work caused them. To dispute this argument puts the inspection team in the uncomfortable position of arguing that they missed something significant in the pre-blast survey, and that perhaps some or all of the other pre-blast surveys in the area are similarly flawed.

2. The pre-blast and post-blast surveys should be close in time to the actual blasting work so no other factors can be responsible for any new cracks or fissures that might be noted in the post-blast survey. Some newly constructed houses settle rapidly during their first year or so. Heavy rains can cause high rates of settlement, as can droughts, significant changes in water table, local-dewatering operations near construction sites, changes in run-off patterns due to nearby construction, seismic movement, soil creep and slump, and so on. If there is too much time between the pre-blast survey and the actual blasting work, or between the blasting work and the post-blast survey, other factors having nothing to do with blasting can possibly cause cracks or fissures to form in a structure. The pre-blast and post-blast surveys lose much of their effectiveness and value when they are not sufficiently timely to exclude other factors.

3. Suitable and believable means of documentation should be used. Pencil drawings and sketches are sometimes suspect because people may believe that a new pencil drawing of crack locations and sizes can be easily substituted for the original one in some kind of conspiracy to defraud, even if they are dated and signed. Photographs and videotapes, especially ones with the time and date embedded in the corner are usually very credible. Since the essential value of pre-blast and post-blast surveys is as a "before and after" argument, it is important to firmly establish the dates on which the "before and after" surveys were made, and prepare the surveys so that they are easy to compare.

## 4.3   Types of Damages Caused by Blasting

There are four types of physical damage that can be caused by blasting operations at construction sites.

1. Flyrock debris impact.
2. Air concussion damage.
3. Air shock wave damage.
4. Ground displacement due to blasting vibrations.

## 4.4  Flyrock Damage

Flyrock debris damage is relatively easy to spot and assess. It occurs when the blaster does not adequately bury or cover the blast charges to prevent blowout from the hole. Rocks and soil materials may become airborne missiles and be hurled away from the point of detonation.

Flyrock debris damage occurs relatively close to the blasting site, and will consist of impacts of rock and soil material on the side or top of the structure toward the blasting site. In most cases, it will appear like a person has thrown rocks and dirt at the structure. The impact speeds will be similar.

If the damage is severe and the structure has a wood or other soft material exterior, some of the rocks may penetrate and become embedded in the structure. Window glass may be cracked or broken. If window glass is cracked from rock impacts, impact craters in the glass will be apparent where the fracture initiated. If the glass has broken and the debris is still undisturbed, the glass and the impacting rock material will be scattered inside the house in a direction consistent with a trajectory from the blasting site. The rock material can then be visually verified as being of the same kind as was in the blasting hole.

Flyrock debris damage does not occur in isolation. Corroborating flyrock debris and perhaps other collateral flyrock damage will be observed in and around the general area of the blast hole. The flyrock debris that caused the damage, and the flyrock debris scattered about the area can be visually checked to ensure that they match the material that was in the blasting hole, and that they have landed in a location consistent with a trajectory from the blasting hole.

This point is brought up because occasionally windows are broken by other means, and then are blamed on flyrock from nearby blasting operations. For this reason, when a blast has created undue flyrock, it is recommended that the blaster immediately examine neighboring areas for damage, perhaps even documenting the examination with time and dated photographs. Even if no damage is noted, it is useful to photographically document the nondamaged items in case some damage is reported later.

If needed, the ejection velocity of the flyrock from the opening of the blast hole can be readily estimated. This is done by locating the flyrock debris that has traveled farthest from the blast hole without encountering a vertical

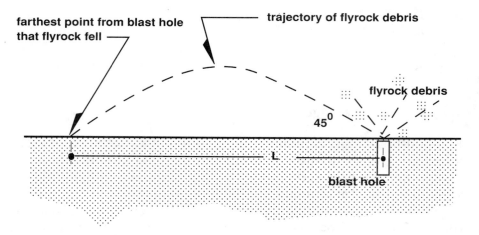

**Figure 4.1** Trajectory of flyrock debris.

obstacle and then measuring its lateral "throw" distance from the blast hole, as depicted in Figure 4.1.

Assuming that the greatest lateral "throw" distance of flyrock is obtained when the flyrock exits the blast hole at a trajectory angle of 45 degrees with respect to the horizontal, then the following vector trajectory equations hold. For simplicity, air resistance is ignored, and it is assumed that the final position of the flyrock is more or less at the same elevation as the blast hole.

horizontal velocity    vertical velocity

$$V = V(\cos 45°)i + V(\sin 45°)j \tag{i}$$

$$t = |V|(\sin 45°)/g$$

$$L = |V|(\cos 45°)\,(2)\,(t) = [|V|(\cos 45°)\,(2)\,|V|(\sin 45°)]/g = |V^2|/g$$

$$|V| = [Lg]^{1/2}$$

where $V$ = exit velocity vector of debris from blasting hole, $t$ = time required to reach maximum elevation during trajectory, $L$ = lateral distance traveled by debris, and $g$ = acceleration of gravity.

In the detonation of construction explosives like ANFO or dynamite, where the explosion is basically open and not properly "smothered," the scatter of debris items can be roughly related to the explosive yield by the following relation.

$$W_E = [L^3]/K \text{ or } [W_E\,K]^{1/3} = L \tag{ii}$$

where $W_E$ = amount of explosive yield in equivalent kilograms of TNT, L = distance from blast epicenter to farthest scatter of debris, and K = scaling factor, 91,000 m³/kg.

Combining Equations (i) and (ii) indicates that the initial velocity of an item hurled from a construction blast is approximately:

$$|V| = [gL]^{1/2} = [g(W_E K)^{1/3}]^{1/2} \qquad \text{(iii)}$$

where V = exit velocity vector of debris from blasting hole, t = time required to reach maximum elevation during trajectory, L = lateral distance traveled by debris, g = acceleration of gravity, $W_E$ = amount of explosive yield in equivalent kilograms of TNT, L = distance from blast epicenter to farthest scatter of debris, and K = scaling factor, about 91,000 m³/kg.

Using Equation (iii), it is found that a 5 kg (11 lbf) charge of ANFO will cause unrestrained flyrock to have an initial velocity of about 27.5 m/s or about 61.5 mph. By Equation (i), flyrock could, therefore, land up to 77 meters or 252.6 feet from the blast hole if the blast hole is uncovered and the flyrock does not meet an obstacle. If the flyrock chunk is a considerable size, the combination of size and speed can be dangerous, and perhaps even lethal.

## 4.5  Surface Blast Craters

Occasionally, an accident may occur where the detonation of an explosive occurs at the surface. When this occurs, a crater is formed. In general, a crater that has been created by a dynamite charge detonated at the surface will have a surface diameter about four times its crater depth. Further, the relationship between crater diameter and surface charge weight has been empirically determined to be as follows.

$$d = (0.8 \text{ m/kg}^{1/3})(W)^{1/3} \qquad \text{(iv)}$$

where d = crater diameter in meters, and W = explosive yield in equivalent kilograms of TNT.

Assuming that the crater formed from the surface explosion is a portion of a sphere, then the volume of material thrown up by the surface explosion can be estimated. If the depth of the crater is presumed to be equal to 1/4th its diameter, then by applying some analytic geometry, it is found that the radius of the sphere is:

$$R = (17/2)h \qquad \text{(v)}$$

where h = depth of crater, and R = radius of sphere.

Applying the formula for the volume of a spherical "cap" and incorporating the previous assumptions, the following is generated.

$$V = (1/3)\pi h^2(3R-h) = (25.65)h^3 = (0.40)d^3 = (0.205 \text{ m}^3/\text{kg})W \qquad \text{(vi)}$$

where V = volume of the crater, in cubic meters, h = depth of the crater, in meters, d = diameter of the crater, in meters, and W = equivalent amount of TNT in kg.

If the blast occurs above the surface, the above equations can be used to estimate the amount of equivalent charge required at the surface to create the same cratering effects. Knowing the distance above the surface the actual charge was placed and an appropriate scaling factor, the amount of charge can then be determined by proportionality: the intensity of a blast decreases by the cube of the distance. An excellent treatment of such scaling factors is contained in the text, *Explosive Shocks in Air*, which is cited at the end of this chapter.

## 4.6 Air Concussion Damage

The usual purpose of blasting at a construction site is the fracturing of rock. To do this in the most cost-efficient manner requires that all the energy from the explosive be directed at fracturing rock. Energy expended in producing air concussion, or air blast, is wasteful and is avoided by the blasting contractor.

For this reason, air concussion effects at blasting sites are generally limited to a relatively small area in and around the blasting hole. For air concussion effects to significantly affect nearby structures, the blast usually has to be unrestrained and located either at the surface, perhaps generating a crater, or elevated above the surface. When air concussion effects are significant, there is often some kind of associated accident or serious blasting mistake.

Even in an efficient blast, however, some fraction of the total charge is expended by the rapid expansion of gases. Essentially, some of the energy contained in the charge is converted into heat and air shock effects that travel radially away from the blast hole and dissipate in intensity as the distance increases.

A measure of the intensity of the air concussion produced by the rapid expansion of gases associated with a detonation is the overpressure. Overpressure is the amount of pressure in excess of the usual ambient pressure, which is sometimes called the gage pressure. An overpressure front of 0.5 lbf/in², for example, means that the pressure front was 0.5 lbf/in² more than the ambient atmospheric pressure of 14.7 lbf/in².

The expansion of gases associated with a detonation blast is so rapid, that the corresponding air concussion or overpressure front is generally

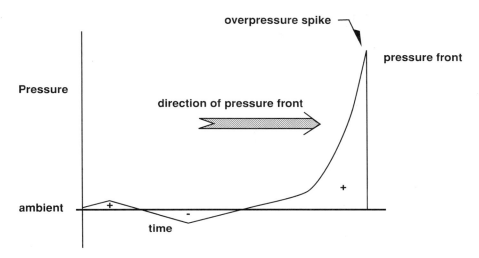

**Figure 4.2** Overpressure front.

considered to be a sharply defined wave front or pressure spike. The pressure on one side of the spike is the ambient pressure. The pressure just at, or slightly behind the spike, is the maximum overpressure that will occur at that particular distance from the blast. The primary destructive power of an air concussion from a blast site is basically contained in the overpressure front.

As the overpressure front passes by, the overpressure behind the front will rapidly diminish. It will drop down to ambient again, and then will even bounce slightly below ambient, similar to the overcorrection bounce in a spring-mass damped vibration system.

In motion pictures showing the passage of pressure fronts from large military bomb explosions, the initial overpressure appears like an invisible, outward pushing surge, knocking things down and dragging loose debris along with the pressure front. After the pressure front passes by, things momentarily stop moving as the local pressure level returns to ambient. When the "negative" overpressure passes through, it appears that there is a slight reverse surge: small items appear to be dragged somewhat backward toward the explosion epicenter. After the negative pressure subsides, the pressure equalizes itself back to ambient, and everything comes to rest.

Figure 4.2, shows a typical overpressure front caused by a blast and the low pressure bounce that follows it.

A reasonable model for calculating the overpressure at the front is as follows.

$$P_r = K[I / r^3] \qquad\qquad\qquad \text{(vii)}$$

where $P_r$ = pressure at overpressure front located "r" distance from blast epicenter, K = arbitrary constant for conversion of units and efficiency of explosion, I = explosion intensity, amount of explosive, or amount of energy release, and r = distance from blast epicenter.

When the overpressure at one point from the blast epicenter is known, it is possible to use Equation (vii) to calculate the expected overpressures for other locations.

$$P_1[r_1^3] = KI = C \qquad\qquad\qquad (viii)$$

$$P_2[r_2^3] = KI = C$$

$$P_2 = P_1[r_1^3/r_2^3]$$

where $P_1$ = known pressure at distance "$r_1$," $P_2$ = unknown pressure at distance "$r_2$," and C = constant.

Damage "markers" can often help in the solution of Equation (vii). For example, it is known that poorly set, single pane glass windows break out when the overpressure exceeds 0.5 to 1.0 lbf/in$^2$, and properly set window glass typically do not break until an overpressure of 2.0 lbf/in$^2$ is exceeded. Thus, if the air concussion from a blast breaks a poorly set, single pane window at "$r_1$" distance from the blast epicenter, then pressure and distance factors are established and the overpressure versus distance function can be estimated.

Structurally, poorly set single pane glass windows facing the blast epicenter are usually the most fragile items that can be damaged by the air concussion from a blast. For this reason, it is usually valuable to determine and map out the greatest distance from the blast epicenter at which single pane glass windows were noted to be broken. This establishes the maximum distance at which the air concussion *could* have caused any damage. The lack of broken windows at a certain distance indicates that no structural damage from air concussion occurred at that distance.

Extensive tests performed by the U.S. Bureau of Mines have consistently shown that internal plaster damages or other such structural damage do not occur unless at least the windows have been damaged. Because the overpressure from a blast decreases with the cube of the distance, overpressure levels beyond the single pane window radius will obviously be insufficient to cause any structural damage.

The above discussion presumes that the explosion has taken place more or less in the open so that the overpressure envelope is in the shape of a hemisphere. The presence of hills, valleys, and even other buildings can occasionally complicate the normally hemispherical propagation of the over-pressure front by causing reflections, refractions, and shadowing effects.

The overpressure front of an explosion can bounce off or be deflected by buildings and geographic features somewhat like waves in a pond bouncing off the side of a boat gunwale. Like waves approaching a dock piling, there can be a zone immediately behind a building or structure where there are no overpressure effects. However, given sufficient distance and blast intensity, this shadow area can be filled in again by overpressure. The overpressure fronts that have gone around either side of the structure spread back into the shadow zone. In some unusual cases, there can even be destructive and constructive interference where the two wave fronts meet and overlap.

There are a number of computer programs that are useful in the calculation of air concussion effects from a blast. Some of these can take into account the effects of nearby buildings and localized geographic effects. Two such programs are *HEXDAM*® (Enhanced High Explosive Damage Assessment Model) and *VEXDAM*® (Vapor Cloud Explosion Damage Assessment Model). Both programs are available from Engineering Analysis, Inc., in Huntsville, Alabama.

## 4.7   Air Shock Wave Damage

In the context of normal construction or quarry type blasting, damage by long distance air shock waves doesn't happen even though it is often alleged.

In very large explosions, the type usually associated with major accidents and terrible calamities, airborne shock waves have been known to travel considerable distances from the epicenter of the blast and cause minor damage to building windows. This type of damage is distinguished from regular air concussion damage by the fact that the damaging shock wave has skipped over a large expanse of area causing no damage to windows in those areas.

For air shock wave damage to occur, two factors have to be present. First and foremost is the fact that a major explosion has to have occurred, something on the order to the 1947 Texas City explosion that killed 561 people and injured 3,000. This is needed because the blast intensity must be very large for the overpressure front to remain significant over a long distance.

The second factor required is an atmospheric inversion layer located a significant distance from the epicenter. An inversion layer is where a cold layer of air is trapped below a warm layer. Normally, air is layered the other way around. When airborne shock waves are generated, even the shock waves parallel to the ground eventually curve upward into the atmosphere. However, when an inversion is present at some distance from the epicenter, the airborne shock waves that curved upward can be bent or reflected downward to the surface.

What does sometimes occur, however, is that people located a long distance from the blast may hear a loud blast noise that perhaps rattles loose glass in the same manner as a sonic boom. While it does not require structurally damaging overpressure to produce a loud noise, people often presume that if the "boom" was loud to their ears, it must have been destructive.

For example, a 135 db noise that can cause pain to the listener has an overpressure of about 114 Pascals or 0.016 lbf/in². However, as was noted before, even poorly set windows are not damaged unless the overpressure is about 0.5 lbf/in² or more. The minimum threshold reported in the literature for occasional glass breakage in poorly set, prestressed, tempered windows with large exposed areas is 0.103 lbf/in². This overpressure corresponds to an extreme noise level of about 151 db. Thus, unless there is broken glass, there will be no structural damage even though the boom may have been perceived as being quite loud.

## 4.8   Ground Vibrations

As was noted before, most of the energy expended in a blast is used to fracture rock and displace earth. However, a significant fraction of the energy released in the blast will be transmitted through the earth in the form of vibrational waves that radiate away from the blast site.

There are two main types of ground vibrational waves generated by a blast: longitudinal waves, sometimes called "P" waves, and shear waves, sometimes called "S" waves. "P" waves cause sinusoidal-like displacements in the direction of wave travel. "S" waves cause sinusoidal-like displacements transverse to the direction of wave travel. Taken together, "S" and "P" waves are sometimes called body waves.

The "P" wave front propagates radially away from the blast hole. It consists of alternating elastic compression and tension of the material through which the wave front passes, sort of a "push-pull" effect. A "P" wave travels at the speed of sound through a material, and is the fastest of the two types of waves. Through the earth, the average speed of a "P" wave is approximately 5,600 m/s. However, of the two main types of waves, "P" waves are the least significant with respect to causing damage to surface structures outside the immediate area of the blast. Only about 5% of the blast energy received by a surface structure is transmitted by "P" waves. Most of the "P" waves travel downward from the blast site, toward the center of the earth.

Shear or "S" waves travel through the ground at a velocity of slightly more than half the speed of "P" waves, approximately 3,000 m/s. Shear waves are similar to the up-and-down wave motion of a rope that has been tied at one end and then whipped at the free end. Shear waves cause similar up-and-down displacements along the surface of the earth and are responsible for most of the damaging effects to structures. Approximately 95% of the blast energy received by a surface structure is transmitted to it by shear or "S" waves.

In addition to causing up-and-down motion along the surface, certain types of shear waves can also cause side-to-side horizontal motions. Sometimes the side-to-side shear waves are called Rayleigh waves, Love waves, or "L" waves, to differentiate them from the up and down shear waves. In most texts, however, the different directions that shear waves may take are simply differentiated by "x-y-z" or "r-θ-φ" coordinates, and all shear waves are simply called "S" waves.

Assuming that the "S" wave ground displacement vibrations from a blast can be modeled with a sinusoidal function, then the displacement, maximum displacement, velocity, maximum velocity, acceleration, and maximum acceleration can all be modeled by the following relations.

$$x = A \sin(\omega t) \qquad\qquad x_{max} = A \qquad\qquad \text{(ix)}$$

$$v = dx/dt = \omega A \cos(wt) \qquad\qquad v_{max} = w A$$

$$a = dv/dt = d^2x/dt^2 = -\omega^2 A \sin(\omega t) \qquad |a_{max}| = w^2 A$$

where x = displacement, v = velocity, a = acceleration, A = displacement amplitude, ω = angular velocity, and t = time.

## 4.9   Blast Monitoring with Seismographs

A seismograph is a device that measures ground vibration levels. Most modern seismographs can measure peak ground displacement, velocity, and acceleration simultaneously in each of the three orthographic coordinates. They typically can also print out a graph of the vibrational wave as displacement, velocity, and acceleration versus time. Some units can simultaneously measure the air concussion overpressure and provide a graph of it correlated with the ground vibration measurements.

By placing a seismograph as close as practical to a structure on the side facing blasting operations, the actual ground vibrations to which the structure has been subjected can be measured and quantified. As long as the ground vibrations and air overpressures are less than some recognized minimum threshold value, it can be said that the building was not damaged during blasting operations. If the vibrations and air overpressures exceed the recognized minimum threshold value, it is said that the structure may have been damaged. If the vibrations then exceed some recognized maximum threshold value, it is presumed that the structure was damaged.

Many contractors routinely monitor ground vibrations in this way during blasting operations. The contractor may own his own equipment and monitor the blast, or the contractor may have an independent third party provide the equipment and monitor the blasting work. The latter method usually has greater credibility when there is a dispute.

During blasting operations, federal law requires that contractors maintain a log of what transpires. Routinely, the log contains the following information:

- location of blasting, charge pattern, and charge depth.
- location of nearest structures to blast site.
- size of individual charges and delay times.
- type of explosive, detonator, and delay devices.
- total amount of charge detonated.
- time and day of blasts.
- weather conditions, miscellaneous items, or problems.
- any seismograph measurements and the location of the seismograph at various times during operations.
- the person supervising blasting.

When seismographs are used, the seismograph automatically records the vibrational data and prints it out. The data will either be noted on the blasting log, or the print out will be attached to the log. When the blast monitoring is done by a third party, there are two records of the seismograph measurements: one maintained by the blast monitoring company, and one maintained by the blasting contractor, perhaps entered directly onto the blasting log at the time the measurement was made. This arrangement is an important hedge against data tampering or fraud.

When disputes occur as to whether a particular structure has been damaged by nearby blasting operations, the matter can often be quickly resolved by consulting the blasting logs, noting the actual vibrational levels that occurred, and comparing the measured vibration levels to accepted standards. Of course, this presumes the blasting logs are complete and accurate.

## 4.10  Blasting Study by U.S. Bureau of Mines, Bulletin 442

An extensive study was conducted by the U.S. Bureau of Mines between 1930 and 1940 to determine the effects of quarry blasting on nearby structures. This study culminated in the publication of *Seismic Effects of Quarry Blasting*, by Thoenen and Windes, Bureau of Mines, Bulletin 442, in 1942.

This study concluded that when the ground acceleration at a residential structure was 1.0 g or more, damage would result. Accelerations between 0.1 g and 1.0 g resulted in slight damage, and accelerations of less than 0.1 g resulted in no damage occurring to the residential structure.

A follow-up study to Bulletin 442 was published in 1943 by the U.S. Bureau of Mines, which was concerned about concurrent air concussion effects from quarry blasting. This study, entitled, *Damage from Air Blast*, by Windes, Report 3708, established the first overpressure thresholds for glass damage. This report indicated that an overpressure of less than 0.7 lbf/in² would result in no window damage, while overpressures of 1.5 lbf/in² or more would. This report also concluded that air concussion damage from blasting operations at quarries was not a significant problem.

## 4.11   Blasting Study by U.S. Bureau of Mines, Bulletin 656

Because of the increased need for objective and verifiable blasting damage assessment criteria, a ten-year study was undertaken by the U.S. Bureau of Mines in 1959 to determine the effects of blasting on nearby structures. Among other things, the study included the establishment of blast damage criteria for residential structures, the determination of blasting parameters that most affect ground vibrations, empirical "safe" blasting limits, and human response to blasting work.

The study included data from 171 blasts at 26 separate sites. Many rock types were included at the sites including limestone, dolomite, granite, diabase, schist, and sandstone, as well as sites with and without overburden. Simple as well as complex geological formations were included. Blasts included the detonation of charges as small as 25 pounds per delay to 19,625 pounds per delay with peak velocity amplitudes ranging form 0.0008 to 21 in/sec. Vibrational frequencies at peak velocities ranged from 7 to 200 hertz.

When the study was completed in 1971, the results were published by the United States Department of the Interior, Bureau of Mines, in Bulletin 656, *Blasting Vibrations and Their Effects on Structures*, by Nicholls, Johnson, and Duvall. This study has formed the basis for most blasting damage assessment guidelines, and is the basis for many legal standards relating to safe blasting practices. Bulletin 656 concluded the following.

- Damage to residential structures from ground vibrations correlate better to the peak velocity than to either peak displacement or peak acceleration.
- "Safe" blasting occurs when the peak velocity in any coordinate direction does not exceed 2.0 in/sec. This insures that the probability of damage from blasting is less than 5%.

**Table 4.1  Peak Velocity Ground Vibration Ranges**

| | |
|---|---|
| Safe blasting level | 2.0 in/sec or less |
| Minor damage level | 5.4 in/sec |
| Major damage level | 7.6 in/sec and higher |

- When seismographs are not being used, a scaled distance of 50 ft/lb$^{1/2}$ may be used as a "safe blasting limit" for ground vibration per the safe blasting formula.
- Air concussion does not contribute to the damage problem in most blasting operations. An overpressure of no more than 0.5 lbf/in$^2$ will prevent glass breakage.
- Except in extreme cases where there is a lack of proper stemming, the control of blasting procedures to limit ground peak velocity levels below 2.0 in/sec also limits overpressures to safe levels.
- Human response to ground vibrations, noise, and air concussion are significantly below those required to cause damage to a residential structure.

Bulletin 656 also established the following standards noted in Table 4.1 for evaluating the effects of peak velocity vibration ranges.

## 4.12  Safe Blasting Formula from Bulletin 656

When seismographs are not used to monitor ground vibrations from a blast, Bulletin 656 recommends the use of the "safe blasting formula."

$$D = K(W^{1/2}) \qquad (x)$$

where D = safe blasting distance from structure to blast hole, K = scaling factor, and W = charge weight per delay (in equivalent dynamite).

According to Bulletin 656, to insure that 2.0 in/sec is not exceeded, a scaling factor K = 50 ft/lb$^{1/2}$ is recommended.

As was noted before, it is the "S" waves, or surface shear waves that transmit the most blast energy through the ground and cause the most damage. These waves can be imagined as concentric rings traveling along the surface of the earth, similar to surface waves in a pond. Because of this, the intensity of the peak velocity ground vibration diminishes with the square of the distance. In other words, if the peak velocity is no more than 2.0 in/sec at 100 feet, it will be no more than 0.50 in/sec at 200 feet.

$$I = k/r^2 \qquad (xi)$$

where I = peak velocity intensity of vibration, k = arbitrary constant, and r = distance from blast hole to location of structure.

Using Equations (x) and (xi), it is possible to determine from the blasting logs whether the blasting was harmful to nearby structures even if seismographic monitoring was not done.

## 4.13   OSM Modifications of the Safe Blasting Formula in Bulletin 656

Since the publication of Bulletin 656, the original "safe" blasting formula and the "safe" blasting vibrational levels have both been modified by the Office of Surface Mining (OSM). The modifications are contained in the publication, *Structure Response and Damage Produced by Ground Vibration from Surface Mine Blasting*, by Siskind, Stagg, Kopp, and Dowding, and published by the United States Department of Interior Bureau of Mines in 1980. The modifications were done to further reduce complaints due to human perception, and to recognize that there is a higher potential for damage in residential structures when the low-frequency ground vibrations dominate.

At relatively close distances to the blast site, it has been found that the higher ground vibrational frequencies dominate. Because of attenuation effects, at greater distances, however, the higher frequencies drop out and the lower frequencies dominate.

For this reason, the allowable peak particle velocity vibration levels measured by seismographs at close range are actually higher than those allowed farther away, as noted in Table 4.2. This is also true of the "K" scaling factor, which is 50 ft/lb$^{1/2}$ for distances less than 300, and 65 lb$^{1/2}$ for distances greater than 5,000 ft.

Similarly, OSM recommends that the following peak ground particle velocity limits for specific frequency ranges be used when using seismographs to monitor blasting work. In general, the allowable peak particle velocities follow a step function related to frequency: low frequencies have low peak particle velocity thresholds.

However, whenever ground vibrational frequencies are verifiably 30 Hz or more, the original peak particle velocity criterion of 2.0 inches per second established in Bulletin 656 is allowed.

**Table 4.2   Modified "Safe" Blasting Criteria**

| Distance | Maximum Peak Velocity | "K" Scaled Distance Factor |
|---|---|---|
| 000–300 ft | 1.25 in/sec | 50 ft/lb$^{1/2}$ |
| 301–5,000 ft | 1.00 in/sec | 55 ft/lb$^{1/2}$ |
| 5,000 and more | 0.75 in/sec | 65 ft/lb$^{1/2}$ |

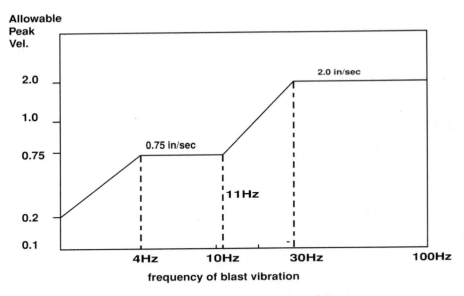

**Figure 4.3** OSMRE blasting level chart for seismograph monitoring.

When blasting ground vibrational levels are kept below the thresholds noted in Table 4.2, Figure 4.3, or below 2.0 in/sec when the frequency is verifiably higher than 30 Hz, or when the charge weight per delay does not exceed that indicated by the various applicable safe blasting formulas, the blast is usually considered safe to nearby structures.

## 4.14  Human Perception of Blasting Noise and Vibrations

One important point already alluded to with respect to public complaints about blasting work is that humans can readily sense vibrations at an order of magnitude or more lower than the level necessary to affect or damage buildings. People often erroneously presume that a noise loud enough to hurt their ears or a ground vibration large enough to allow them to feel the shake is indicative of a blast powerful enough to cause damage to a structure. Many claims of blasting damage are solely predicated upon the homeowner hearing or feeling blast vibrations that felt "uncomfortable" or "really large" on a personal basis.

As Bulletin 656 pointed out, the human body is capable of detecting vibrations that are no more than 1% of that required to cause damage to a building. Human perception of vibration is dependent upon both the frequency and the peak displacement. To a human, a vibrational displacement

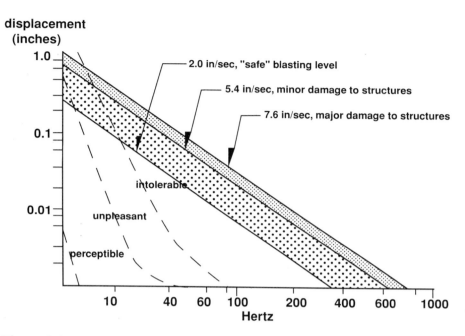

**Figure 4.4** Human perception and ground vibration levels.

of 0.06 inches at 1 Hz is perceptible; at 4 Hz it is unpleasant; and at 10 Hz and greater, it is intolerable. To a building, however, none of these circumstances will produce damage because all are well below the safe blasting criterion vibrational established by Bulletin 656.

Figure 4.4 depicts an approximate comparison between human perception of ground vibrations and the various levels at which they affect buildings. Note that the human perception of ground vibrations being "intolerable" does not cross the line where damage to a structure actually occurs unless the ground vibration level is below 4 Hz.

When an air blast sound or ground vibration feels unpleasant or intolerable to a person, it is likely that the person will fear and presume that his residence has been damaged. He may then examine the residence in great detail, perhaps for the first time in many months or years. When he finds cracks that he does not recall, which is likely if he has not inspected the structure at regular intervals, he presumes that the cracks were caused by the blasting work which he personally felt and heard to be quite severe.

Thus, it is predictable that when blasting operations are loud, or when they produce ground vibrations with displacement and frequency combinations that are unpleasant or intolerable to humans, the number of complaints and damage claims will be high.

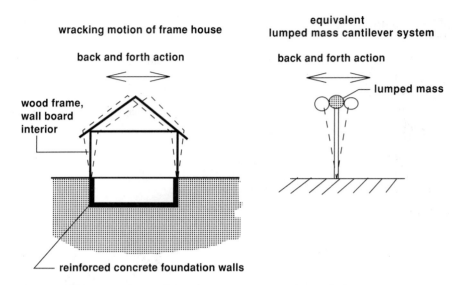

**Figure 4.5** Equivalent lumped mass cantilever to frame house.

## 4.15   Damages Typical of Blasting

Most typical residences in the Midwest have wood frame upper structures set on reinforced concrete foundation walls, slabs, or basements. With respect to back-and-forth wracking motion, this type of residence can be modeled by a lumped mass at the end of a cantilever beam, as shown in Figure 4.5.

If the top of the house in Figure 4.5 were pushed by an arbitrary force to the right a small amount, it would respond as follows:

$$F = kx \text{ or } x = F/k \qquad \text{(xii)}$$

where F = small force, parallel to ground, applied to top of house, k = spring constant of house at the top, and x = amount of displacement in response to applied force.

If it is assumed that the cantilever beam is continuous, that it is firmly fixed into the foundation, and that the force can be applied anywhere along its length, then standard engineering handbook equations indicate that the deflection is a function of the following:

$$x = [F/EI][(lc^2/2) - (c^3/6)] \qquad k = EI/[(lc^2/2) - (c^3/6)] \qquad \text{(xiii)}$$

where $l$ = full length or height of the cantilever, c = is the height at which the force is applied, E = Young's modulus for the material in the cantilever, and I = second moment of area of the beam cross-section.

If the force is only applied at the end or top of the cantilever, then the deflection is given as follows:

$$x = [F][(l^3)/(3EI)] \qquad k = 3EI/l^3 \qquad\qquad (xiv)$$

If it is assumed that the house structure has no damping, and the top of the house is displaced slightly to the right and released, then the top of the cantilever will have the following motion:

$$x = A \sin(\omega t) \qquad\qquad\qquad (xv)$$

where x = displacement, A = maximum displacement, t = time, and $\omega$ = natural angular velocity of the cantilever.

$$\omega = [k/m]^{1/2} = 2pf$$

where m = the lumped mass at the end of the cantilever, k = spring constant, and f = natural frequency of the cantilever in Hz.

It is notable that the maximum back-and-forth motion occurs at the end or top of the cantilever, and not at the middle or bottom. At the bottom, where the cantilever embeds into the foundation and is fixed, there is no relative motion between the cantilever and the ground.

In a frame structure with a rigid roof truss, if the walls are considered to be fixed to both the foundation and the roof truss, it can be shown by the application of Rayleigh's energy method (cited at the end of the chapter) that the lower 3/8 of the walls are not even kinetically involved in the back-and-forth movement of the roof. This means that in a frame house, if any damage has occurred due to ground vibrations, nearly all of it will have occurred in the upper half of the house.

Assume now that the ground motions produced by a nearby blast are a sinusoidal forcing function for the above cantilever system, at least for the first "wave." Assume also that due to damping, the first vibrational wave will produce the largest deflection at the end of the cantilever, and therefore will cause the primary damage to the structure. Under these circumstances, the response of a frame structure, as shown in Figure 4.4, to destructive ground vibrations from nearby blasting work can be modeled as follows.

$$m[d^2x/dt^2] + kx = Q\sin(\theta t) \qquad\qquad (xvi)$$

$$x = Q\sin(\theta t)/m[\omega^2 - \theta^2] \qquad x_{max} = Q/m[\omega^2 - \theta^2]$$

where Q = force applied to the cantilever by the forcing function, i.e., the ground vibrations, $\theta$ = angular velocity of the forcing function, i.e., the

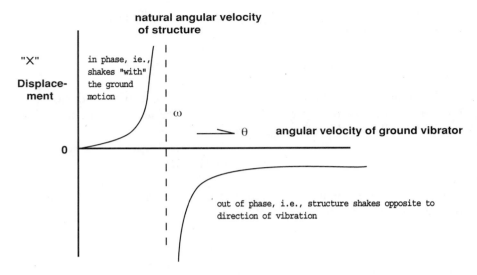

**Figure 4.6** Relationship of displacement of structure to blast frequency.

dominant angular velocity of the ground vibrations, x = movement at the top of the house, i.e., at the end of the cantilever.

Simply put, the above analysis indicates that when there are lateral ground vibrations from nearby blasting work, i.e., shear waves, vibration levels in a structure will be highest when the dominant vibration frequency of the blast is close to the natural frequency of the house walls (i.e., the cantilevers). The closer the two frequencies are to each other, the greater the displacement at the top of the walls will be. Figure 4.6 depicts this.

From experimentation in the field, it has been found that the natural frequency of one- and two-story residential structures varies from about 3 to 12 Hz. Two-story structures, of course, have lower natural frequencies than similarly made one-story structures.

As noted before, because of attenuation factors, low frequency vibrations from blasts tend to dominate at longer distances from the blast site. Thus, the need for OSM's modifications to the original safe blasting criteria contained in Bulletin 656 is evident in the solution of Equation (xv). Low-frequency ground vibrations near the natural frequency of a residential structure are more damaging than high-frequency ones.

The foundations of most modern homes are usually made of thick concrete slabs with steel reinforcement bar embedded into the concrete. The upper areas of frame houses, however, are structurally composed of braced 2 × 4 wood beams, window frames, exterior siding, and gypsum board wall panels.

Thus, if a frame structure has been subjected to damaging ground vibrations, theoretical considerations indicate that damage to the uppermost areas of the structure should be greatest, especially in relatively fragile materials like plaster and glass. Broken knickknacks that have fallen from shelves, broken or cracked windows, or pictures hanging on walls that have fallen should be the first things affected in the upper areas of a residential structure if there are ground vibrations. The absence of these affects indicates the absence of ground vibrations.

Similarly, damage to the lowest level of a residential structure by ground vibrations should generally be zero or very limited, especially in relatively strong materials like reinforced concrete. In point of fact, this expectation was borne out in field tests and observations as reported in Bulletin 656.

Thus, actual wracking damages caused by blasting ground vibrations generally follow a "top-down" pattern of severity; the most severe damages occur at the top of the structure, and the least severe damages occur at the bottom. If the claimed damages do not follow this "top-down" pattern of severity, it is likely that they were not caused by nearby blasting work.

## 4.16  Types of Damage Often Mistakenly Attributed to Blasting

The most common deficiencies that are blamed on nearby blasting work are basement floor and wall cracks, and cracks in plaster or gypsum board walls, especially at corners and headers. The second most common deficiencies blamed on nearby blasting work are sunken floors, stuck doors and windows, and nail pops. Occasionally, problems in septic tanks and sewer lines, leaking well casements, nonfunctioning well pumps, deficient soil percolation in septic systems, and water table changes are also blamed on blasting work.

If the blasting work has been recent, a visual inspection inside the cracks in question may be useful in determining the age of the cracks. Cracks that have been present for some time may have paint or plaster drippings inside the fissure. These paint or plaster drippings may have flowed into the fissure by capillary action when the room was painted or plastered. For example, if both red and white paint is found in a basement wall crack and the wall is presently painted white, it is obvious that the crack has been present since before the wall was painted white, when it was still red.

Dead bugs, accumulated dirt and debris, tree roots and plant tendrils that have worked in from the outside, and other such items that could not have come to be inside the crack in the relatively short time span since the blasting work was completed can also be useful in dating the crack.

Both concrete and plastic will oxidize and absorb moisture after fresh fracture surfaces are exposed. This will usually cause the surface of an old fracture to appear darker and less bright than a fresh one. It is often useful to deliberately make a small chip in the concrete or plaster near the crack in question to directly observe the color of a fresh fracture. This fresh fracture surface can then be visually and photographically compared to the exposed fracture surfaces in the crack in question. The use of a Munsell® color chart, the same chart used by the USDA to classify the color of soils and gravels, can be very useful in distinguishing and documenting the variations in color of fracture surfaces in concrete, plaster, and gypsum board.

It may be useful for the inspector to document the color of a fracture surface within a crack during the initial inspection. During the same initial inspection, he can also deliberately make a fresh crack surface. Allowing about the same amount of time to pass as occurred from the end of blasting operations to the initial inspection, a follow-up inspection can then be made to see if the color of the fresh crack has changed over time to that observed in the crack in question during the initial inspection.

A pattern of cracking in walls and ceilings that appears most severe at the lowest level of the house and then diminishes in severity toward the top of the house is often indicative of foundation or settlement problems. When the foundation settles or shifts position, the whole weight of the house bears down on the structural elements at the lowest level, while the uppermost levels bear the least weight. The widest and most severe cracks observed in the exterior or interior of a residential structure often occur nearest the causative displacement.

Because most residential structures have foundations made of concrete or similar nonflexible building materials, the greater bearing loads impressed on the lower level components often create excessive bending or shear stresses, even when differential settlement is slight. Concrete and cement blocks just don't bend very well. Their resistance to shear is also relatively poor, especially when it is generally intended that they carry their loads in bearing.

Cracks in a residential structure caused by blasting work ground vibrations will usually exhibit characteristic bending or shear fracture features due to the wracking motion of the house. Cracks that exhibit pure compressive mode characteristics are not due to ground vibrations. They are caused by bearing load inadequacies: perhaps too much bearing load, or not enough bearing load support. Ground vibrations do not cause a structure to significantly increase its bearing load. In other words, shaking a house doesn't cause it to gain weight.

The following is a brief, and certainly not exhaustive, list of deficiencies in a residential structure that are commonly mistaken for having been caused by blasting damage by the owners.

- Shrinkage and expansion of low grade lumber that was not properly kiln dried. When used for structural components, it causes wall cracks and nail pops. Wood beams, joists, and trusses intended for structural uses will be stamped with their grade.

- Foundation settlement problems. This includes poor foundation soil preparation, poor foundation design, and differential settlement. Often, settlement problems will not be evident until a drought occurs, an unusually wet period occurs, or there is a change in the local drainage pattern. This can be many years after the house has been constructed.
- Substandard concrete in the foundation walls, floors, and slabs. Cracks in concrete walls are sometimes due to a low strength concrete, a concrete with too much water in it, a concrete with contaminants, or a low density concrete that has too much entrained air in it. To guard against this in new homes, reputable contractors will do slump tests of the concrete when it is delivered and provide records of such slump tests to the owners when the house is purchased. When there has been too much water in the concrete mix of a floor, the floor surface will appear "alligatored," crazed, or mottled.
- Basement cracks and water leakage due to the drainage water from roof gutters not being carried away from the building by splash blocks or down spout conduits. The excess water piling up against a foundation wall can cause settlement, heaving, and freeze/thaw damage. If there is any sort of crack and there is enough water available, eventually the water will find a way in.
- Significant changes in soil moisture content around the foundation footings. Footings too shallow. Footings in frost zone. Footings set on fill dirt that was not properly compacted.
- A higher amount of weight in the house than the floors and supports of the house can carry. Sometimes homeowners overload floors and foundations, such as when a house is used to store excessive amounts of office filing cabinets, or home business equipment.
- Water damage that has caused damage to plaster, caused wall beams, floor joists, or roof trusses to warp or swell.
- Poor water/plaster mixtures in the finishing "mud" that cause shrinkage cracks. Poor water mixtures also cause shrinkage cracks in stucco and other cementitious materials.
- Wall or ceiling plaster applied too thin.
- Not enough nails or glue nails used to properly affix structural items.
- Floor joists improperly spaced or rafters improperly spaced.
- No lintels over windows. Lintels the wrong size, allowing undue flexing. No headers over doors or windows.
- Leaking water pipes. Water from leaking pipes in crawl spaces can often undermine foundation footings and pillars causing them to shift position.

- Improper application of joint tape and mud to wallboard joints and corners.
- Use of "green" concrete blocks that have not properly cured. When set in position, the blocks will continue to cure and shrink. If the shrinkage is sufficient, cracks will appear through the blocks or the blocks will pull away from their mortared joints.
- Significant changes in the local runoff and drainage around the foundation footings.
- Excessive passive lateral pressure exerted by expansive clay soil formations on basement walls and floors. This causes the walls to bulge inward, and sometimes causes the floors to heave upward. This is a common problem in block foundation walls, especially those with inadequate drainage to prevent hydrostatic pressure buildup.

## 4.17 Continuity

As was noted earlier in the chapter, ground vibrations emanating from a blast site and traveling to nearby structures are primarily transmitted by shear waves, or "S" waves, which propagate along the surface. Therefore, for a structure to have been affected by such shear waves, it is necessary that the surface between the structure and the blast site is continuous.

For a surface to be continuous, there should be no significant breaks in the surface or significant changes in composition. Examples of what constitutes a significant break in surface continuity include the following:

- ravines, gulches and canyons.
- rivers and deep creek beds.
- deep ditches.
- cliffs.
- buttes.

In general, when an "S" wave encounters a break in the surface, it will partially reflect at the surface of the break or discontinuity. This is similar to the way that when a rope is attached to a door handle and a wave at the held end is created, the wave travels down the rope and "goes around" the rope at the door handle and travels back to where it is held.

A portion of the wave energy may also be expended by dislodging loose materials at the surface of the break. Dirt and small rocks may be seen to "pop off" the side of a ravine when a blasting ground wave impinges on it with sufficient intensity.

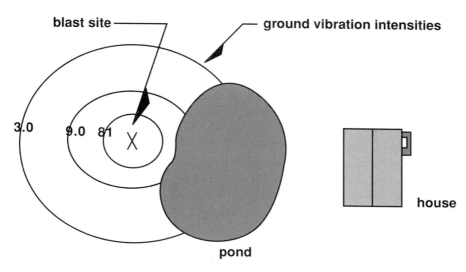

**Figure 4.7** Intervening deep pond shielding house from ground vibrations.

"S" waves cannot "jump" through the air from one side of a discontinuity to the other. Significant discontinuities effectively stop the propagation of "S" waves. Thus, if a significant discontinuity exists, such as a ravine or cliff, between the blast site and the structure in question, the structure will not be subject to ground vibrations.

It is possible for an "S" wave to go around a relatively small discontinuity. However, small discontinuities in the surface medium weaken the intensity of the vibration and disrupt propagation. The portion of the wave that directly impinges on the small discontinuity reflects back, leaving the rest of the wave front to bend and "fill in" the portion of the wave front lost by reflection. Many such small discontinuities clustered together can effectively act as a large discontinuity and wholly block transmission of ground vibrations.

Significant changes in surface composition can also change the way in which "S" waves are transmitted along the surface. In general, fluid or fluid-like materials rapidly dampen out the transmission of "S" waves. For all practical purposes, bodies of water like lakes or rivers can be considered to be nontransmitting media for "S" waves. Likewise, significantly deep saturated soils, wet sand, bogs, swamps, or marshes quickly dampen out "S" waves.

Figure 4.7, shows an example of a house that is shielded from nearby blasting vibrations by an intervening deep pond.

For this reason, it is usually a good idea to walk or reconnoiter the ground surface between the point where blasting occurred and the structure in question.

## Further Information and References

"Air Blast Phenomena, Characteristics of the Blast Wave in Air," reprint of chapter II from, *The Effects of Nuclear Weapons*, by Samuel Glasstone, Ed., April 1962, U.S. Atomic Energy Commission, pp. 102–148. For more detailed information please see Further Information and References in the back of the book.

*An Introduction to Mechanical Vibrations*, by Robert Steidel, Jr., John Wiley & Sons, New York, 1971, pp. 65–70, 83. For more detailed information please see Further Information and References in the back of the book.

*Blasting Damage — A Guide for Adjusters and Engineers*, American Insurance Association, 1972. For more detailed information please see Further Information and References in the back of the book.

*Blasting Vibrations and Their Effects on Structures*, by Nicholls, Johnson, and Duvall, United States Department of the Interior, Bureau of Mines, Bulletin 656, 1971. For more detailed information please see Further Information and References in the back of the book.

*Damage from Air Blast*, by Windes, Bureau of Mines Report of Investigation 3708, 1943. For more detailed information please see Further Information and References in the back of the book.

*Engineering Analysis of Fires and Explosions*, by R. Noon, CRC Press, Boca Raton, FL, 1995. For more detailed information please see Further Information and References in the back of the book.

*Explosive Shocks in Air*, by Kinney and Graham, Springer-Verlag, 2nd ed., 1985. For more detailed information please see Further Information and References in the back of the book.

*Investigation of Fire and Explosion Accidents in the Chemical, Mining, and Fuel Related Industries — A Manual*, by Joseph Kuchta, United States Department of the Interior, Bureau of Mines, Bulletin 680, 1985. For more detailed information please see Further Information and References in the back of the book.

*NFPA 495: Explosive Materials Code*, National Fire Protection Association, 1990 Edition. For more detailed information please see Further Information and References in the back of the book.

*Seismic Effects of Quarry Blasting*, by Thoenen and Windes, Bureau of Mines, Bulletin 442, 1942. For more detailed information please see Further Information and References in the back of the book.

# Building Collapse Due to Roof Leakage

# 5

## 5.1  Typical Commercial Buildings 1877–1917

In the period from post-Civil War Reconstruction to about World War I, many of the small and medium towns that now dot the Midwest were settled and built. In general, the sequence of town building in each case was similar. Homesteads were built first, which congregated around crossroads, river fords or ports, depots, mines, sawmills, or other natural points of commercial activity. This was followed by the construction of temporary tent and wood-frame commercial buildings. When business activity was sufficient, the temporary wood-frame buildings were replaced by more permanent masonry commercial buildings. Many of these masonry buildings still populate the original downtown areas.

In general, these permanent masonry commercial buildings followed a similar structural design. A typical front elevation is shown in Figure 5.1. Most were made of locally quarried stone or locally fired brick. It was com-

**Figure 5.1** Typical two-story commercial building with load bearing, masonry side walls.

mon for the interior wall to be stone and then faced with brick on the exterior side. Sandstone and limestone were commonly used. The mortar for both the stonework and brickwork was generally quicklime from local kilns, and sand from local riverbeds or deposits. The front width of the building was usually 1/2 to 1/4 the length of the building. The long side walls carried most of the structural load.

The roof was simply sloped at a low angle. The high side of the roof was at the front and the low side at the back, as shown in Figure 5.2. Usually, the front of the building had a large facade parapet wall for advertising, as is shown in Figure 5.1. The sides also usually had parapet walls. However, the rear portion of the roof usually did not have a parapet wall. Thus, rainwater

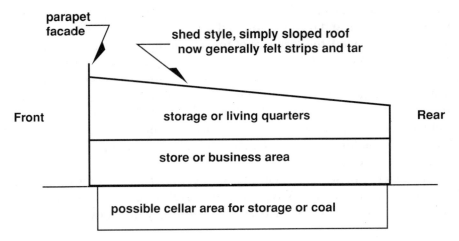

**Figure 5.2** Cutaway view showing roof slope and floor usage.

drained from the front and sides of the building toward the rear. When the buildings were first built, there would often be rain barrels at the rear corners of the building to catch the runoff.

The original roof was usually a wooden board deck over simple wood beam roof joists. The deck would be covered with various layers of felt and bitumen, roofing rolls, or in some cases, overlapping galvanized metal sheeting. In recent years, many of these roofs have been converted to conventional tar and gravel built up roofs (BURs), or rubber membrane roofs.

The floor and roof decks were generally supported by simple wood joist beams. When it was necessary to splice joists together to span the distance across the side walls, the joists would be supported in the middle by a beam and post combination. Splices were often accomplished by overlapping the two pieces where they set over the support post, and nailing or bolting them together. In many cases, however, the joists were simply overlapped and set side-by-side on top of the post with no substantial fasteners connecting them. It was presumed that the decking or flooring nailed to their upper surface would hold them in place. Additionally, the joists were usually, but not always, side braced to ensure they would stay vertical.

The floor and roof joists were supported at the ends by the side walls. A bearing pocket would be created in the side wall, and the end of the wood joist was simply set into the bearing pocket. Often, the end of the joist was mortared into the bearing pocket so that it would be rigid and vertical.

Usually the bearing pocket only extended about halfway through the thickness of the wall. In some buildings, however, the bearing pocket would go all the way or nearly all the way through the side wall, such that the ends of the floor and roof joists could be seen from the outside. To keep the ends of the joists from weathering, the butts would be covered with mortar, tarred, or painted. Roof decking and floor decking were nailed directly on top of the joists. Figure 5.3 shows the basic structural support system as described.

## 5.2 Lime Mortar

Lime or quicklime for the mortar in these buildings was generally obtained by roasting calcium carbonate in kilns. The calcium carbonate may have come from local limestone deposits, chalk deposits, or even marble deposits. The fuel for the roasting process was usually wood and charcoal, although coal and peat were occasionally used when it was convenient.

The chemical process for converting calcium carbonate to quicklime is simple. The raw material is heated in a kiln to above 1000°F, usually 1400 to 2000°F. At this temperature, the calcium carbonate decomposes into calcium oxide and carbon dioxide. The liberated carbon dioxide may then sequen-

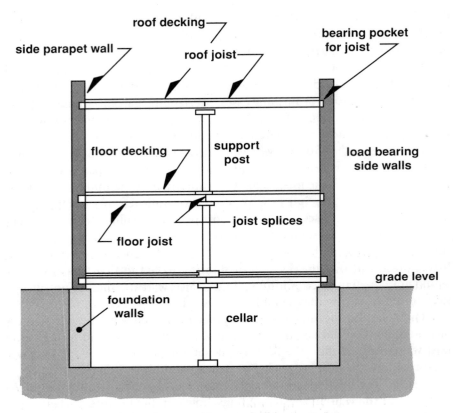

**Figure 5.3** Basic structural support system.

tially react with residual moisture, i.e., steam, in the kiln's flue gas atmosphere to form carbonic acid. The carbonic acid precipitates out as a liquid when the flue gases cool below the dew point.

In the environmentally unsophisticated lime kilns built prior to World War I, flue gases were discharged directly into the air without much thought. Typically, a kiln flue stack was not very tall and could not "punch" the flue gases very high to disperse the precipitates. Thus, the carbonic acid vapor in the flue gases usually precipitated out within a short distance downwind of the kiln. Such precipitated carbonic acid often caused significant damage to nearby plant life, especially if the kiln were operated for several years in the same spot.

The basic chemical equation for the conversion process is given in Equation (i).

$$\underset{\text{heat}}{\overset{\text{calcium carbonate} \quad\quad \text{calx} \quad \text{carbon dioxide}}{CaCO_3 \;\rightarrow\; CaO + CO_2}}$$  (i)

The residue product of the process, calcium oxide, is whitish or grayish-white in color with a specific gravity of 3.4. Iron impurities in the raw material charge can cause the calcium oxide to have a yellowish or brownish tint. Other names for the calcium oxide residue from the reaction include calx, quicklime, burnt lime, or unslaked lime.

Calcium oxide readily reacts with water to form calcium hydroxide and in doing so, produces an ample amount of heat. This process is called slaking. When the slaked lime is noted to form steam, water must be added and the resulting quicklime paste must be thoroughly mixed to stop steam from forming.

$$CaO + H_2O \rightarrow Ca(OH)_2 + heat \qquad (ii)$$

Calcium hydroxide, or slaked lime, is a white-colored material with a specific gravity of 2.34, which is a weight density of about 146 pounds per cubic foot. It loses water at 1076°F, and is slightly soluble in water. Most chemical handbooks report that the solubility of calcium hydroxide in cold water is 0.185 grams per 100 ml of water. It reacts readily with most acids, and absorbs carbon dioxide from the air.

The usual recipe for making quicklime mortar in the era being discussed consisted of two to three volumes of sand to one volume of lime paste. Lime paste was generally two parts by weight of water to one part quicklime. Sometimes the lime mortar would be strengthened by the addition of cement, especially Portland.

These mixing ratios and recipes, however, were simply common rules of thumb. Since there were no quality control standards or building codes, conditions and mixture compositions varied greatly. Incomplete decomposition of the limestone in the kiln, or contamination of the quicklime by combustion materials could greatly affect mortar strength. Further, most masons added water as they saw fit for "workability."

The sand used in the mixture was usually not sieved for size consistency. Thus, the sand could be well-graded, gap graded, or uniformly graded and have significant amounts of silts and fines mixed in. Mortars containing sand of which 48% is able to pass through a #100 sieve, for example, have only about 1/10 the compressive strength of one in which only 5% passes through a #100 sieve.

Even when good practice was adhered to, lime mortar is inferior to Portland cement type mortar mixtures. A lime mortar with no cement additives, as described previously, will develop a tensile strength of at most 26 pounds per square inch after 84 days of curing . This is about 1/10 the tensile strength that a modern concrete mixture will develop after 28 days. Because of its slow hardening rate and poor strength, lime-based mortars are not used in standard U.S. construction anymore except for interior, nonload-bearing walls.

## 5.3   Roof Leaks

While some of these buildings have been well maintained over the years by their owners, most of them have not, especially their roofs. Water leakage through the roof membrane is a common maintenance problem. The most common point of leakage in the roof membrane is the parapet wall flashing where the roof decking abuts the parapet walls.

Often, the flashing along the parapet wall simply dries out and cracks due to ultraviolet light exposure, weathering, and age. Sometimes the membrane will have been broken or penetrated by foot traffic on the roof. Sometimes the flashing membrane simply pulls away from the wall due to a combination of age-related material shrinkage, and weather-deteriorated tar, mastic, or sealing compound. Whichever is the case, deficient flashing allows rainwater to enter the building next to a load-bearing wall. The rain water then runs down the interior side of the wall.

Because there is often a stud wall cavity between the interior side of the masonry wall and the interior finish of the building, the leakage can go unnoticed for years if the water simply "hugs" the interior side of the masonry wall. The water may not cause any noticeable staining of the interior walls or ceilings because the interior finish may not even get wet despite extensive leaking. Further, because many of the buildings now have been outfitted with drop ceilings, wall paneling, and fixtures that obscure the original interior finish, any stains on the original walls and ceilings that may have occurred may not be observable. Thus, if regular inspections of the roof are not done to check for roof membrane continuity, leakage can go undetected for years.

This is especially true with the, "if it ain't broke, don't fix it," mentality of some businesspersons. They often do not spend any time or money on preventive maintenance items. They often wait until an item actually fails or creates a crisis before doing anything. Even then, they may only do the most minimal of repairs to keep the immediate maintenance expenditures low. The practice is termed deferred maintenance.

## 5.4   Deferred Maintenance Business Strategy

For the short term, a deferred maintenance business strategy can be beneficial to the current owner. By deferring maintenance expenditures and selling the building before any major maintenance work must be done, the current owner of the business can essentially transfer the costs of major maintenance to the future owner or perhaps an insurer.

For example, a building owner may know that the flashing of his roof is in poor condition and is likely leaking now, or will soon leak. The owner is well aware that such leakage can cause structural damage to the building over time. However, the owner may decide to do nothing about the problem because it is not causing any immediate unsightly staining of the building's interior, it is not damaging any stored inventory, and because the owner does not wish to spend money on preventive maintenance. The owner defers the maintenance and pockets the money that would have otherwise been spent.

If the roof flashing leaks as expected, and the leakage damages the material integrity of the wooden roof joists, the ability of the roof to carry the required dead and live loads diminishes. If the integrity of the wooden joists diminishes sufficiently, the roof may eventually collapse when the next heavy snowfall, heavy storm, or freezing rain occurs.

At that point, the owner of the building may be able to make a property loss claim on his insurance under the provisions for sudden collapse. Any inventory damaged in the collapse and any damages to the interior finish resulting from the collapse may also be covered by the insurance policy. Further, the loss of business activity that would have occurred if the business had not sustained the collapse may be claimed if the owner carried business interruption insurance.

If the owner is successful in collecting on these claims, he will have essentially transferred the major costs of maintenance of the roof to the insurance company. Further, the owner will have also transferred the inherent risks assumed by such a deferred maintenance strategy to the insurer. In essence, the structural collapse policy on the building becomes a major maintenance contract.

If the building owner is particularly astute, not particularly honest, and the building is located in an area where building code enforcement is poor, the owner may further profit from the above situation. Most insurance companies provide sufficient money, less any deductible, to properly repair the collapse damage. They will have estimates made by reputable contractors who are required to make repairs in accordance with good practice and applicable building codes. The insurer will then issue a check to the owner in an amount commensurate with the estimates so that the owner can have the work done by whichever reputable contractor he or she chooses.

However, if the building owner personally manages the repair work and the repairs are done "on the cheap" rather than as estimated, the owner can pocket the difference. When the owner then sells the building, he can claim that the building has a brand new roof and a remodeled interior despite the fact that perhaps neither was constructed according to code or good practice.

In the long term, the practice of deferred maintenance substantially increases maintenance costs severalfold. In a sense, deferred maintenance is like deferred loan payments. Not only do they accumulate into a harder-to-afford large sum that eventually has to be paid anyway, but they also generate interest. If a roof is promptly repaired, only the roof costs are expended. If a roof is not promptly repaired, money must still be spend on repairing the roof itself, and additionally on all the items that have become water damaged as a result of the roof work being deferred. Sometimes the resulting water damages severely reduce the structural integrity of the building, and therefore its usable life. This, of course, is another additional cost directly attributable to the deferred maintenance.

Unfortunately, in the short term, the practice of deferred maintenance can make economic sense to the owner when the costs and risks can be transferred to other parties. Thus, even though deferred maintenance increases the overall costs in the long run, if the current owner doesn't have to pay them and can find an unsuspecting "patsy" who will, the owner can save or even profit by employing the strategy.

## 5.5   Structural Damage Due to Roof Leaks

With respect to the type of building being discussed, water leakage through the roof causes two main types of structural damage:

- weakening of wooden roof and floor joists.
- weakening of the load-bearing walls.

Most people are familiar with the usual ways in which leaking water damages wood. It provides needed moisture for colonies of bacteria and fungi. The bacteria and fungi initially establish themselves on the surface of the damp areas, and then go on to digest the organic materials within the wood itself. The resulting wood rot, or rust, is often black, brown, or white in color.

Water leakage can also cause structural wood timbers to swell and soften. Structural-grade wood that has been properly kiln dried usually has a moisture content of 8–12%. Green wood, which is significantly weaker, less dense, and less stiff, can have a moisture content as high as 40%.

Wood is hygroscopic, which means it readily incorporates moisture into its cellular structure when such moisture is available. If structural timbers are carrying load when they absorb excess moisture, they can permanently deform in response to the loads. This is, of course, the basic woodworking method for shaping straight wood pieces into curves.

These deformed timbers can cause structural load and deflection problems due to second moments associated with eccentric loading. Second moments are those bending moments created when a beam or column is misshapen and the applied forces are no longer centered or symmetric. The eccentrically placed loads cause the beam or column to twist or flex. Trusses with significant second moments, for example, might no longer form a plane but instead flex into a saddle-shaped curve.

Furthermore, wood that has absorbed excess moisture and become softened may allow nails and bolts that are carrying load to loosen or pull out. The loss of fastener integrity can also cause a significant share of structural headaches.

Both of these effects, wood rot and distortion, are readily observable by inspection, and would be expected to occur where the water directly impinges on the wood over a period of time. These areas will also likely be stained by minerals and dissolved materials carried by the water. These minerals and dissolved materials are picked up by the water as it percolates through the building, and are then deposited on the wood during the various wetting and drying cycles.

A less familiar way that water leakage damages roof and floor joists in these older buildings is by chemical attack. Water that makes its way to the bearing pockets reacts with the lime mortar surrounding the wood. The calcium hydroxide in the mortar is soluble in water, and forms an aqueous, caustic solution. The pH of the solution at room temperature may be as high as 12.4, which is more than sufficient to attack the wood embedded in the bearing pocket.

When this occurs, the affected surfaces become discolored, lose mass, and appear dimensionally reduced. Material around the bearing surface, usually on the bottom side of the joist, loses strength and flattens out due to compressive failure. Because of this, the joist will often drop down in the bearing box or become loose within it. This may allow the beam to twist or tip.

With respect to the load-bearing walls, structural weakening by leakage from the roof is accomplished in several ways. The primary way is the leaching of calcium hydroxide, the primary binding ingredient of the mortar, from the mortar by water that percolates down the wall from the roof.

As noted before, about 0.185 grams of calcium hydroxide will dissolve in 100 ml. of water at room temperature. When a roof leak has been present for many years, the water running down the interior side of the wall steadily dissolves the slaked lime from within the mortar. Since the interior side of the wall is not seen by anyone, the mortar is usually rough, unfinished, and unpainted. This makes it easy for the water to wet the mortar, as compared with the exterior side. On the exterior of the building, the mortar is finished and has minimal surface area for water to contact. The exterior wall surface may even be painted, which further protects the mortar from leaching damage.

With the calcium hydroxide dissolved from the mortar, the mortar becomes a very weak, porous material. If leaching damage is sufficiently severe, the wall essentially becomes a pile of loose bricks or stones held together by a slightly sticky sand with a high voids ratio.

It is easy to verify in the field if a wall has been damaged in this way. First, the water stains on the wall will be readily apparent. Secondly, mortar that has been damaged in this way can be easily removed from between the stones or bricks by a penknife, metal key, or even a person's fingers. The mortar will flake and crumble easily, like porous sandstone. In severe cases, it is even possible for a person to remove whole stones or bricks from the wall by digging them out with his fingernails.

## 5.6  Structural Considerations

The deleterious effect that calcium hydroxide leaching from the mortar has on the structural stability of the wall can be assessed mathematically. This is done by considering the equation that describes the elastic stability of a thin plate under compression, and presuming that it is analogous to the masonry wall under consideration. According to Roark (see references at end of chapter), the critical stress at which buckling occurs when a thin plate is uniformly loaded in compression along two parallel edges is as shown in Equation (iii).

$$\sigma_{crit} = K[E/(1 - v^2)][t/b]^2 \qquad \qquad \text{(iii)}$$

where t = thickness of wall, b = length of wall, a = height of wall, E = Young's modulus, v = Poisson's ratio, K = factor to account for ratio of height to length and end conditions of plate, and $\sigma_{crit}$ = compressive stress at which instability occurs.

The thin plate modeled by Equation (iii) is presumed to be homogenous and isotropic throughout, while the load-bearing walls under consideration are composed of discrete units of brick, stone, and mortar. Despite these differences, however, Equation (iii) suffices to show the overarching principle that applies in this case, especially if the weakest material values in the wall are presumed to apply in the formula.

It is apparent by inspection of Equation (iii) that significant changes in Young's modulus, "E," greatly affect the critical stress at which buckling in the thin plate occurs. If we compare the condition of the wall before leaching occurs, to that after leaching occurs and presume that the dimensions and end clamping conditions of the wall have not changed, then the following is true.

**Table 5.1  Young's Moduli for Brick, Stone, and Soil**

| Material | Typical "E" Value |
| --- | --- |
| Limestone | 7 to $8 \times 10^6$ lb/in$^2$ |
| Sandstone | $\sim 3 \times 10^6$ lb/in$^2$ |
| Brick | 3 to $4 \times 10^6$ lb/in$^2$ |
| Concrete | $\sim 2 \times 10^6$ lb/in$^2$ |
| Soil (unconfined) | 0 to $10 \times 10^3$ lb/in$^2$ |

$$\sigma_{crit}/\sigma_{crit}' = [E/(1 - v^2)]/[E'/(1 - v'^2)] \qquad \text{(iv)}$$

where the prime mark (') denotes the condition of the mortar after leaching has occurred.

Since Poisson's ratio for most masonry materials is approximately 0.25, Equation (iv) simply reduces to the following ratio.

$$\sigma_{crit}/\sigma_{crit}' = E/E' \qquad \text{(v)}$$

Table 5.1 shows some representative values of Young's modulus for brick, stone, and soil.

In relating the values in Table 5.1 to Equation (v), it is obvious that if the mortar loses significant stiffness by leaching and essentially becomes equivalent to something between a soil and porous sandstone, the wall will be structurally weakened, perhaps becoming sufficiently unstable for buckling to initiate.

This structural problem with the walls is further exacerbated if there has been chemical attack of the wooden joists in the bearing pockets. Since most of these walls were constructed without formal tie-ins, the wooden roof and floor joists act as tie-ins to the rest of the structure and help stabilize the walls. If the wall is considered analogous to a Euler column with respect to buckling, tie-ins divide the wall into smaller columnar lengths. This strengthens the wall against buckling. The joist tie-ins also improve the end conditions of the column sections, which further strengthens the wall against buckling.

Euler's formula for column buckling is given below in Equation (vi). The reader will likely note the basic similarity of Equation (vi) to Equation (iii) taken from Roark's text, *Formulas for Stress and Strain*.

$$\sigma_{crit} = \pi^2 E\, I\, C/L^2 \quad \text{Euler's formula for column buckling} \qquad \text{(vi)}$$

where $\sigma_{crit}$ = Euler buckling stress, stress at which buckling could occur, E = Young's modulus, C = factor for end conditions of column, L = effective length of column, and I = moment of inertia.

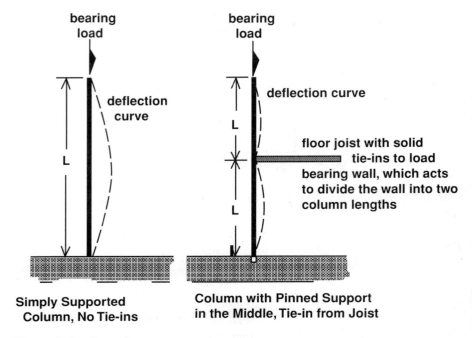

**Figure 5.4** Effect of joist tie-ins.

When one end of a column is rounded, and the other end is fixed, C = 2. When the column is pivoted at both ends, C = 1. When both ends are fixed, C = 4. When one end is fixed and the other free, C = 1/4.

Thus, with no effective tie-ins, the wall can be modeled as a column of length "L," with a free end and a fixed end. However, if there is a rigid tie-in in the middle of the wall, the lower portion becomes a column with a fixed end and a pivoted end, and the upper portion of the wall becomes a column with a free end and a pivoted end. The effective tie-in afforded by the joist not only shortens the length of the column, which reduces the critical stress point, but also improves the end conditions, which further reduces the critical stress point.

With respect to the bottom of the wall being assumed to be a fixed end, some care should be made when applying this assumption. Often when water leakage is sufficient, water will run down the wall and pool at the base of the wall. If significant damage has occurred at the bottom, it should be considered a rounded end condition.

Thus, when water leakage from the roof has both leached significant amounts of calcium hydroxide from the mortar and caused rot in the bearing boxes, the combined structural effects can very pronounced. The effective value for Young's modulus is reduced, the overall length of the wall as a column can be doubled, and the applicable factors for column end conditions are lowered.

## 5.7 Restoration Efforts

Some of the buildings that fit the category under discussion have become icons of the city or town in which they were constructed. Because they were built when the town was first established, they have historical and sentimental value to the residents. Unfortunately, the repair of such buildings is neither easy nor cheap.

If the damage to the wall is localized and confined to a small area, repairs may entail removal of the stone or brick in the leached areas with complete resetting with a modern cement based mortar. Since removal of such brick or stone may leave a large hole in the load-bearing wall, such repairs may require significant structural shoring. To do this safely requires the expertise of experienced rehabilitation contractors.

Sometimes when such water damage is finally discovered in a turn-of-the-century building, leaching of the mortar has progressed to the point where entire sections of load-bearing walls are affected and are at risk. Simple tuck-pointing repairs or remortaring *in situ* are the usual first attempts to deal with such damage because they are the cheapest option. While these efforts can temporarily hide the damage, they can not reverse it. In fact, such repairs often will not even be cosmetically effective for long because the new mortar will not be able to adhere to the crumbling old mortar.

Meaningful structural repair and salvage of the wall in such cases requires more than superficial tuck pointing. It may require more resources than are available, or at least more than the owners are willing to spend. This creates the following dilemma: the building is too sentimentally significant to the town to be torn down, but costs too much to be properly fixed by the owners. Because of the indecision that results, the building usually sits for a long time, decays more, and is eventually razed when decay is obvious and collapse is imminent.

## Further Information and References

*Formulas for Stress and Strain*, by Raymond Roark, 2nd ed., McGraw-Hill, New York, p. 300. For more detailed information please see Further Information and References in the back of the book.

*Mechanical Engineer's Pocket Book*, by William Kent, 9th ed., John Wiley & Sons, New York, 1915, p. 372, "Strength of Lime and Cement Mortar," and p. 370, "The

Strength of Brick, Stone, Etc." For more detailed information please see Further Information and References in the back of the book.

*U.S. Bureau of Standards, Technical Paper 123*, published in 1919. For more detailed information please see Further Information and References in the back of the book.

# Putting Machines and People Together

# 6

## 6.1   Some Background

Ergonomics is a branch of engineering that is concerned with the way machines and people work together. Sometimes called the science of work, it is the application of psychology, physiology, engineering, industrial management, and other information about how people work and function to the design of machinery and work environments. The purpose of ergonomics is to improve productivity, safety, and the general working environment.

Until about the middle of the 19th century, manufacturing equipment and machinery were generally designed with few concerns about the human operator. Employers were virtually unrestrained by any safety regulations, legal, or social responsibilities. Humans were subordinate in importance to the equipment they were employed to operate, and were simply expected to adapt to the needs of the equipment. If a person refused or could not adapt to the needs of the equipment, another person would be found who was willing to do so.

In some cases, people were hired solely for their size or shape because it suited the equipment. Young boys, for example, were often hired to work in coal mines so that the horizontal mine shafts could be made smaller. Children were also the favorite in textile operations. Because of their small size and nimbleness, they could get around the tightly packed loom equipment better than adults.

Some of the equipment and working conditions also caused new types of diseases and maladies. Brown lung, which was unknown prior to the invention of textile factories and flour mills, resulted from the long-term inhalation of textile particles or flour particles. The more famous cousin disease, black lung, resulted from the inhalation of fine coal and dirt particles. Disfiguring injuries were so common with some types of equipment that it was possible to guess a person's trade by the location of their scars, the number and placement of gnarled joints and crooked fingers, or the pattern of lost limbs.

In response to these and other problems, labor organizations were developed. The first labor union in the U.S., the Knights of Labor, was organized in Philadelphia in 1869. Perhaps predictably, its membership was initially drawn from the garment manufacturing industry. By 1886, it had 700,000 members.

Various other labor organizations were formed concurrently with the Knights of Labor. Some, unfortunately, used violence in an attempt to achieve their goals. The terrorist labor organization, the Molly Maguires, was an organization of Irish coal miners in the Scranton, Pennsylvania, area. In an attempt to gain better working conditions, they kidnapped and murdered various mining officials and law officers who protected the mining company interests. In 1877, eleven Molly Maguires were hanged and the organization was effectively broken up.

From 1869 to the turn of the 20th century, ergonomics, such as it was at that time, was mostly a mixture of politics, business economics, labor law, and theoretical social movements. However, starting around 1900 there was an attempt in the U.S. to apply psychology to the workplace. This new discipline was called industrial psychology. In 1901, for example, Walter Dill Scott applied psychology to the selection of salespersons and to the use of advertising. He developed profiles for screening people who would likely be the more successful salespersons, and developed advertisements that incorporated psychological principles to make them more appealing and effective.

In 1911, the text, *Principles of Scientific Management,* by Frederick Taylor was published. Taylor was an inventor and engineer, and is sometimes called the father of scientific management. In 1881, while working for the Midvale Steel Company, Taylor studied how individual tasks were being performed by timing each task. He then rearranged work stations and how materials were being processed through the mill to minimize the overall work time. After graduating from Stevens Institute of Technology in 1883, Taylor went on to apply these same techniques to other industries, primarily in the steel industry, as a private consultant.

*Principles of Scientific Management* was the first systematic text about industrial psychology and its application to the management of a factory or workplace. The term "Taylorism" originated from the success of the text. The moniker "efficiency expert" was often used later on to describe those who practiced Taylor's theories and principles.

Also published in 1911 was the text, *Motion Study*, authored by Lillian and Frank Gilbreth, a husband and wife engineering team. *Motion Study* was then followed by a series of books on industrial productivity and efficiency. Like Taylor, the Gilbreth's published body of work grew out of their practical industrial consulting work, first started in Providence, Rhode Island, and then later in Montclair, New Jersey.

Their work included the assessment and role of job skill, the effects of work fatigue, and the design of workstations for the handicapped. Their study and recommendations concerning hospital surgical teams resulted in many procedures still in use, especially the procedure where the surgeon calls for an instrument instead of looking for it himself.

Many people are familiar with the Gilbreth's home life which was written about by two of their children in the popular books, *Cheaper by the Dozen*, and *Belles on Their Toes* by Frank Gilbreth, Jr. and Ernestine Gilbreth Carey. Both books were also made into motion pictures under same titles respectively in 1950 and 1952.

During World War I, industrial psychology was given a great boost by the rapid mobilization for war in the U.S. For the first time on such a large scale, psychologists were employed to write job specifications, develop job skill screening tests, develop officer rating forms, and develop psychological counseling programs. Over 1,700,000 men were rapidly classified and assigned military duties using industrial psychology techniques.

Like World War I, World War II was also a great impetus in the development of ergonomics. Engineers were pressed by the urgency of the war to come up with equipment that could be easily and quickly put into service. "Operator skill" was designed into the equipment so that an inexperienced person could become productive with the equipment in short order. Psychological screening was also used to a greater degree than before in the selection and training of special-purpose crews, especially in the Army Air Corps.

By the end of World War II, the "efficiency expert" had become a "human factors engineer." In the 1970s, due to the expansion of the field to include more and diverse subject matter, the term "human factors engineering" gave way to the term ergonomics, which had been the preferred term in Europe.

The third great impetus to ergonomics in the U.S. in the 20th century to date has been the Williams-Steiger Occupational Safety and Health Act, passed by the U.S. Congress in 1970. This act established the Occupational Safety and Health Administration, or OSHA, and required OSHA to codify comprehensive industrial and construction safety standards by the end of 1972. Since a number of excellent standards already existed at that time, OSHA simply incorporated many of them into the OSHA code. As stated in the act, the purpose of OSHA is "… to assure so far as possible every working man and woman in the Nation safe and healthful working conditions and to preserve our human resources."

Why was there a need for the Williams-Steiger Act? In 1970 when the act was being debated, the following statistics were cited.

- Job-related accidents accounted for more than 14,000 worker deaths per year.
- Nearly 2,500,000 workers were disabled, about 10% of the population of the U.S. at that time.
- Ten times as many person-days were lost from job-related disabilities as from strikes.
- The estimated new cases of occupational diseases totaled 300,000 per year.

The passage and implementation of the Williams-Steiger Act is a landmark in ergonomics in the U.S., especially the safety aspects of ergonomics, because it established nationally uniform standards. Prior to 1972, design and safety regulations were a patchwork. They varied by state, labor union jurisdiction, industry, and civil law precedent. Importantly, the act also established an orderly method for the modification and deletion of standards, and the adoption of additional standards. The method takes into consideration new research findings, and employs an open hearing review process to consider industrial, labor, and public concerns.

Nearly all the major engineering disciplines incorporate some ergonomic principles into their professional practice. However, the fields of mechanical engineering and industrial engineering are more likely to be concerned with ergonomics than the others. This is because mechanical engineers and industrial engineers are usually responsible for the actual equipment configuration and placement.

Two primary themes are at work in the modern field of ergonomics that sometimes seem to be in conflict with each other. The first is the modification of equipment and work environments to improve productivity and optimize the work task. The second is the improvement of the work process and work environment to enhance worker satisfaction and welfare.

## 6.2   Vision

It is known that within the population of adult men in the U.S., about 8% have red-green color recognition deficiencies. Medically, this condition is known as anomalopia and is an inherited condition. Similar color recognition deficiencies can also result from injury or degenerative disease. Thus, in a statistically normal population of grown men, about 8% will have trouble distinguishing red from green in certain circumstances through no fault of their own.

It is traditional in Western countries that green is the color for go, okay, or start, and red is the color for danger, stop, or hazard. For example, U.S. traffic lights use green for go and red for stop. Further, red and green lights are often used as status indicators on machinery. A green light indicates the a machine is running or is ready to start. A red light indicates that a machine has stopped, or that something is wrong.

With respect to signs used around machinery, OSHA regulations (CFR 1910.145[f] Appendix) require that red be used for the background color on signs that warn of danger, yellow for signs that caution, yellow-orange for warning signs, and orange-red for biological hazard signs. Some of these background colors are colors commonly confused by persons with anomalopia.

Because of this, it is possible that if a person with color deficient vision is put into a situation where it is very important that he recognize the meaning of a machine indicator light or sign by its color, a mistake may occur. Does this mean, then, that people with anomalopia should not be allowed to drive a car or operate machinery? Should employers give prospective employees color vision tests to screen out people with anomalopia? Not necessarily.

One way to alleviate the possibility of color-related recognition mistakes among persons with anomalopia is to provide secondary visual cues. For example, not only are the traffic lights red, yellow, and green, but they are always positioned the same way. The red light is always on top, the yellow light is always in the middle, and the green light is always on the bottom. In this way, even if a person cannot distinguish red from green, he can observe the position of the light that turns on and know what to do. Similarly, in painted traffic signs, the red stop sign is distinguishable from the red "wrong way" sign by the fact that the stop sign is octagonal, and the "wrong way" sign is rectangular.

The use of position and shape provide an alternative to color-only recognition. Because of this, people with anomalopia are not barred from driving. Similarly, on machinery, red and green lights can be grouped in certain positions. When a certain indicator light comes on, the operator can know by its position whether it is a "good" green light, or a "bad" red one. Further, the shape of the light can also be a visual cue. A green light on a machine control panel might be made to always be round in shape, while a red one might be required to be square.

## 6.3   Sound

Suppose a new machine were being installed in a factory. The machine is big and powerful, and could injure a person if warning alarms are not recognized by the operator. One of the features of the machine is that when the machine malfunctions, a warning buzzer actuates and sounds off. The sound level of the warning buzzer is 90 decibels, which to most people is startling. As a comparison, a diesel truck engine will often be as loud or louder than 90 decibels.

However, where the machine is installed at the factory, the general noise level is enough, 95 decibels, that earplugs or "Mickey Mouse" ears are required for all employees working in the area. In this particular work environment, the effectiveness of the warning buzzer has been negated. The operator of the machine cannot hear the warning buzzer of 90 decibels when the general din of the facility is at 95 decibels or his ears are plugged up for protection from the sound.

Further, consider what would occur if several of these same machines were installed near each other in a production line and all were equipped with the same warning buzzer. If the warning buzzer for one machine went off, how could an operator immediately determine which machine was malfunctioning?

In fact, if the warning buzzer for one machine went off with regularity due to malfunction or an unusually high sensitivity, the operators of the other nearby machines might come to ignore their warning buzzers, since there would be a pattern established of the one going off for no apparent reason. Under these circumstances, an operator could be lulled into complacency by repeated warning buzzers from other machines and not recognize when his specific machine's warning buzzer was actuated.

Thus, the above example indicates.

- How an otherwise good audible warning alarm system can be negated by the work environment.
- Why it is important that warning devices have specificity: that is, a specific warning signal for a specific problem or machine.

A similar effect occurs in vehicles at railroad crossings. Train engineers usually sound an ear-splitting horn when the train approaches intersections with highways and roads. The idea is to warn drivers of the approaching train, especially at crossings where there are no automatic warning lights or barrier gates. The sound level of the horn at close range can be 95 db.

When the windows are rolled up in most modern cars, the exterior sound is attenuated about 20 to 25 db. Most car manufacturers have spent considerable effort to accomplish this. Because of this, however, a locomotive horn at close range would only have a sound level of about 70 to 75 db inside a car with the windows rolled up. If the train were further away, the sound level of the train's horn would drop off to even lower levels.

Since most cars are equipped with tape players, radios, air conditioners, ventilation fans, and many other sound-producing devices (like small children), it is not unusual for the noise generated within the car to be 75 to 80 db or more. Driving at moderate to moderately high speeds on gravel roads in rural areas also causes the interior sound in a car to be 70 to 75 db or loader.

To readily distinguish one sound over another, the sound to be distinguished should be at least 3 db greater, if not 5 db greater, than the other sound. To hear the warning horn of a train, the sound of the horn inside the car should be at least 3 to 5 db louder than the sounds generated within the car itself.

In short, with the windows up and the radio operating, it is likely that some drivers will not hear the warning horn of a locomotive at an intersection. Unfortunately, without automatic lights flashing directly in front of most drivers to warn them, or better yet, automatic barrier gates to prevent crossing the tracks when a train is approaching, many drivers will not look left and right to visually verify that it is safe to cross railroad tracks.

## 6.4   Sequencing

One of the more visible applications of ergonomics has been to computer software. When "user-friendly" software was first sold, many programs had single stroke "delete" or "erase" commands. Because of this, many a computer operator has contemplated suicide when he has inadvertently erased an entire data file or program. One wrong keystroke and it was all over.

However, most contemporary software requires at least a two-step keystroke sequence to initiate a delete or erase command. By incorporating a two-step sequence, it ensures that no mistake is being made. While the operator might hit a single keystroke by mistake, the operator would not hit two or perhaps even three specific keystrokes in an ordered sequence unless he really means to activate that particular command. In this way, inadvertent or random mistakes are greatly reduced.

## 6.5   The Audi 5000 Example

One of the more famous examples of automotive ergonomics relates to the Audi 5000. In a *60 Minutes* television expose in 1986, it was reported that the Audi 5000 was prone to sudden acceleration. In a typical such incident, the driver of the car would report that when he put his foot on the brake to put the car into drive, the car accelerated forward at great speed. In many instances, the driver of the car also reported that he was pushing the brake pedal to the floor as hard as possible, but the car continued to accelerate.

The *60 Minutes* report was picked up by the general press and generated great public interest and consternation for several years. Many high-dollar lawsuits were filed. There was even an Audi sudden acceleration victim support group organized called the Audi Victims Network, or the AVN.

Because of the *60 Minutes* report and complaints of sudden acceleration incidents in other brand vehicles, the National Highway Traffic Safety Administration conducted a thorough study of ten vehicle types with above-average complaints about sudden acceleration events. The results of that study were published in 1989. They indicated that, in general, there was nothing mechanically wrong with any of the cars. The drivers were simply putting their feet on the wrong pedals.

This finding, of course, was contrary to many theories fostered in the public press. The more popular of these were: cruise control problems, problems in the new computer-controlled fuel system components, and problems in the transmission control system. It may have been no coincidence either that the movie *Christine* had been prevously released. *Christine*, based on the Stephen King novel of the same title, is about an evil car with a mind of its own.

Notwithstanding the public's conjectures, the sudden acceleration problem was significantly improved by applying simple ergonomic principles. First, the gas pedal was made to look and feel distinctly different from the brake pedal. Secondly, a new device called an automatic shift lock was installed in subsequent Audi 5000 cars. An automatic shift lock is a device that requires the driver to put pressure on the brake pedal, and not the gas pedal, before the gear shift selector can be shifted out of park.

Since the introduction of the automatic shift lock, the number of sudden acceleration incidents has dramatically dropped. During the period September 1986 through November 1988, for example, Audis so equipped had 60% fewer sudden acceleration incidents than identical models not so equipped.

What are the lessons of the Audi 5000 example? For one, it demonstrates that what people *believe* they are doing with a machine is not always what they are *actually* doing with the machine. With all due sincerity, many of the drivers believed they had been pressing down on the brakes when, in fact, they had been pressing down on the accelerators.

Secondly, it points out what can be done by design to eliminate the chance of the operator of the machine making the wrong decision. By incorporating an interlocking device between the transmission and the brake, the driver could not put his car in drive without having his foot first placed firmly on the brake. This removed some of the opportunity for error by the operator.

## 6.6  Guarding

The basic premise of guarding is to prevent exposure to hazardous things in the workplace. The regulations and requirements for machinery guarding are primarily contained in OSHA regulations 1910.211 through 1910.222, and 1910.263(c), revised as of July 1, 1990. Additionally, OSHA has published a booklet, *Concepts and Techniques of Machine Safeguarding* (OSHA 3067, 1980), which more simply explains and illustrates the various ways of guarding machinery.

Often related to guarding matters, OSHA also requires certain signs and markings to designate hazards. These are contained in OSHA regulations 1910.144 and 1910.145.

In general, any machine component or operation that could cause injury must be effectively guarded. Most mechanical hazards are associated with the following three situations.

- **Point of Operation:** The point in the machinery where cutting, shearing, shaping, material removal, or forming operations are performed.
- **Power Transmission Equipment:** The machinery components that transmit power through the machinery. This includes sheaves, belts, connecting rods, cams, couplings, chains, gears, cogs, flywheels, spindles, sprockets, and shafts.
- **Other Moving Parts:** This includes the moving parts of the machinery not directly associated with the first two categories such as feed mechanisms, augers, flyball governors, moving fixtures, control mechanisms, and so on.

An easy rule to consider with respect to whether guarding is needed is could a person put his face in it without harm occurring? If not, protective guarding is needed.

A machine guard can take many forms. The most common is the direct barrier. A barrier simply prevents contact being made with the dangerous item while it is operating. A common example is the chain guard on a bicycle. It prevents the inadvertent contact of a person's leg or clothing with the bicycle chain and sprockets. Without the chain guard, clothing or skin could become entangled at the pinch point between the chain and sprocket, or snagged by the irregularities of the moving chain links.

For a barrier to be effective guard, a person should not be able to easily remove it and it should be durable. In some industries where pay is based on piecework, an operator may be tempted to remove a guard in order to save time and increase his pay. Of course, such removal may substantially increase the risk of injury. Such an easily removable guard may constitute an invitation to circumvent the guard. Similarly, a guard that easily breaks apart or wears out is not considered effective. The guard must stand up to normal wear and tear, including bumps and impacts from tooling that might be associated with routine work in that area.

To protect the operator from his own inadvertent action, or those of others, guards that might be removed occasionally by the operator for legitimate reasons can be equipped with trip switches or interlock devices. This will disable the operation of the machinery when the guard is removed. Thus, a person who is engaged in resetting some mechanism and is obscured from sight will not accidentally be injured by another person who unknowingly attempts to operate the machinery.

A barrier guard should also cause no new hazard and should allow the operator to work unimpeded. Guards made of sheet metal with sharp edges may create a new laceration hazard where there was none before. Guards that get in the way of work or aggravate a worker often are circumvented and workers will connive to get rid of them if possible.

Guarding by the use of direct barriers is often considered the preferred method of guarding. This is because no special training or actions by the operator are needed. The hazard is eliminated at the source. However, some situations do not easily lend themselves to this method and require alternative guarding techniques, such as protective clothing or equipment.

Protective clothing or equipment is especially useful where there is a danger of chemical splash, high intensity noise or light, noxious vapors, or similar hazards that cannot be fully eliminated by good practices or design. In essence, instead of putting a guard around the hazardous item, the guard is put around the person.

Items such as face shields, ear plugs, helmets, protective cloaks and coats, respirators, goggles, and steel-toed boots are in the category of protective clothing and equipment. Whenever possible, however, the root source of the hazard should be eliminated or reduced by design and engineering.

In instances where a machine's hazardous equipment cannot be effectively safeguarded by a fixed barrier, various electromechanical safety devices can be utilized to prevent contact. For example, photoelectric sensors can be used to determine if a person's hands or body is in a danger zone. While they are in this zone, the machine is disabled. Such a device is often used in power brake presses, where the duration of action is very short, but the force of the forming action is sufficient to severely maim or sever body parts.

Pullback devices are often used on machines with stroking actions. A pullback device allows an operator to manipulate a workpiece at the point of operation when the punch or shear blade is in the "up" position. When the machine begins its downstroke, cables attached to the operator's wrists automatically ensure that his hands are pulled clear of the work area.

Alternately, a two-handed trip switch can often accomplish the same effect. In such a system, the operator must have both hands and sometimes both feet simultaneously pressing independent switches before the machine will operate. This ensures that the operator's hands and feet are out of harm's way.

Trip wires and trip bars are devices that disable a machine when depressed or activated. The devices are used when operators have to work near equipment pinch points or moving machinery and there is a risk of falling into the equipment.

While not strictly guards, other safety devices that prevent inadvertent injury by machinery should not be overlooked. Such devices include safety blocks and chocks, lock-outs on power disconnect boxes, lock-outs on pneumatic and hydraulic valves, push sticks and blocks forefeeding materials, and so on (see OSHA 1910.147).

A machine can also be safeguarded by location: that is, it can be so placed that a person cannot normally get to it or to its hazardous equipment. Access to hazardous machinery can be denied by fence enclosures or by fixed wall barricades. Safety can even be further assured by placing a safety switch in the door or gate that will stop the machinery when the enclosure is entered.

## 6.7  Employer's Responsibilities

In a nutshell, it is an employer's responsibility to ensure that all equipment used in his shop meets current OSHA safety standards. Older machines bought or built before current OSHA standards were enacted must be brought up to current standards. Similarly, imported equipment that does not meet current OSHA standards must be modified to do so. Unlike an older car or building, for safety reasons, manufacturing equipment is not "grandfathered" in.

Because of this, employers must regularly inspect equipment as per OSHA regulations, and maintain records of such inspections. Worn out guards and safety devices must be promptly repaired or replaced.

Consider the following: a belt guard was also equipped with an interlock switch so that when the guard was removed, the belt and pulleys would not operate. The switch wore out. To expedite production, a "jumper" wire was run to bypass the switch and keep the machine going. The switch was then forgotten. Two months later, an operator wished to adjust the belt tightness. He turned off the machine and removed the guard. While he was making adjustments, another person inadvertently turned on the machine and the operator was injured. In such a situation, it is likely that the employer will be held accountable for the accident rather than the employee who mistakenly turned on the machine without first checking.

Employers must also be watchful for employees who circumvent guards and safety equipment. If it can be shown that the employer knew about or failed to reasonably correct such equipment abuses, the employer may be held accountable for any ensuing accidents.

Work rules are not a substitute for guarding. An employer cannot simply promulgate a work rule that employees must keep their hands and limbs away from dangerous equipment, or that employees should work safely. An employer cannot depend on notions of common sense by his employees to avoid accidents and injuries. Thus, an employer cannot assume a passive role with respect to shop safety. He must become informed of standards and requirements and be actively involved. In the legal sense, since he is inviting people to his workplace, they have a reasonable expectation that they are being provided with reasonably safety equipment and working conditions.

## 6.8　Manufacturer's Responsibilities

The manufacturer of equipment has, of course, the responsibility to design and construct its equipment so that it is in compliance with current OSHA standards. Also, the equipment should be in keeping with current industry standards and practices. A commonly used legal expression is "state-of-the-art design" which is usually taken to mean that the item under consideration is as good and as bad as everyone else's like product currently being sold in the marketplace.

There are persons, primarily plaintiff attorneys, who will argue that this is still not sufficient. They argue that the various published standards are just a minimum, and that a machine is not deemed safe simply because it meets all published or required standards and regulations.

There is, of course, some merit to this argument. As new machines are designed and built with innovative features, faster feed rates, greater power, and so on, new hazards appear that must be dealt with. For example, the original OSHA safety standards published in 1972 had not yet fully addressed the use of high-powered lasers in manufacturing, which were just being introduced at that time. It would be specious to argue that a patently unsafe laser sheet metal cutter met all industry standards when there weren't any.

In this context then, it is incumbent on the manufacturer to at least recognize the spirit or underlying intent of safety standards and regulations, and to make reasonable efforts to ensure that a machine is safe and properly guarded, even if official standards or regulations don't yet exist.

The designer of machinery would do well to consider that even the best operators make mistakes when they are tired, bored, or simply asleep. Given enough operating hours, a mistake that would be considered "one in a million" might occur several times. Is it good design if a person could lose a hand or eye because of a simple or even stupid operating mistake?

## 6.9 New Ergonomic Challenges

With the advent of the Americans with Disabilities Act, or ADA, the problems of matching people to machinery and equipment will become even more challenging. For example, in the past it was assumed that a person who operated a punch press might be standing or perched on a shop-type stool. The work platform on the machine was then sized accordingly. Now it is possible that the work platform, or some work platforms, may have to be sized for persons in wheelchairs.

Similarly, many shop processes like punch presses and sheet metal shearers rely on the operator to use a foot switch to activate the machine. This may require some modifications if the operator has no feet or if the operator's feet are artificial and he has no feeling in them.

## Further Information and References

*All About OSHA*, U.S. Department of Labor Occupational Safety and Health Administration, OSHA 2056, 1995 (revised). For more detailed information please see Further Information and References in the back of the book.

*Code of Federal Regulations*, Parts 1900 to 1910, and Part 1926. For more detailed information please see Further Information and References in the back of the book.

*Human Factors in Engineering and Design*, by Sanders and McCormick, 6th ed., McGraw-Hill, New York, 1987. For more detailed information please see Further Information and References in the back of the book.

*The One Best Way*, by Robert Kanigel, Viking Press, New York, 1997. For more detailed information please see Further Information and References in the back of the book.

# Determining the Point of Origin of a Fire

# 7

To the trained eye there is as much difference between the black ash of a Trichinoply and the white fluff of bird's eye as there is between a cabbage and a potato.

**— Sherlock Holmes, in *The Sign of the Four***
*Arthur Conan Doyle, 1859–1930*

## 7.1 General

Fire is a type of chemical reaction. It is a rapid, self-sustaining oxidation reaction where heat, light, and byproducts are produced. While it may seem almost trivial to repeat, in order for fire to take place there must be three things:

- combustible fuel in sufficient quantity.
- sufficient air or an oxidizing agent to react with the fuel.
- an energy source, agent, or action sufficient to facilitate ignition.

These are the fundamental principles for determining the point of origin of a fire. The point of origin of a fire must be where all three components necessary to initiate fire were present at the same time and place when the fire began.

Admittedly, some types of fires can be self-starting in terms of activation energy. Simply bringing certain materials into contact with one another, like air and white phosphorus at 30°C, can initiate burning. In such cases, the fire is not initiated by a spark or flame, but by some type of action that allows the reactants to come into physical contact with each other.

There are also some materials that can self-ignite when a third chemical or agent is added to the first two, which are already in contact with one another. The third chemical, or catalyst, does not react with either of the other two chemicals to initiate burning. The catalyst does, however, react with them in a way that lowers the activation-energy threshold between the two. Because of this, the oxidation-reduction reaction between the two materials can initiate at a lower temperature, or with significantly less initial energy input. Thus, two chemicals that would ordinarily not self-ignite with one another, could self-ignite if a certain type of catalyst or agent is added to the mix.

However, returning to the basic point, there is ample oxygen in the air to support combustion nearly everywhere on earth where there are buildings and people. Thus, having enough oxygen to support burning is usually not a problem except perhaps in vaults, vacuum chambers, or other unusual airtight structures.

Similarly, most buildings, homes, and manufacturing facilities contain ample amounts of combustible materials. At one time, it was popular to advertise that a warehouse or hotel was "fire proof." This claim was made because the building itself was made only of masonry, steel, and other noncombustible materials. However, these buildings usually contained large amounts of furniture, clothing, decorations, papers, boxes, and other combustible materials.

Of course, just because a material does not burn does not mean it is fire proof. Masonry and steel are both adversely affected by fire. It is amazing how many of these so-called fire proof hotels and warehouses have burned down over the years. For this reason, the designation "fire proof building" is no longer used.

Thus, two of the three components necessary for a fire to occur are typically already present in most places, and they likely have been present for a long time without any fire occurring. Consequently, the key to solving the point of origin of a fire and its causation often revolves around determining the nature of the third component necessary for fire, the source of ignition energy.

## 7.2  Burning Velocities and "V" Patterns

From casual observation, it is apparent that fire propagates at different rates depending upon the material and its orientation. With respect to orientation, for example, it is common knowledge to anyone who has used matches that a fire burns up a match faster than it burns down a match. It is also common knowledge that when a match is held horizontally, the burn rate is somewhere between.

Consider the following simple experiment. A common wooden kitchen match is about 50 mm long, excluding the match head, and is about 2.5 mm × 2.5 mm square in cross-section. If timing is begun after the match head has ignited and the wood stem begins to burn, it takes about 30 seconds for the match to burn to the end when held horizontally. Similarly, it takes 6 seconds for the match to burn to the end when held vertically with the ignited end at the bottom. When the ignited end is held at the top, in most cases the match goes out after fire travels 15 mm downward from the head in an average lapsed time of 35 seconds.

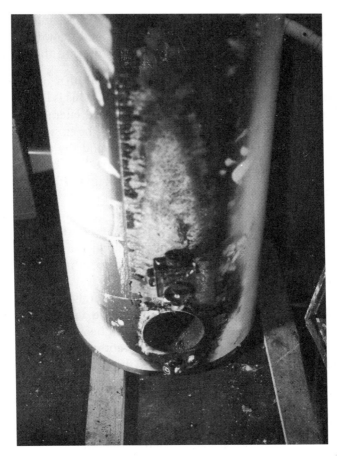

**Plate 7.1** "V" pattern on side of hot water tank emanating from gas apparatus.

Using parametric equations, the above information can be expressed in the following.

$$z = (50 \text{ mm/6 sec})t = (8.67 \text{ mm/sec})t \qquad \text{(i)}$$

$$x = (50 \text{ mm/30 sec})t = (1.67 \text{ mm/sec})t$$

$$zd = (15 \text{ mm/35 sec})t = (0.43 \text{ mm/sec})t$$

where z = fire travel distance in upward, vertical direction, x = fire travel distance in horizontal position, and zd = fire travel distance in downward, vertical direction.

Inspection of the above three equations for fire velocity on a wooden match finds that:

- burn velocity is about five times faster upward than lateral.
- burn velocity is about four times faster lateral than downward.
- burn velocity is about twenty times faster upward than downward.

Because of the above, when fire is allowed to burn on a vertical wall surface, it will generate a "V"-shaped burn pattern. This is because as the fire burns upward, it will, at the same time, also spread laterally, but usually at a slower rate. If the lateral burn rate were equal to the upward burn rate, the angle of the "V" with the vertical would be 45°, and the angle at the notch of the "V" would be a right angle, 90°. It is for this reason that upward moving fires have a "V" angle of 90° or less. Due to convection processes, lateral burn rates do not equal or exceed upward burn rates. If the "V" angle is more than 90°, the fire was not upward burning.

Consider what occurs if the fire is downward burning instead. Let us suppose that a fire begins somewhere in the ceiling, and then spreads downward along a vertical wall. The initial point of burn on that wall begins at the top of the wall, in the middle.

Because the burn rate is several times faster laterally than downward, the shape of the downward burn pattern is an oblique angle, and not an acute angled "V." For example, using the previous numbers for burn rates in thin wooden match sticks, the lateral burn rate is about 4 times faster than the downward burn rate. This would cause the fire to spread laterally 4 length units to the left, and 4 length units to the right, as it descended one length unit downward. This is an angle of 76° with the vertical, and would create a "V" with an angle of about 152° at the notch of the "V."

Thus, downward burning fires do not create "V" patterns that have an acute angle. They create burn patterns that have oblique angles at the notch of the "V."

By measuring the angle of the "V" burn pattern, the ratio of the upward burning rate to the lateral burning rate can be determined with the help of a little trigonometry. For example, if the angle at the notch of the "V" were 45°, the angle with the vertical would be 22.5°. The tangent of 22.5° is 0.414. Thus, the average burn rate in the lateral direction would be 41% that of the burn rate in the upward direction, or the upward burn rate is 2.4 times the lateral burn rate.

Because there will also be some downward burning, the actual point where the fire began is not always located exactly at the notch of the "V." It will often be located slightly above the lowest burned area of the "V" pattern.

For example, using the fire velocity numbers noted previously for the dry, thin wood used in matches, in a vertical wall made of the same material, a fire began at some point on that wall and burned for 5 seconds, the resulting burn pattern would appear something like that depicted in Figure 7.1.

In Figure 7.1, the ratio of upward to lateral fire velocity is 5.2 to 1. This creates an angle of 11° with the vertical, or an angle at the "V" notch of 22°.

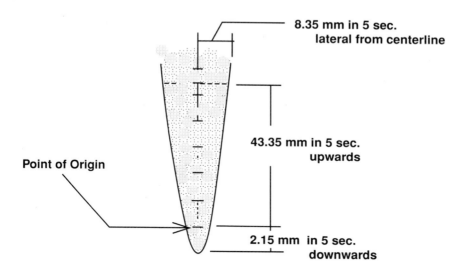

**Figure 7.1** Typical "V" burn pattern on thin, dry wood veneer in vertical position.

Because there was some downward burning, the actual point of origin of the burn is located 2.15 mm above the lowest point of the "V."

Of course, the specific flame velocity parameters will change from material to material. However, the general principle stays the same: fire propagation upward is significantly faster than lateral propagation, and lateral propagation is significantly faster than downward propagation. It is not unusual in some materials for upward burn rates to be as much as 20 times faster than lateral burn, or upward burn rates to be as much as 50 times that of downward burn rates.

It is for these reasons that fire investigators often look for the lowest point of significant burn damage, often called the low point, and observe whether fire damage appears to fan out laterally and upward from that point. It is sometimes the case that the point of origin of the fire is located at the low point.

## 7.3 Burning Velocities and Flame Velocities

There is an important exception to the previous discussion of "V" patterns, and that exception involves the use of liquid accelerants. When a liquid accelerant, especially one whose flash point is below room temperature, is splashed on a wall and ignited, the fire travels quickly throughout the contiguous area wetted by the accelerant: downward, upward, and lateral. With respect to burn within the wetted area, the directional fire velocities are nearly the same, for all practical purposes.

This is because the fire actually spreads by deflagration of the flammable vapors clinging close to the liquid layer of the accelerant. This is why when the vapors of a flammable liquid accelerant are ignited, they often catch fire with a sudden "whoosh" or "whomp" sound as the flame front quickly spreads in an explosion-like way. Unconfined vapor explosions of this type are often called "puffers."

As vapors are given off by the liquid accelerant and mix with the available air, there will be five basic zones of flammability close to the surface of the wetted area:

1. The outermost zone is where the vapors are too thin or diffuse to support combustion. No combustion occurs here even though some flammable vapors are present.
2. The next zone, closer in, is where the vapor concentration is more than the lower limit fuel-to-air ratio necessary for combustion to occur, but is less than stoichiometric conditions.
3. The third zone is where the air-to-fuel ratio is at stoichiometric conditions. Combustion effects are maximum. This is not a continuous line or boundary, but is usually a thin, unstable region subject to the ambiguities of diffusion and convective turbulence between the air and the fuel vapors.
4. The fourth zone is where the fuel-to-air ratio is more than stoichiometric conditions, but not so high as to choke off combustion by exceeding the maximum limit of flammability.
5. The fifth zone, which is right up next to the wetted surface, is where the fuel-to-air ratio is too high for combustion to occur. It is for this reason that in a pool of highly volatile flammable liquid, sometimes the flames seem to burn slightly above the surface of the pool.

Of course, once the vapors are ignited, the heat released during combustion causes the gases to expand and push away from a vertical wall, which expands the zones. The gases first move outward into the room, and then also upward due to convection effects. Thus, the burning vapors of a flammable liquid can extend well beyond the perimeter of the area wetted by the accelerant.

As noted before, the burning velocity of the thin wood used in match sticks is about 1.67 mm per second when the match stick is held horizontally. Under similar horizontal burning conditions, the maximum burning velocity of acetone is 425 mm per second. Similarly, methanol is 570 mm per second, and ethylene oxide is 890 mm per second. Respectively, the burning rates of these accelerants are 254, 341, and 533 times the burning rate of the match wood.

However, a simple comparison of burning velocities is not the whole story. The definition of "burning velocity" with respect to gaseous fuels is:

the speed at which a laminar combustion front propagates through the flammable mixture relative to the unburned mixture. In essence, burn velocity is measured with the fuel itself as the reference point. Expansion or convection effects on the gas mixture itself are ignored.

In other words, the burning velocity of a gaseous fuel does not take into account the fact that the whole gaseous mixture may be expanding or moving upward, downward, or sideways relative to a stationary object, like a wall.

When such expansion or convective effects are taken into consideration, the term used is "flame speed." Flame speed is the velocity at which the fire itself is seen to travel relative to stationary objects.

The fire or flame velocity associated with deflagration of the flammable liquid vapor is much faster than the burn velocity of the underlying solid fuel that has been wetted by the accelerant. This is to be expected since fire velocities are a function of several factors including the activation energy needed to initiate combustion; the amount and rate of heat given off during combustion; and the resulting expansion and convection of hot, burning gases away from the point of ignition. When the activation energy required for combustion ignition is low, when the heat of combustion per gram is relatively high, and the rate of combustion energy release is high, the flame velocity is correspondingly high.

In general, the flame speed is related to the burn velocity in the following way:

$$V_f = V_b + V_g \qquad\qquad (ii)$$

where $V_f$ = flame speed, $V_b$ = burn velocity, and $V_g$ = the velocity of the gas mixture itself relative to a stationary object.

If stoichiometric conditions are assumed, and it is also assumed that the flames are not turbulent, the maximum flame speed of most combustible gases is roughly eight times the burn velocity. The number eight comes from the fact that this is the average amount of expansion of a combustible gas when it burns in air at optimum conditions. (Most deflagration type explosions reach a maximum pressure of eight times atmospheric pressure at stoichiometric conditions.)

Thus, the vertical flame speeds of combustible gases can be as much as eight times their burning velocities. This would mean, for example, that the vertical burning rate of a wooden match of about 8.67 mm per second, would be nearly 392 times less than the vertical flame speed of acetone, which is about 3400 mm per second at stoichiometric conditions.

Essentially, the accelerant burns so much faster than the wood or whatever is used, that the fire spreads throughout the wetted area before any of the underlying solid can significantly ignite and burn on their own. Thus,

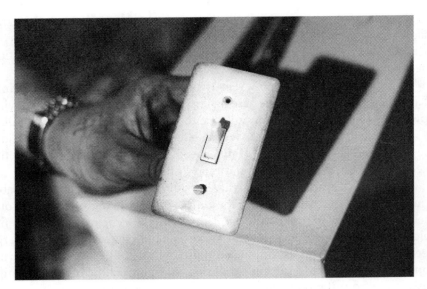

**Plate 7.2** Determining position of switch during fire from smoke stains on exposed portions of toggle.

the fire in the underlying solid seems to have begun basically all at once in the accelerant wetted area. The entire wetted area appears to be the point of origin, which when a person thinks about it, is exactly true.

Once the underlying solid has begun to burn, and the accelerant is largely consumed, the fire will once again follow the "normal" rules with respect to "V" patterns. This is one of the reasons why a fire that does not appear to follow the "normal" rules, may have had human intervention.

## 7.4  Flame Spread Ratings of Materials

The relative burning rates of materials used in buildings are often specified in certain building codes and specifications, especially with regard to interior finish. This, of course, is done to prevent the use of materials that allow unusually rapid and deadly spread of fire into living areas. The terms used in connection with the relative burning rates of materials are "flame spread rating" or "flame spread index."

One of the more common tests used to determine the flame spread rating of a material is the Steiner tunnel test. The specific procedures for this test are contained in ASTM E84, *Standard Method of Test for Surface Burning Characteristics.* Briefly, a sample of material 0.565 m wide by 7.32 m in length is burned and timed. The amount of time required to burn the material is then correlated to relative rating. Some common interior finish materials rated by this method are listed in Table 7.1.

**Table 7.1   Flame Spread Ratings According to ASTM E84**

| | |
|---|---|
| Gypsum plaster in ceiling | 0 |
| Enameled metal in ceiling | 0–20 |
| Wood-based ceiling tile with flame proofing | 20–75 |
| Wood-based ceiling tile with no flame proofing | 75–300 |
| Brick, concrete, or plaster walls | 0 |
| Enameled metal walls | 0–20 |
| Gypsum board | 20–75 |
| 1/2-inch thick wood | 70–200 |
| Cork sheets for walls | 100–250 |
| Untreated fiberboard | 200–500 |
| Plywood paneling | 70–300 |
| Shellac finish on paneling | >500 |
| Concrete floor | 0 |
| Linoleum[a] | 50–600 |

[a]   In general, the use of the Steiner tunnel test is not recommended for floor coverings like carpet, linoleum, and such. For floor coverings, ASTM E648 or NFPA 253, *Standard Method of Test for Critical Radiant Flux of Floor Covering System Using a Radiant Heat Source,* is generally used. (See section F, Radiation, for a discussion of radiant heat flash-overs.)

As noted in Table 7.1, items that do not burn at all are given a "0" rating. Some items that are not ordinarily thought of as being flammable, such as enameled metal surfaces, can surprisingly burn and propagate fire, albeit poorly with respect to items like wood paneling.

Because it is not so easy to obtain about 7.32 meters, or 24 feet, of continuous sample from a fire scene, an alternate type of test is often done with materials taken from fire scenes. ASTM E162, *Standard Method of Test for Surface Flammability of Materials Using a Radiant Heat Source,* only requires a piece of material about 15 cm by 46 cm. ASTM E162 can be correlated loosely with the findings of ASTM E84. However, the fire investigator should be aware that the correlation is not exact. Since some building codes rely specifically on test results according to ASTM E84, there may be some legal problems if a proceeding depends precisely upon whether a particular material meets the fire code flammability spread rating per ASTM E84.

For carpets and rugs there is a specific test, often called the "pill test," which is used to determine their relative flammability. The pill test, is more formally entitled, "Standard for the Surface Flammability of Carpets and Rugs, FF I-70," (Federal Register, Vol. 35, #74, pg. 6211, April 16, 1970). This test measures the ignitability of carpets and rugs from small ignition sources such as dropped cigarettes and matches. Sometimes a carpet material that passes the pill test can still contribute to fire spread because of general chemical decomposition due to exposure to large sources of heat. When this is suspected to have occurred, the carpet should be tested according to ASTM E648.

With respect to the structural elements of a building, the fire resistance of endurance of those materials are given in "hour" ratings in accordance with testing procedures contained in ASTM E119, *Standard Methods of Fire Tests of Building Construction and Materials*. This testing procedure is also contained in NFPA 251.

In this test, structural materials or elements, such as a beam or column, are put into a special test furnace and the temperature of the materials on the protected side or space is plotted against lapsed time. The resulting curve is call the "fire endurance standard time-temperature curve." The test is continued until failure occurs. Failure would be one of the following:

- inability to carry rated load.
- cracking, bending, or other significant dimensional failure in the material.
- loss of surface continuity such that flames or hot gases can pass through the material; it is easily pushed aside by a stream of water.
- loss of insulation capacity such that the temperature on the other side rises to more than 250°F.

It has been found from experimental testing that a one-hour fire rating as per ASTM E119 corresponds to a fire load of about 10 pounds of ordinary combustible materials (like wood) per square foot of floor space in a building, or about 80,000 BTU/ft².

It is emphasized to the reader, however, that the hour fire rating of ASTM E119 is not to be considered as reflecting how long a particular material or structural member will actually endure in a fire. A "two-hour" rated column will not necessarily endure two hours of an uncontrolled fire; it may endure much less than two hours, or perhaps more than two hours. The hour fire rating of ASTM E119 is simply an arbitrary rating system. To avoid this confusion, it would have been better if the officials at ASTM had used a different term. Unfortunately, the term has become ingrained in many building codes and legal descriptions, and is with us to stay.

While some materials are intrinsically flammable, such as wood and natural textiles, often they can be chemically treated to reduce their flame-spread rating. Cellulose and textile fibers, for example, can be treated with a boric acid solution to reduce their rate of burning. Wood is often treated by pressure impregnation with mineral salts to reduce burning rates.

When these treatments are done, the material is often labeled as being fire resistant or fire retardant. Sometimes fire-treated materials can be substituted for all or a portion of gypsum board fire rated barriers when a certain hour fire rating is required by code. For example, in some jurisdictions, the use of fire-retardant treated wood trusses with a one-hour rated ceiling of gypsum board underneath might be substituted for a two-hour rated ceiling of gypsum board with untreated trusses.

However, when fire retardant chemicals are used, it is sometimes the case that the basic properties of the material are altered. For example, when wood is treated with certain fire retardants, the structural strength of the wood may be significantly reduced. This loss of strength must be well-considered by the designer and allowed for in the initial design. A simple substitution of fire retardant trusses or columns for untreated trusses or columns without proper design consideration could result in unacceptable structural weakness in a building. Thus, the use of such fire retardant treated materials should be a design decision by either the responsible architect or engineer.

Due to the effects of some fire-retardant chemicals on wood, there may even be geographic or ambient temperature limitations placed on the use of such treated wood. For example, wood trusses treated with one type of fire retardant might be restricted to areas where the average humidity is low, away from coastal areas, or where attic temperatures do no exceed 140°F.

Lastly, fire-retardant chemicals can sometimes be lost from a material by exposure to water or solvents, repeated washing, or inadvertent exposure to other chemicals that may react with them. When this occurs, it is possible that the protection thought to be present by the use of fire-retardant treated materials has been lost.

With fire-treated wood, for example, high humidity and high ambient temperatures might leach out the impregnated mineral salts. With fire-retardant-treated clothing for children, repeated washing in hot water or in chlorine-containing bleaches might leach out or react with the fire-retardant chemicals. This is why such clothing has tags that caution the owner against improper washing, and provides recommended washing instructions.

Other tests related to flame spread potential and flammability that may come up from time to time include the following.

**Open cup flash point**, ASTM D-1310. To determine the flash point of a liquid.

**Tag closed test**, ASTM D-56. To determine the flash point of liquids with a viscosity of less than 9.5 cSt at 25°C and a flash point below 93°C.

**Closed drum test**, ASTM D-3065. To determine the flammability of aerosols in a confined volume with an open flame available.

**Setaflash closed test**, ASTM D3278. To determine the flash point of paints, enamels, lacquers and the like having a flash point between 0 to 100°C and a viscosity of 150 st at 25°C.

**Pensky-Martens closed test**, ASTM D-93. To determine the flash point of fuel oils, lubricants, oils, suspended solids, liquids that form a surface film, and similar.

**CIPAC (Collaborative International Pesticide Analytical Council Handbook) MT 12 flash point** (Abel method). To determine the flash point of petroleum products and mixtures and viscous products.

**Flammability of aerosol products**, ASTM D-3065. To determine the flame projection length of aerosols.

## 7.5   A Little Heat Transfer Theory: Conduction and Convection

To more fully understand how fire propagates, it is first necessary to understand some basic principles about heat transfer. In general, heat can be transferred from one point to another by only three means: conduction, convection, and radiation. In any of the three modes, there must be a temperature difference or temperature gradient between the two points for heat transfer to occur. Heat will flow from the area or point where the temperature is higher, to the area or point where the temperature is lower.

A useful analogy between heat flow and water flow can be made. Let temperature be analogous to elevation, and heat be analogous to water. In the same way that water will roll down the elevation gradients and collect in the low spots, heat flows down the temperature gradients and collects in the cool spots. Like elevation, temperature is a measure of potential. When there is no difference in potential, there is no flow of either water or heat.

Conduction is where heat is conducted through a material by diffusion of the thermal kinetic energy through the material. The equation that describes one-dimensional conduction through a material is given in Equation (iii) below.

$$q = -kA[dT/dx] \qquad \text{(iii)}$$

where q = rate of heat conduction (energy per time), k = thermal conductivity of the material (energy/time × distance × temp), A = area of the path perpendicular to the direction of heat flow, and dT/dx = temperature gradient in the "x" direction, that is, the direction of heat flow.

From Equation (iii), it is seen that:

- when the area "A" is increased, the heat transfer increases.
- when the thermal conductivity "k" increases, the heat transfer increases.

- when the temperature difference "dT" increases, the heat transfer increases.
- when the thickness of the material "dx" decreases, the heat transfer increases.

In general, materials that have high electrical conductivity also have high thermal conductivity. Thus, materials like copper and aluminum, which are excellent electrical conductors, are also excellent thermal conductors. Similarly, materials like concrete or porcelain, which are poor electrical conductors, are also poor thermal conductors.

The second mode of heat transfer, convection, is actually a special type of conduction which takes place between a solid material and a moving fluid. As with regular conduction, in order for heat to be transferred, there must be a significant temperature difference between the solid material and the moving fluid. The fluid heats up or cools down as it passes over the surface of the solid material. By its motion, the fluid then transports the heat or cold to another point downstream. New fluid then replaces the old fluid that has moved on, and the process is repeated.

In natural or free convection, the fluid is set into motion by buoyant effects. The density of nearly every fluid is dependent upon its temperature. When a small portion of the fluid is either heated or cooled, that small portion becomes either more buoyant or less buoyant with respect to the rest of the fluid. Buoyant forces then cause the fluid to move away from the surface, and the old portion of fluid is replaced by a new portion of fluid.

When a portion of fluid is passed over the surface of the solid material, flows away, and then is returned to the surface to repeat the process over and over, the repeating pattern is called a circulation cell. If the system is in equilibrium so that the temperatures at specific locations in the system stay about the same over time, stable convective fluid circulation patterns develop.

For example, consider an old-fashioned steam radiator in a closed room when it is cold outside. When the radiator is hot, the air that comes into contact with the radiator heats up and rises. As it rises to the ceiling, the air begins to cool. When it meets the ceiling, it moves laterally across the ceiling, spreads out, and cools some more. When it has sufficiently cooled, it gets heavy and begins to sink to the floor. As it sinks, it is drawn into the current of air flowing towards the radiator. The air then contacts the radiator again, and the process repeats.

In forced convection, the natural convection effects are enhanced by fans or pumps that force the flow of fluid over the surface of the solid. In forced convection, usually the velocity of the fluid is such that free or natural convection effects are not a factor and are neglected. In other words, the velocity of the fluid as set into motion by a fan or pump, is usually many times faster than the velocity of the fluid as set into motion by natural convection effects.

Since fire in a house or business structure generally spreads more by convection than the other two modes, sometimes an arsonist will turn on fans, blowers, or strategically open windows to deliberately set up strong drafts to aid in the spread of the fire. In essence, he is attempting to set up forced convection rather than depending upon free convection. Forced convection will not only cause the fire to spread more rapidly, but can be used to direct the spread of the fire to specific areas in the building.

An important characteristic of free convection is that there is always a thin layer of the fluid next to the surface of the solid that which does not move. Essentially, this thin layer is stuck to the exterior surface of the solid. This thin layer is called the boundary layer. Heat from the solid surface is actually conducted through this thin boundary layer into the stream of moving fluid. In forced convection, however, this boundary layer is usually in a constant state of removal and replacement, especially when the flow is extremely turbulent.

The equation that describes convection-type heat transfer is as follows:

$$q = hA \, (\Delta T) \tag{iv}$$

where q = rate of heat flow, A = area involved in heat transfer, $\Delta T$ = temperature difference between solid surface and fluid, and h = convection heat transfer coefficient.

The heat transfer coefficient "h" for forced convection is given by the following equation:

$$h[D/k] = K[Du\rho/\mu]^{\alpha}[C_p\mu/k]^{\beta} \tag{v}$$

where D = dimension associated with the length of the heat transfer area (with pipes, it is diameter; with plates it is the length of contact across the flat surface), k = conductivity of fluid, K = an experimentally determined constant, $\rho$ = density of fluid, u = velocity of fluid, $C_p$ = heat capacity of fluid, $\mu$ = viscosity of fluid, $\alpha$ = experimentally determined exponential coefficient, and $\beta$ = experimentally determined exponential coefficient.

It is noteworthy that the above equation has been arranged in groups of dimensionless parameters in the manner consistent with the Buckingham pi theorem for similitude modeling of fluid systems. For those readers who may be interested in simulating fires and fire damage by the use of laboratory models, knowledge of similitude and the Buckingham pi theorem is a must.

In fluid mechanics and heat transfer, each dimensionless parameter in Equation (v) has its own name. This is because these same groupings of variables come up routinely in the mathematical description of fluid flow. The names of the parameters are noted below:

$$h[D/k)] \quad \text{Nusselt number.}$$

$$[Du\rho/\mu] \quad \text{Reynolds number.}$$

$$[C_p\mu/k] \quad \text{Prandtl number.}$$

In like fashion, the heat transfer coefficient "h" for free or natural convection is given by the following equation:

$$h[D/k] = K[D^3ga(\Delta T)/\mu^2]^\alpha [C_p\mu/k]^\beta \qquad \text{(vi)}$$

where g = gravity and a = density temperature coefficient.

As before, Equation (vi) has been arranged in groups of dimensionless parameters. The new parameter, the one in the middle of the right term which replaced the Reynolds number in the forced convection equation, is called the Grashof or free convection number. In comparing Equations (v) and (vi), only the middle parameter is different.

In general, the above equations can be reduced to the following typical relationships.

$q = C_1A(\Delta T)u^{6/5}$      Forced convection.      (vii)

$q = C_2A(\Delta T)^{5/4}$      Free convection when the Reynolds      (viii)
number is less than 2000
(laminar flow).

$q = C_3A(\Delta T)^{4/3}$      Free convection where the Reynolds      (ix)
number is more than 4000
(turbulent flow).

Here $C_1$, $C_2$, and $C_3$ are arbitrary constants determined by experiment and A is the heat transfer area.

It is left up to the reader to solve for "$C_1$," "$C_2$," and "$C_3$" in terms of the other variables listed in Equations (v) and (vi).

Examination of equations (v), (vi), (vii), (viii), and (ix) finds the following general principles at work.

1. Increasing the temperature difference increases the heat transfer.
2. Increasing the heat transfer area increases the heat transfer.

3. Increasing the fluid flow in forced convection increases the heat transfer.
4. Increasing the turbulence in free convection increases heat transfer.

In most fires, convection effects are responsible for much of the fire spread. As the reader may have already surmised, the reason why the burning velocity up a wall is several times greater than down a wall is because the primary mechanism for upward fire travel is convection, while the primary mechanism for downward fire travel is conduction.

Since most walls are made of insulating-type materials, the conduction of heat through the wall material is poor. In fact, most of the common materials that burn in a fire like wood, paper, textiles, and such are poor conductors of heat. This is why downward fire travel is generally so slow, except when accelerants or similar faster burning materials are involved.

## 7.6  Radiation

The third mode of energy transfer, radiation, is where heat is transferred from one point to another by electromagnetic waves, typically in the wavelength range of 0.1 to 100 microns. No intermediary material is needed. In fact, when there are intermediary materials, such as gases or suspended dust, the amount of heat transferred by radiation is decreased due to absorption of heat by the gas or dust.

In this method of heat transfer, the amount of energy radiated away into space is proportional to the fourth power of its surface temperature. The amount of heat or energy emitted this way is governed by the Stefan-Boltzmann law, given in equation (x) below.

$$q = \sigma \varepsilon A T^4 \tag{x}$$

where q = the amount of heat or energy emitted, $\sigma$ = 5.67 × $10^{-8}$ watts/m$^2$ ($^{\circ}$K$^4$), the Stefan-Boltzmann constant, A = area of the radiating surface, T = surface temperature in absolute units, and $\varepsilon$ = emissivity, that is, the efficiency of the surface with respect to a black body or perfect radiator.

For perfect radiators or black bodies, the "$\varepsilon$" value is equal to one. This means that the radiating surface is 100% efficient in allowing the energy to be radiated from the surface. Alternately, another definition of a black body is one that absorbs all the radiant energy that falls on it; it does not reflect away any radiation.

Absorptivity and emissivity are essentially the same thing. The only difference is the direction of the flow of radiation. Emissivity typically refers to radiation leaving the surface while absorptivity refers to radiation being

**Table 7.2 Absorptivities of Some Common Materials**

| Material | Absorptivity |
|---|---|
| Water | 0.96 |
| Rough red brick | 0.93 |
| Concrete | 0.63 |
| Aluminum foil | 0.087 |
| Polished silver | 0.02 |
| Hot tungsten filament at 6000°F | 0.40 |
| Lampblack | 0.97 |
| Well-rusted steel | 0.95 |
| Enamel paints, any color | 0.90 |
| Fire brick | 0.70 |
| Molten steel | 0.30 |
| Polished sheet steel | 0.55 |
| Hard rubber | 0.95 |
| Roofing paper | 0.92 |
| Steam at 540°F and 1 at., 1 ft. thick | 0.28 |

received by the surface. Numerically, the two parameters are the same, which is known as Kirchoff's identity. More formally, it is stated: the emissivity and absorptivity of a body are equal at the same temperature.

When radiation strikes a surface, there are only three things that can happen to it. It can be absorbed; it can be reflected away, as occurs with a highly polished, mirror type surface; or it can be transmitted through the material, as occurs with glass.

Table 7.2 lists some representative values for the absorptivity of some common materials at room temperature except where noted. The specific emissivity of a material does change with temperature.

When two radiating bodies "see" one another, they will transfer energy to one another. Assuming no energy losses due to absorption of energy by air, water vapor, etc. the net energy exchange between the two bodies is given by the solution of the Stefan-Boltzmann equation for both radiating bodies:

$$q_{net} = \sigma\varepsilon CA[T_1^4 - T_2^4] \tag{xi}$$

where C = factor which accounts for relative position and geometry of the two surfaces relative to one another, A = area of the surface at which heat flow measurements are being made, and $\varepsilon$ = relative emissivity/absorptivity of the surface at which heat flow measurements are being made.

For example, consider a small electric heater used to provide spot warmth in a chilly room. The heater uses 1500 watts of electrical power. If we assume that the area of the heating coils is 0.01 m$^2$, the emissivity is 0.55, and the surface temperature of the resistance coils is about 1000°K when operating, then the proportion of input energy converted to radiant energy can be estimated by the application of the Stefan-Boltzmann relationship.

$$q = \sigma\varepsilon AT^4$$

$$q = [5.67 \times 10^{-8} \text{ watts/m}^2 \text{ (°K}^4)][0.55][0.01\text{m}^2][1000°\text{K}]^4$$

$$q = 312 \text{ watts.}$$

Thus, 312 watts or $312/1500 = 21\%$ of the energy input to the heater ends up as radiant heat. The rest of the energy is dissipated convectively.

Consider also the following example of radiant heat transfer from the ceiling of a room to its floor. Let's say that the whole ceiling is on fire such that the flames form a deep, luminous layer across the ceiling. The ceiling is 7 m by 10 m and the distance from the floor to the layer of flames is 2.25 m. Under these conditions, factor "C," which accounts for geometry and position, is equal to 0.7.

If the luminous flames are relatively hot and are burning at a temperature of about 1350°K, and it is noted that the paper items on the floor (which burn at a temperature of 550°K) have ignited, then the heat transfer from the ceiling to the floor is as follows:

$$q_{net} = \sigma\varepsilon CA[T_1^4 - T_2^4].$$

$$q_{net} = [5.67 \times 10^{-8} \text{ watts/m}^2 \text{ (°K}^4)][0.65][0.70][70 \text{ m}^2] \\ [(1350°\text{K})^4 - (550°\text{K})^4].$$

$$q_{net} = 6{,}163{,}000 \text{ watts.}$$

$$q_{net}/A = 88{,}100 \text{ watts/m}^2.$$

Thus, in this case, in order for fire in the ceiling to flash over and cause items on the floor to ignite by radiant heat, it is necessary for the fire in the ceiling to be at 1350°K, which is about 1970°F, or almost as hot as is needed to melt copper, and the rate of radiant heat transfer 88,100 watts/m². 

This is a heat rate roughly equivalent to burning 120 grams of liquid octane gasoline per minute per square meter of ceiling and having all the released energy be in the form of radiant energy. Of course, in actuality, it would take 5 to 25 times as much fuel as this because not all of the released energy becomes radiant heat. Only a small fraction will become radiant heat. Most of the released energy will be soaked up by convection and conduction effects.

An interesting point to note with respect to radiation flashovers is that they more often occur when there is a luminous fireball or layer of luminous flames at the ceiling rather than on the floor. In other words, flashovers more often occur when radiant heat transfer is occurring from the ceiling to the floor rather than vice versa. Why is this? A check of Table 7.1 provides the answer.

When most fires burn, the flue gases contain ample amounts of water vapor and unburned particles. The unburned particles are mostly soot. Both of these materials are very good absorbers of radiant energy. Thus, when viewed from above, a hot, luminous fire on the floor is cloaked by its flue gases of water vapor and soot particles, which soak up much of the radiant energy before it can reach the ceiling.

On the other hand, when the fire is located in the ceiling, the "butt" of the flames can poke through the cloud of flue gases. This is, of course, because the flames are actually hot expanding gases that are pushing away from the ceiling. When the expansion is exhausted, the fumes then simply turn and begin their convective rise back toward the ceiling.

When radiation leaves a hot, luminous object, unless it is guided, confined, or otherwise directed in some fashion, it will lose intensity according to the inverse square law.

$$I_1/I_2 = r_2^2/r_1^2 \text{ or } I_1 r_1^2 = I_2 r_2^2 = K \tag{xii}$$

where $I$ = radiant energy intensity in watts/m$^2$, $r$ = distance from source of radiant energy, and $K$ = constant equal to the radiant energy output at the source.

Thus, if the radiant energy intensity measured at 1 meter from the source is 100 watts/m$^2$, then at 4 meters, the radiant energy intensity is:

$$[100 \text{ w/m}^2]/I_2 = 16 \text{ m}^2/1 \text{ m}^2$$

$$I_2 = 6.25 \text{ w/m}^2 \text{ and } K = 100 \text{ w}.$$

Luminous fire is actually a cloud of burning flammable particles. The temperature of a burning cloud and its emissivity can be measured at a distance by the use of an instrument called an optical pyrometer. In general, when temperatures are more than 600°C, pyrometers are generally used to measure the temperature.

In optical pyrometers, the instrument works by comparing the color of the flames with that of a hot incandescent filament. The filament is viewed against the luminous cloud and the current through the filament is adjusted until the color of the filament matches the color of the cloud. The current through the filament is calibrated to correspond to the various temperatures of the filament.

In total radiation type pyrometers, the radiation is directly measured by a solar cell type semiconductor. The amount of radiant heat that falls on the semiconductor is converted to a current, and the current is calibrated to indicate the amount of incident radiation being absorbed and the temperature of the radiating body.

When steel is heated to a temperature of about 1000°F, it begins to glow with a very dull red color. As the steel is heated up, the color shifts from red to orange, and then to yellow. Near the melting point at about 2800°F, steel will radiate with a nearly whitish color. Thus, given the right circumstances, the temperature of some substances, like steel, can simply be estimated from their apparent, visible radiant color.

## 7.7   Initial Reconnoiter of the Fire Scene

In observing a fire scene for the first time, it is often best not to rush directly into the fire-damaged areas in search of the origin. Most fire investigators will first reconnoiter the fire scene to observe which areas did not burn.

This is important for two reasons: first and obvious is the fact that the areas that did not burn do not contain the point of origin. Often knowing what was not burned by the fire allows the elimination of many theoretical point-of-origin possibilities. The second reason is to determine the extent of fire damage to the building and to examine it carefully for structural weakness before entering it. A crippled building can be a death trap for the unwary investigator.

This last point cannot be emphasized enough. It is very prudent to first examine the fire-damaged structure before entering it to see where it has been structurally weakened, or is in peril of collapsing. Then, if needed, the proper equipment can be secured to do the job safely. Generally, with a little thought and imagination, every building, no matter how damaged, can be safely examined without putting people's lives in jeopardy. No fire diagnosis is worth dying for.

As noted previously, many investigators will search for the "low point" of the fire. This may be evident by conspicuous two-dimensional "V" patterns on walls, or perhaps less recognizable three-dimensional "V" patterns (or cone patterns) in stacks of materials. As discussed, fire spreads generally more by convection than conduction. Thus, it is often the case that the lowest point of burn in a building will be the point of origin.

However, as noted in the previous section, when a flashover from radiant energy transfer occurs, it can produce a false low point. For example, if the ceiling were to fill up with hot, luminous flames sufficient to radiate heat downward onto the tops of furniture and rugs, fires could break out in those places.

Fortunately, radiant heat flashovers are usually easy to recognize. They have the following characteristics.

1.  There will be a source point or area for the radiant heat. This will be an area that has sustained very hot fire usually over an extended area, like a ceiling or wall.
2.  There will be several radiant heat low point areas of ignition, and they will be positioned such that they can "see" the source point or area for the radiant heat.
3.  Other evidence indicates that these areas began burning well after other parts of the building or structure had already burned.

Other false low points can be caused by fall down debris. "Fall down" is the general term used to indicate materials that have become displaced to new locations, generally to floor areas, because of fire damage.

For example, in a particularly severe fire, burning roof shingles might fall into the basement space because fire has consumed the supporting structures between the roof and basement. The burning shingles may ignite small, secondary fires in the basement, perhaps even causing obvious "V" patterns. These secondary burn patterns, even though they may technically be the lowest burn patterns in the structure, are not the point of origin of the fire. They are simply secondary points of fire origination caused by burning fall down debris. However, sometimes they can mistakenly be taken for the primary point of origin.

In general, such secondary points of origin can be distinguished from the primary point of origin by the following.

1.  Since they occur later in the fire, the lateral spread from them is smaller and more limited. The fire spread from the secondary source may not even be contiguous with the main portion of the fire damage.
2.  Sometimes the piece of fall down that caused the secondary fire is located nearby and is recognizable.
3.  The fire at that point will have no other source of ignition except for the piece of fall down, which would not normally be associated with ignition. For example, it is not logical that timbers found in the attic could be the primary source of ignition for a fire that started in the basement.
4.  The secondary fire pattern does not meet the conditions necessary for it to be the point of ignition. The point of origin must be where all three components necessary to initiate fire were present at the same time and place when the fire began.

## 7.8 Centroid Method

Imagine that a large piece of cardboard is held horizontally. Now imagine that some point in the plane of the cardboard near the center is ignited. The resulting fire will then burn away from the initial point of ignition more or less equally in all directions in the plane. If the fire is extinguished before it reaches the edges of the cardboard, the fire will burn a circular hole in the cardboard. Of course, at the center of the hole is the point of origin of that fire.

The above is an example of the underlying principle of the centroid method for determining the point of origin of a fire. This method is particularly useful in single-story structure fires, where the building is made of more or less the same materials and has the same type of construction throughout. This homogeneity of materials and construction means that lateral fire spread rates will be similar in all directions.

Basically, the method works as demonstrated in the above example. The extent of burn damage in the structure is noted and perhaps even drawn to scale on a plan drawing. Then the center of damage is determined. This can be done by "eyeball," for those with calibrated eyeballs, or by more sophisticated methods including graphic integration and mathematical analysis.

Graphic integration involves the use of a drafting instrument called a planimeter, along with a detailed scaled sketch of the fire-damaged areas. The sketch is done on a Cartesian "x-y" plane and a convenient point of origin is selected. The sketch of the fire-damaged area is then sectioned off into smaller areas. The planimeter is used to measure the plane area encompassed by fire damage in the various sections. The distances from the center of these smaller areas to the respective "x" and "y" grid origins are then measured.

The position of the fire damage centroid is calculated using the equations for determining the center coordinates for plane centroids. The origin of the fire is then near or at the location of the centroid of the fire-damaged area.

$$x_c = \frac{\int(x)dA}{\int dA} \quad y_c = \frac{\int(y)dA}{\int dA} \tag{xiii}$$

where $dA$ = differential of burned area, $(x)$ = distance from "y" axis, $(y)$ = distance from "x" axis, $x_c$ = "x" coordinate of centroid, and $y_c$ = "y" coordinate of centroid.

Since most fire-burned areas are not regular and easily modeled by an integratable function, a plot of the burn pattern can be made over a grid system, and the areas can be determined by counting grid squares, or portions thereof. The integrals can be approximated by numerical evaluation as follows:

$$xc = \frac{\Sigma x_i(\Delta A_i)}{\Sigma(\Delta A_i)} \qquad yc = \frac{\Sigma y_i(\Delta A_i)}{\Sigma(\Delta A_i)} \qquad \text{(xiv)}$$

where $x_i$ = distance from "y" axis to area section, $y_i$ = distance from "x" axis to area section, and $\Delta A_i$ = incremental area section.

Some discretion must be used in applying the centroid method to allow for fires that start near an edge or corner. In the example using a piece of cardboard, the fire had plenty of material to burn in any direction of travel in the plane, and was put out before it reached an edge. Thus, the fire was able to move laterally away from the point of origin in all directions.

However, if the fire had begun at an edge or corner, it would have burned away from that point in a skew symmetric fashion. The edge or corner would prevent the fire from burning further in that direction; the fire then could only burn in directions away from the edge or corner. For example, a fire that begins at an edge will burn away from that edge in a semicircle pattern. A fire that begins in a corner burns away from the corner in a quarter-circle pattern. In short, the radial symmetry produced by a more or less equal fire travel rate in the lateral direction is affected when the fire must stop at a boundary for lack of fuel.

This shortcoming can be overcome by the use of mirror symmetry. For example, suppose that a fire began at the center of the east wall of a square-shaped building, and was extinguished before it spread more than halfway to the west wall. The fire pattern in the building would then appear roughly half-moon-shaped in the plan view. By placing a mirror alongside the east edge of the building plan, the burned area reflects in the mirror and a round burn pattern is then observed. The centroid method is then applied to both portions: the "real" side and the "virtual" or mirror side. For a corner fire, two mirrors or two mathematical reflections are needed.

When a fire begins near a wall or corner, but not exactly at the wall or corner, the mirror symmetry technique described above has shortcomings. In such an instance, the "virtual" portion of the fire damage must be estimated. The "virtual" fire damage is that portion that would have burned if the building or material was infinitely continuous in the horizontal plane. In essence, the fire damage that would have occurred had the fire continued is estimated, and a centroid is determined taking into consideration both the actual and the virtual damage.

## 7.9  Ignition Sources

When the centroid method has been completed, and the point or area of fire origination has been determined, that area must then be examined and inventoried for potential ignition energy sources. Such energy sources commonly include pilot lights, space heaters, electrical appliances, fluorescent light fixtures, fireplaces, chimneys, smoking materials, cooking equipment, lamps, electrical wiring and outlets, and so on.

Occasionally, the ignition of combustibles can be spontaneous, but this usually involves organic oils or materials that can readily undergo decomposition. It is commonly assumed that a pile of oily rags can catch on fire by itself. This is not true when the oil is a petroleum-based product. In fact, it is only true when the oil comes from an organic source, like linseed oil, tung oil, spike oil, and the like. These organic oils contain large amounts of organic acids that can react with air at room temperature. Motor oils and lubricants generally do not ignite spontaneously and are very stable.

Decaying fecal matter covered with straw and without ready access to air can spontaneously ignite. Also, wet hay or moist vegetative material can also spontaneously ignite when stored in bulk. However, spontaneous combustion is a multistep process initially involving bacterial decay. It typically occurs in barns, feed silos, or animal pens that have not been cleaned out in a while. The occurrence of such spontaneous combustion in one- or two-family dwellings is relatively rare compared to other causes.

The flaring up of smolders is often mistaken for spontaneous combustion. In a smolder, the combustible materials are sometimes starved for air, and thus the combustion reaction proceeds slowly. This effect usually occurs in a woven or porous material.

However, some types of smolders are simply a function of the material. Despite sufficient air, the rate of combustion for some materials proceeds very slowly. This is typical of some plastics and manmade organic materials and fibers.

In general, a smolder is any type of combustion process where the leading edge of the combustion zone moves only about 1 to 5 cm per hour. Because of this, a smolder produces modest heat and smoke, and is hard to detect when in progress. Smolders can be dangerous not only because of their obvious fire hazard, but also because while going undetected, they can release toxic gases such as carbon monoxide or even cyanide, which can slowly poison the atmosphere of an occupied room.

When an oxygen-starved smolder reaches a place where there is ample air, it may burst into flames. Similarly, when the leading edge of fire in a smoldering material reaches a more readily flammable material, it also may suddenly flare up. In both cases, the resulting fire may appear to have spontaneously combusted where the smolder finally burst into flames.

Some materials when heated to relatively low temperatures, decompose and then react to produce more heat, which eventually causes ignition of the material. This effect is also mistaken for spontaneous combustion. Polyurethane foam is a type of material that exhibits this behavior. When heated, perhaps by hot wiring or a hot flue pipe, the polyurethane foam undergoes chemical decomposition. The chemicals present after decomposition occurs, and react with one another and give off heat. If the heat cannot escape, it may accumulate sufficiently to ignite the rest of the foam.

## 7.10 The Warehouse or Box Method

Figure 7.2 shows a typical fire damage pattern that often occurs in single-level warehouse type buildings. The fire spreads in a radial pattern in the ceiling area and reaches one perimeter wall first, and then another. The perimeter wall that the fire reaches first will exhibit the most damage along its length. The wall that the fire reaches second will exhibit the secondmost fire damage, etc. The virtual fire-damaged areas, where fire would have spread had the building been larger in the horizontal plane, are also shown in Figure 7.2.

With respect to Figure 7.2, if a person simply bisects the wall damage on the right side and extends a line perpendicular to the wall into the building, the line will intersect the point of origin. By doing this also on the left side and on the bottom, three lines can be generated that intersect the point or area of origin. If the fire had spread to the top wall, a fourth line could also have been generated.

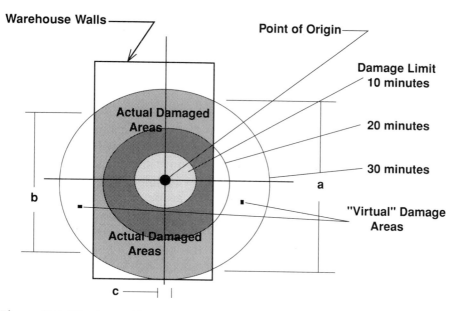

**Figure 7.2** Warehouse fire: actual and "virtual" damaged areas.

In warehouses and similar rectangular buildings, the point of origin can be quickly, and usually very accurately, estimated by this method. This technique is sometimes called the "box" method or the "warehouse" method for determining the point of origin. It works especially well in buildings where the fire has reached the ceiling area, and the ceiling area is generally open to fire spread.

In essence, the method simply requires observation of the fire damage along the exterior of the perimeter walls and the generation of perpendicular lines that bisect the damaged area. This is especially easy to do when the exterior walls are metal sheeting, and the paint has blistered away where the fire has directly impinged on the interior side.

The warehouse method of determining the point of origin of a fire and the centroid method are essentially the same technique. In both techniques, it is presumed that the lateral burn rate is about the same in all horizontal directions, and that the fire would burn out in a circular pattern of damage with the point of origin at the center. It is left up to the reader to prove the equivalency of the two methods.

## 7.11   Weighted Centroid Method

The weighted centroid method is similar to the basic centroid method except that this method assumes that the point subject to the longest duration of fire, and consequently the most severely damaged, is the point of origin. To this end, areas more severely damaged by fire are given more weight in determining the centroid than areas with less severe damage. As before, the method presumes a generally homogeneous structure in terms of material flammability and construction.

For example, the various fire-damaged areas can be divided into severity zones, with the most severe areas of burn damage given a weighted factor of "10." Areas with less severity of fire damage are assigned lower weighted factors as shown in Table 7.2. Areas with no fire damage are assigned a weighted factor of "0" and drop out of any equations.

A sketch of the building can be made drawing in the various areas that have been damaged. When the weighting factors are applied to the various areas, the resulting burn damage sketch will appear somewhat like a topographic map with contour lines. Whereas the contour lines in a "topo" map indicate elevation, the contour lines in the fire sketch indicate fire severity.

The centroid of the fire-damaged area can then be found by applying the weighting factors to the various areas affected. The equations for the determination of the centroid coordinates would then be as follows:

$$X_c = \frac{\Sigma x_i (\Delta A_i) f_w}{\Sigma (\Delta A_i)} \quad y_c = \frac{\Sigma y_i (\Delta A_i) f_w}{\Sigma (\Delta A_i)} \tag{xv}$$

where $f_w$ = weighting factor, $x_i$ = distance from "y" axis to area section, $y_i$ = distance from "x" axis to area section, and $\Delta A_i$ = incremental area section.

As with the basic centroid method, this method is often done by "eyeball" by an experienced fire investigator. However, the method also lends itself to more sophisticated computational analysis. In fact, there are several inexpensive computer-assisted drafting and design (CADD) programs available that will allow the building plan to be laid out on a computer screen and the fire-damaged areas drawn in as an overlay. Some of the CADD programs can then automatically find the centroid of an area.

It is also possible to scan in actual photographs of the fire scene and work directly from them in determining the point of origin, burn severity areas, etc. This, of course, requires that the photographs be taken from strategic vantage points. However, the use of actual photographs in conjunction with computer-generated overlays indicating the significance of the damage patterns visible in the photographs can be a very powerful demonstrative tool.

The "10" point system listed in Table 7.2 is somewhat arbitrary. Actually, any reasonable method of rating fire damaged areas can be used as long as the area of severest damage is assigned the highest weighting factor, and the area of least damage is assigned the lowest weighting factor. A simpler system using "1-2-3" is more practical for hand calculations. In that system, "3" is severe, "2" is moderate, "1" is light, and "0" is a no damage area.

One note of caution should be observed when using this method. Not all buildings burn homogeneously, of course. For example, if a fire had started in a kitchen and then spread to a storage area where an open 55 gallon drum of oil was stored, the most severely damaged fire area would be the storage

**Table 7.2  Weighting Factors for Fire Severity**

| Factor | Description |
|---|---|
| 10 | Materials are gone, wholly burned away |
| 9 | Materials are mostly gone, some residual |
| 8 | Materials are partially gone, recognizable residual |
| 7 | Materials are burned all over but shape intact |
| 6 | Materials are mostly burned |
| 5 | Materials are partially burned |
| 4 | Materials are slightly burned |
| 3 | Materials are heat damaged from nearby fire |
| 2 | Materials are heavily smoke damaged |
| 1 | Materials are slightly smoke damaged |
| 0 | Materials exhibit no significant fire damage |

**Plate 7.3** Fire in garage that originated in dashboard short in the van. Note the greater fire damage on the right side of the roof.

area, not the kitchen. Thus, the initial reconnoiter of the damage should note where there were fuel concentrations that might skew the weighting.

## 7.12   Fire Spread Indicators — Sequential Analysis

The point of origin of a fire can also be found by simply following in reverse order the trail of fire damage from where it ends to where it began. Where several such trails converge, that is the point of origin of the fire. In essence, the method involves determining what burned last, what burned next to last, etc. until the first thing that burned is found.

In the same way that a hunting guide interprets signs and markers to follow a trail of game, a fire investigator looks for signs and markers that may lead to the point of origin. For example, a fast, very hot burn will produce shiny wood charring with large alligatoring. A cooler, slower fire will produce alligatoring with smaller spacing and a duller-appearing char. Fire breakthrough or breaching of a one-hour rated firewall will likely take longer to occur than fire breakthrough of a stud and wood-paneled wall.

As noted, wood char and alligatoring patterns are useful indicators. The reader may recall that commercial wood is about 12% water by weight. As heat impinges on a piece of wood, the water in the surface material will evaporate and escape from the wood. The rapid loss of the water at the surface is also accompanied by a rapid loss of volume, the volume formerly occupied by the water. The wood surface then is in tension as the loss of water causes the wood to shrink. This is the reason why wood checks or cracks when exposed to high heat or simply dries out over time.

Of course, if the heat is very hot, more of the water "cooks" out, and the cracking or alligatoring is more severe. When the heat is quickly applied and then stopped, there is only sufficient time for the surface to be affected. When the heat is applied for a long time, there is sufficient time for the wood to be affected to greater depths.

Other indicators of temperature and fire spread include paint, finishes, coatings, and the condition of various materials (e.g., melted, charred, burned, warped, softened, oxidized, annealed, etc.).

For example, the paint finish on a furnace is a valuable indicator of the temperature distribution on the furnace. As the temperature increases to perhaps 250 to 400°F, the first thing to occur is discoloration of the paint. As the temperature increases over about 350 to 400°F, the paint will bubble and peel off, exposing the underlying primer. As the temperature increases again to more than perhaps 400 to 450°F, the primer will come off, and the undercoat, often zinc, will be exposed and oxidized. And as the temperature increases still more, perhaps beyond 786°F, the undercoat will melt away leaving only the bare steel, which itself oxidizes.

It is often easier to visually determine the hottest point on the furnace or appliance several days after the fire. The areas where all the paint, primer, and galvanized undercoating have been removed by high heat exposure will be bright red where the bare sheet steel has rusted.

Another example is metal ventilation ductwork. Metal ventilation ductwork is often made of galvanized steel, that is, steel with a thin exterior coating of zinc. The galvanization, which is normally shiny, will first dull and darken upon exposure to heat. When temperatures get above 500°F, the zinc will begin to oxidize significantly and will become whitish. As the zinc is heated past 500°F, it will whiten more and more. However, when the temperature approaches 786°F, the unoxidized zinc will melt and slough off, leaving bare steel exposed. (The zinc oxide itself will not actually melt until a temperature of about 3600°F, but it usually sloughs off with the unoxidized zinc underneath it.) Thus, the hottest spots on ventilation ductwork are also the red spots, where the exposed steel has oxidized to rust.

The interpretations of many such markers, "V" patterns, etc. are then combined into a logical construct of the fire path. One of the favorite tests used in the sequential analysis method is the question: "Which is burned more, the material on this side or that side?" The answer to this question then supplies a directional vector for the fire spread, and the "vector" is backtracked to another position where the question is again posed. In a sense, following a trail of indicators is like playing "Twenty Questions."

**Plate 7.4** Lawnmower was stored next to furnace in garage. Furnace pilot ignited fumes from spilled gas around mower fuel tank.

In order to avoid a "false" trail due to fall down, it is common to backtrack several fire trails from finish to start. When several such trails independently converge to a common point of origin, the confidence in the answer is greatly increased.

The advantage to the sequential method is that no special assumptions need to be made concerning structural homogeneity. The disadvantages are twofold. First, it relies upon the individual skill and knowledge of the investigator to find and properly interpret the markers. Not all fire investigators have the same knowledge about materials, fire chemistry, heat transfer, etc. One fire investigator may spot an important marker that another also saw, but ignored.

Secondly, it assumes that sufficient markers are present to diagnose the fire, and can be found. This is not always true. Sometimes the severity of the fire or the fire-fighting activities themselves destroy significant markers and indicators. Also, sometimes the markers may be present but are lost in the jumble of debris. Thus, there are gaps in the evidence, and the resulting sequential analysis is discontinuous.

## 7.13   Combination of Methods

Few fires lend themselves to complete analysis by only one of the methods described. Many require a combination of methodologies. For example, it is common to determine a general area where the fire began using one of the centroid methods, and then determine a specific point of origin using a combination of fire spread indicators and an examination of available ignition energy sources.

## Further Information and References

*ATF Arson Investigation Guide,* published by the Department of the Treasury. For more detailed information please see Further Information and References in the back of the book.

*Code of Federal Regulations 29,* Parts 1900 to 1910, and Part 1926 (two volumes). For more detailed information please see Further Information and References in the back of the book.

*The Condensed Chemical Dictionary,* revised by Gessner Hawley, 9th ed., Van Nostrand Reinhold Company, 1977. For more detailed information please see Further Information and References in the back of the book.

*Fire and Explosion Investigations*-1992 ed., NFPA 921. For more detailed information please see Further Information and References in the back of the book.

*Fire and Explosion Protection Systems,* by Michael Lindeburg, P.E., Professional Publications, Belmont, CA, 1995. For more detailed information please see Further Information and References in the back of the book.

*Fire Investigation Handbook,* U.S. Department of Commerce, National Bureau of Standards, NBS Handbook 134, Francis Brannigan, Ed., August 1980. For more detailed information please see Further Information and References in the back of the book.

*General Chemistry,* by Linus Pauling, Dover Publications, New York, 1988. For more detailed information please see Further Information and References in the back of the book.

*Handbook of Chemistry and Physics,* Robert Weast, Ph.D., Ed., 71st Edition. CRC Press, Boca Raton, FL. For more detailed information please see Further Information and References in the back of the book.

*Investigation of Fire and Explosion Accidents in the Chemical, Mining, and Fuel Related Industries — A Manual,* by Joseph Kuchta, United States Department of the Interior, Bureau of Mines, Bulletin 680, Washington DC, 1985. For more detailed information please see Further Information and References in the back of the book.

*NFPA Life Safety Code (NFPA 101),* National Fire Protection Association, February 8, 1991. Also listed as ANSI 101. For more detailed information please see Further Information and References in the back of the book.

*A Pocket Guide to Arson and Fire Investigation,* Factory Mutual Engineering Corporation, 3rd ed., 1992. For more detailed information please see Further Information and References in the back of the book.

*Principles of Combustion,* by Kenneth Kuo, John Wiley & Sons, New York, 1986. For more detailed information please see Further Information and References in the back of the book.

*Rego LP-Gas Serviceman's Manual,* Rego Company, 4201 West Peterson Avenue, Chicago, IL, 60646, 1962. For more detailed information please see Further Information and References in the back of the book.

*Standard Handbook for Mechanical Engineers,* by Baumeister and Marks, 7th ed., 1967. For more detailed information please see Further Information and References in the back of the book.

*Thermodynamics,* by Ernst Schmidt, Dover Publications, New York, 1966. For more detailed information please see Further Information and References in the back of the book.

# Electrical Shorting

# 8

Properly applied fuses and circuit breakers protect branch circuit conductors from reaching ignition temperatures of ordinary combustibles, even under short circuit conditions. However, lamp cords, extension cords, and appliances of lower rating than the branch circuit conductors may reach higher temperatures without blowing the fuse or tripping the breaker if there is a malfunction in the appliance or the cord.

— **Section 2-3.3, NFPA 907M**

## 8.1  General

A significant proportion of all fires occurring in structures are caused by electrical shorting. The building's wiring system, lighting fixtures, appliances, installed machinery, and extension cords are some of the more common items in which shorting occurs.

The shorting components themselves usually do not directly catch fire. Most electrical components contain insulated metal conductors. Except in unusual circumstances, the metal conductors themselves are not flammable. The insulation around the metal conductors do not ordinarily ignite or burn except in some older types of wiring where the insulation material may be flammable.

Shorting causes electrical conductors to excessively heat up. This may cause the plastic insulation material coating the conductors to melt and slough off, leaving the conductors bare. Fire can then ensue when flammable materials come into direct contact with the hot conductors.

If the short is sufficiently hot, the metal conductor itself may melt, flow, and drip onto flammable materials located below. The melting temperature of conductor materials such as copper and aluminum is usually higher than the ignition temperature of common construction materials such as wood, paper, and textiles. Sufficient heat transfer between the conductor drippings and the flammable materials during contact can result in the initiation of a fire.

Fire can also ensue when there is high-voltage electrical arcing. In such arcing, molten conductor droplets may spatter onto nearby flammable materials. Since spattering can cause molten material to be thrown off above,

135

below, and to the sides of the short, it is possible that fire from arcing can be initiated in places other than directly below the short.

Lastly, fire can also ensue when a flammable material is sufficiently close to the shorting component that it ignites due to radiative, conductive, or convective heat transfer directly from the electrical arc plasma. The arc plasma created by electrical shorting is the same type of plasma created during arc welding, which can have a temperature ranging from 2500°F up to 10,000°F. In fact, a significant number of accidental fires associated with arc welding occur each year. Usually, they occur because the welder did not adhere to proper safety procedures. One common situation is when flammable vapors drift over to and collect in an area where arc welding is being done.

While shorting is indeed a common cause of fires, unfortunately some fires not caused by shorting are conveniently blamed on electrical shorting. It is the case that electrical shorting is the "cause of last resort" for some investigators. This is because nearly all inhabited buildings in the U.S. have electrical wiring of some type. When a building catches fire and burns, it is probable that the fire will cause something electrical to short out, no matter what actually caused the fire. Thus, an investigator who cannot determine the specific cause of the fire, can always find some evidence of shorting to blame as being the cause, and close out his paperwork.

Because of this, it is necessary to discriminate between primary shorting and secondary shorting. Primary shorting is shorting that causes the fire. Secondary shorting is shorting that is caused by the fire. Of course, both types can and do occur in the same fire. Primary shorting can occur at one location and cause a fire to ensue. The fire can then burn up energized electrical equipment at another location, resulting in secondary shorting within that equipment.

In general, primary shorting has the following characteristics.

- It occurs at or close to the point of origin of the fire. There are indications of fire spread away from the short, and the point of shorting is often in the area of severest burn damage.
- Heat damage to the conductor is more severe at the interior than at the exterior (inside-to-outside damage pattern).
- Significant movement or travel of the short has occurred, i.e., the short appears to have been active for a relatively long time. Ample beading may be present.
- The severest damages in the electrical item that shorted are limited to a small area proximate to the short, rather than being spread over a large, general area.
- In consideration of the whole body of evidence, it is the short that must have occurred first in the timeline of the fire.

**Plate 8.1** Classic "V" pattern emanating from a short in the wire where it abraided on the hole and grounded out.

Secondary shorting has the following general characteristics.

- It occurs in locations away from the point of origin of the fire. There are indications of fire spread to the short from other locations. The short may be in a general area where, except for the short itself, there is little difference in fire damage severity. This indicates that the area had been approached by a fire front spreading from another area.
- The conductor interior may not be as severely damaged as its exterior (outside-to-inside damage pattern).
- Little movement or travel of the short has occurred. The short appears to have been active for only a short time. Beading effects may be limited.

**Plate 8.2** Shorting at switch such that the switch no longer operates the equipment. The short turned on the equipment.

- In consideration of the whole body of evidence, it is a short that may have occurred at any time during the course of the fire.

It should be noted that the above characteristics cited for both primary and secondary shorting are generalizations. They may not all apply to a specific case.

For example, consider the situation where a fire begins in one short extension cord due to shorting, and the ensuing fire then engulfs a second short extension cord plugged into the same outlet, causing secondary shorting to it. Both shorts would have occurred near the point of origin of the fire, and both might be located in the area of severest burn.

Similarly, consider the situation where a short causes a fire, which then spreads to where flammable fuels are stored. When ignited, the fuels cause very severe damage in their general area, including damage to some electrical wiring. More severe, in fact, than what occurred at the point where the fire began.

## 8.2   Thermodynamics of a "Simple" Resistive Circuit

Consider a simple circuit that only has resistance and an alternating voltage source, such as shown in Figure 8.1. Like most household circuits, the voltage varies sinusoidally.

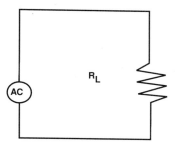

**Figure 8.1** Simple resistive circuit.

In this circuit, the instantaneous current flowing through the resistive load follows Ohm's law.

$$E_{max}(\sin \omega t) = I(R_L) \qquad \text{Ohm's law} \qquad (i)$$

where I = current, $R_L$ = resistive load in circuit, $E_{max}$ = maximum voltage amplitude, t = time, and $\omega$ = frequency of alternating current given in radians per second. 1 Hertz = 2p radians/sec.

Given that the resistive load is constant, rearranging Equation (i) shows that the current through the circuit varies as follows:

$$I = (E_{max}/R_L)(\sin \omega t) \qquad (i\text{-}a)$$

If it is assumed that the root-mean-square values for current and voltage are substituted for the instantaneous values, then Equation (i) becomes simply:

$$E = I R_L. \qquad (ii)$$

The amount of energy consumed by the resistance in the circuit per unit of time, i.e., the power consumed by the circuit, is given by:

$$P = EI = I^2 R_L = E^2/R_L \qquad (iii)$$

where the E and I terms are understood to mean root-mean-squared values for voltage and current, respectively, and the circuit is either DC or single phase alternating current.

The above is usually referred to as the power of the circuit and the common units for electrical power are watts. One watt is the product of one volt of potential and one ampere of current. In other units, one watt is equivalent to one joule of energy per second, 3.413 BTU per hour, or 0.7376 lbf-ft/sec. A kilowatt, 1000 watts, is equivalent to 1.341 horsepower in English units.

In the circuit shown in Figure 8.1, since the load in the circuit is purely resistive, the power consumed by the load becomes heat. As power is consumed, the temperature of the resistive load will increase until its cooling rate, or heat transfer from the load, equals the electrical power being consumed. When that occurs, the circuit is in thermal equilibrium with the environment and the temperature of the resistive load will stabilize.

In less technical terms, the above situation is analogous to a large holding tank that has a water input valve and a drain valve. The input valve is like the power being consumed by the resistive load, the cooling rate is like the drain valve, and the holding tank is like the mass associated with the resistive load. If the drain is closed and the input valve is open, the tank will fill and the water level (or temperature) will rise. If the drain is opened a little, the tank will fill more slowly. If the rate of drain is set equal to the rate of water input, the tank will neither fill nor empty, and the water level in the tank will stabilize.

Mathematically, the above situation is described by the following linear differential equation.

$$P - (dQ/dt) = mC_p(T - T_o)/t \qquad (iv)$$

where P = power input to the component, $dQ/dt$ = rate of heat transfer from the circuit to the environment, m = mass of the resistive load, $C_p$ = specific heat at constant pressure of the resistive load, $T_o$ = ambient temperature (in absolute units), T = temperature of resistive load (in absolute units), and t = lapsed time.

In the above equation, the first term, "P," is the amount of energy coming into the resistive load that is converted into heat. The second term, "$dQ/dt$," is the amount of heat removed from the resistive load by cooling. The right-hand term in the equation, "$mC_p(T - T_o)/t$," is the rate at which heat is being stored in the resistive load. As more heat is stored, the temperature of the load or resistance increases.

In Equation (iv), the mass in the wires connected to the resistive load has been ignored to keep things simple. It is assumed that the only item heating up is the mass of the resistive load.

Despite the above simplifying assumption, Equation (iv) is still fraught with complications. First, the resistance of the load changes with temperature. For most of the common materials used in electrical conductors, as temperature increases, resistance increases. However, there are notable exceptions, such as carbon film resistors. They actually lose resistance with increasing temperature within certain ranges. In either case, the relationship between resistance and temperature is typically described mathematically as follows:

$$R = R_o(1 + \alpha(T - T_o)) = R_o + R_o\alpha(T - T_o) \tag{v}$$

where R = resistance, $R_o$ = resistance at ambient temperature $T_o$, $\alpha$ = temperature coefficient of resistance, and T = temperature.

When the material resistance increases with increasing temperature, "$\alpha$" is positive. When the material resistance decreases with increasing temperature, "$\alpha$" is negative.

To complicate matters just a bit more, the coefficient of resistance, "$\alpha$," also varies with temperature. With common conductor materials, like copper or aluminum, it diminishes slowly in absolute value as temperature increases. For example, for aluminum at 0°C, "$\alpha$" has a value of +0.00439/°K. At 25°C, the value decreases to +0.00396/°K, and at 50°C the value decreases further to +0.00360/°K. However, for this simple model of a resistance circuit, it is assumed that the average coefficient of temperature between the two end point temperatures is used.

Another complication of Equation (iv) is that the rate of heat transfer is also dependent upon temperature. As the temperature difference between the hot resistive load and the ambient increases, the rate of heat transfer increases. This is shown in Equation (vi) below.

$$dQ/dt = UA(T - T_o) \tag{vi}$$

where U = heat transfer coefficient for convection and conduction around resistive load, and A = heat transfer area.

And now, to be really ornery, the "U" term in Equation (vi) is also dependent upon temperature. In fact, between some regions, especially the transition zone between laminar and turbulent convective flow, "U" can be decidedly nonlinear. However, again to keep things simple, it will be assumed that in the range under consideration, the "U" value is constant.

If Equations (v) and (vi) are substituted back into Equation (iv), then an expression is obtained that approximately depicts the relationship of temperature and electrical power consumption in a "simple" resistive circuit.

$$P - (dQ/dt) = mC_p(T - T_o)/t \tag{vii}$$

$$I^2R - UA(T - T_o) = mC_p(T - T_o)/t \qquad \text{Substituting Equations (iii) and (vi)}$$

$$I^2[R_o(1 + \alpha(T - T_o))] - UA(T - T_o) = mC_p(T - T_o)/t \qquad \text{Substituting Equation (v)}$$

Collecting terms and simplifying gives the following.

$$I^2R_o + I^2R_o\alpha T - I^2R_o\alpha T_o - UAT + UAT_o = mC_p(T)/t - mC_p(T_o)/t \quad \text{(viii)}$$

$$T[I^2R_o\alpha - UA - mC_p/t] = T_o[I^2R_o\alpha - UA - mC_p/t] - I^2R_o$$

$$T = T_o + [I^2R_ot]/[(mC_p) + UAt - I^2R_o\alpha t]$$

The foregoing expression allows the increase in temperature of the resistive load to be calculated provided that the following factors are known: the applied current, the heat capacity and mass of the resistive load, the initial resistance at ambient temperature, ambient temperature, the heat transfer coefficient, and the area of heat transfer.

In the derivation of Equation (viii), the expression "$I^2R$" was substituted for "$P$," the power term, in Equation (iv). However, an alternate substitution, "$E^2/R$," can be made for "$P$" the power term, as follows:

$$P - (dQ/dt) = mC_p(T - T_o)/t \quad \text{(ix)}$$

$$E^2/R - UA(T - T_o) = mC_p(T - T_o)/t$$

$$[E^2/R_o(1 + \alpha(T - T_o))] - UA(T - T_o) = mC_p(T - T_o)/t$$

Collecting terms and simplifying gives the following:

$$E^2 - UA(T - T_o)[R_o + R_o\alpha(T - T_o)] = \quad \text{(x)}$$
$$[mC_p(T - T_o)/t][R_o + R_o\alpha(T - T_o)]$$

$$[E^2]/[R_o(mC_p/t + UA)] = (T - T_o) + \alpha(T - T_o)^2$$

Equation (x) allows the calculation of the temperature of the resistive load if the following is known: the applied voltage, the initial resistance, the ambient temperature, the heat transfer coefficient and area, and the heat capacity and mass of the resistive load. Unfortunately, Equation (x) is not as easy to work with as Equation (viii). While Equation (viii) is a linear equation in "$T$," Equation (x) is a quadratic in "$T$." Thus, Equation (x) could have two solutions: one real and one extraneous.

Of course, it is possible to take Equation (viii), and substitute "$E^2/R_o$" for the "$I^2R_o$" terms, and obtain the following:

$$T = T_o + [E^2t/R_o]/[(mC_p) + UAt - \alpha E^2t/R_o] \quad \text{(xi)}$$

**Table 8.1  Some Temperature Coefficients of Resistance at 20°C, or 293.15°K[a]**

| Material | $\alpha$, Given in Dimensionless Units per Degree Kelvin |
|---|---|
| Aluminum | 0.00403 |
| Brass | 0.0036 |
| Copper wire | 0.00393 |
| Steel | 0.0016 |

[a] *Standard Handbook for Mechanical Engineers,* 7th ed., Marker and Baumeister, McGraw-Hill, New York, 1967, pp. 15–8.

Equation (xi) is useful in that it does not have the quadratic form of Equation (x) and closely resembles the linear Equation (viii).

In Equations (viii), (x), and (xi), there are terms included to account for the variation of resistance with temperature. Table 8.1 lists some common values for "$\alpha$," the coefficient of resistance with respect to temperature.

To determine how important these terms might be, consider the following. Given a 40°K rise in temperature, from 0°C to 40°C, a copper wire has an increase in resistance of

$$[1 + (0.00393/°K)(40°K)] = 1 + 0.157 = 1.157 \text{ or } 16\%.$$

A similar 100°K rise in temperature, from −30°C to 70°C would result in a 39% increase in resistance. Thus, the increase in resistance caused by increased temperature is a significant factor and should not be neglected when significant temperature increases are involved.

In some electrical circuits the applied voltage can be considered constant within certain ranges. Consider what theoretically occurs in that situation due to the change in resistance with temperature.

$$E = IR = I[R_o + R_o\alpha(T - T_o)] \tag{xiia}$$

$$P = EI = E^2/[R_o + R_o\alpha(T - T_o)] \tag{xiib}$$

As the resistance increases due to increased temperature, in order to maintain a constant voltage, the current must reduce proportionally. Likewise, as the resistance increases with increasing temperature, the power expended diminishes. This is sometimes called "heat choke" of the current.

Similarly, in some electrical circuits the applied current can be considered constant within certain ranges. In that case, the following applies.

$$I = E/[R_o + R_o\alpha(T - T_o)] \tag{xiiia}$$

$$P = I^2[Ro + R_o\alpha(T - T_o)] \tag{xiiib}$$

As the temperature increases, the resistance increases and the voltage must also increase to maintain a constant "I." In turn, this causes the amount of power being consumed to increase linearly. Thus, in constant current applications, heating of the resistive load will cause the applied voltage to increase, and the power to increase.

Sometimes, equipment is designed to ensure that the power of the system remains constant within certain operating ranges. In those types of circuits, the product of voltage and current will be constant. Thus, in such a circuit, if there was a low voltage condition, the current would correspondingly increase to maintain constant "P." Similarly, if the current were to drop, the voltage would increase accordingly. These circuits, theoretically, will neither increase nor decrease in power as the temperature of the resistive load increases.

While very idealized, the basic principles noted in the foregoing analysis of a simple resistive circuit have wide general application. They apply to many types of electronic equipment, lighting equipment, and resistance-type heating equipment. These appliances for the most part convert electricity to heat like a simple resistive load.

In fact, in determining the internal heat load of an office space or building, it is common practice to simply sum the various power consumption ratings of the appliances in a given space. This includes copy machines, lights, telephones, computers, coffee makers, etc. Whatever electricity they consume is eventually converted to heat and is released into that same space.

However, when electricity is converted into mechanical work, such as in an electric motor, a modification of the basic thermal energy equation is required. An additional term is added to account for the conversion. When electricity is converted into mechanical work, significant amounts of energy can leave the component without causing the component itself to heat up. Eventually, the mechanical work is converted to heat also. But, it is possible that the conversion to heat will not be associated with the electrical component that originally generated the work. In other words, the mechanical work produced by the component can cross the boundary of the system, and the eventual conversion to heat can occur somewhere else. For this reason, mechanical work is considered an energy loss term in the same way as heat transfer from the component.

To account for mechanical work, Equation (vii) is modified as follows:

$$P - (dQ/dt) - W/t = mC_p(T - T_o)/t \tag{xiv}$$

where W = work output of the equipment.

By substituting as was done in Equation (vii) to obtain Equation (viii), the following is obtained:

$$T[I^2R_o\alpha - UA - mC_p/t] = \\ T_o[I^2R_o\alpha - UA - mC_p/t] - I^2R_o + W/t \tag{xv}$$

$$T = T_o + [I^2R_ot - W]/[(mC_p) + UAt - I^2R_o\alpha t]$$

Similarly, by substituting as was done in Equation (ix), the following is obtained:

$$E^2 - [UA(T - T_o) + W/t][R_o + R_o\alpha(T - T_o)] = \\ [mC_p(T - T_o)/t][R_o + R_o\alpha(T - T_o)] \tag{xvi}$$

$$[E^2t - WR_o]/[R_o(mC_p + UAt + W\alpha)] = (T - T_o) + \alpha(T - T_o)^2.$$

It is left up to the reader to incorporate a mechanical work term into Equation (xi) and derive the results.

The efficiency of a motor or machine, that is, the conversion of electrical energy to useful mechanical work, is given by:

$$\eta = W/Pt \tag{xvii}$$

This is what typically occurs when a motor wears out. As it wears, it becomes less efficient; more of the input energy becomes heat instead of useful work. The motor then runs hotter. Given time and further deterioration, the motor may become hot enough for general failure to occur. The bearings may overheat, or the dielectric in the windings may fail allowing internal shorting to occur. If there is no internal thermal protection, such as a bimetallic thermal switch in the windings, the motor may eventually burn out and catch fire. Often, the dielectric insulation covering the windings is flammable and can constitute a significant fire load. If there is internal thermal protection, the motor will turn off if the local area around the switch reaches the set point temperature. This may prevent the insulation from catching fire.

Some thermal protection switches will reset when the motor cools down, and some may stay tripped. If it is the former, the motor may restart when it cools and run a short time until the temperature again builds up. This start-stop-start behavior is often encountered in air conditioner compressors that are in the last stages of operational life. Eventually, the motor will draw

**Plate 8.3** Electrical arcing damage to sheet metal in switch gear.

enough current to cause the breakers servicing the unit to trip. This, of course, assumes that the breakers have been properly sized.

## 8.3   Parallel Short Circuits

In an electrical circuit, conductor pathways are provided so that electricity can flow to and from an appliance or component to perform some function. A short circuit is when the electrical current flows through an unintended pathway, i.e., a "shortcut" electrical pathway, between the conductors.

There are two basic types of short circuits: shorts that create parallel pathways for the electricity, and shorts that create series pathways for the

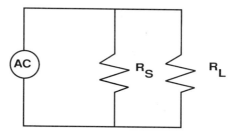

**Figure 8.2** Parallel short circuit.

electricity. In the former, the short can be considered as a parallel resistive load, and in the latter, the short can be considered as a series resistive load.

Consider the parallel resistive load first. A simplified schematic of a parallel, short circuit resistive load is shown in Figure 8.2.

The equivalent resistance for two resistive loads in parallel is given below.

$$[1/R_{Eq}] = [1/R_S] + [1/R_L]. \qquad (xviii)$$

$$R_{Eq} = [R_S][R_L]/[R_L + R_S].$$

When there is no short, the value for "$R_S$" is infinite and "$R_{Eq} = R_L$."

Consider the following example of a short circuit in the cord of a toaster. The resistance of the heating elements in the toaster is 20 ohms. The electric cord for the toaster is rated for 15 amperes of current. If the toaster is plugged into a 120 VAC source of electrical power and the toaster operates normally, the current and power to the toaster is:

$$I = 120 \text{ v}/20 \text{ ohm} = 6 \text{ amperes.}$$

$$P = (120 \text{ v})(6.0 \text{ a.}) = 720 \text{ watts.}$$

Assume now that an internal break has developed in the toaster's electrical cord due to the cord being bent in an excessively tight radius. The break is such that the insulation is much thinner there than elsewhere. Assume also that because of this break the resistance across the insulation to the other wire is 100,000 ohms instead of the typical 10,000,000 ohms.

In this case then, the equivalent resistance of the toaster and the short circuit in parallel is 19.996 ohms. Assuming that a constant voltage is being supplied to the circuit, that is, there is a constant voltage applied across the equivalent resistance, then the current and power consumed by the equivalent resistance is given by:

$$I = (120 \text{ v}/19.996 \text{ ohm}) = 6.001 \text{ amperes.}$$

$$P = (120 \text{ v})(120 \text{ v}/19.996 \text{ ohm}) = 720.14 \text{ watts.}$$

Now, within the parallel circuit, the toaster resistance is consuming the same 6 amperes of current and 720 watts of power. However, the resistance in the short circuit is consuming 0.001 amperes of current and 0.14 watts of power. Since the short circuit is a pure resistance, this power is being converted to heat and is concentrated in the area of the insulation break. The concentrated heat then causes further deterioration of the insulation. The deterioration may involve melting, charring, or just general degradation of the dielectric properties of the insulation. This results in further breakdown of the insulation.

Because of the damage caused by heating effects, the short resistance now drops to 1,000 ohms, a factor of 100 less than before. In this case, the equivalent resistance is now 19.608 ohms. This causes a current consumption of 6.12 amperes, and a power consumption of 734.4 watts. Again, the toaster is still consuming 720 watts, but the short is consuming 14.4 watts. As before, the heat produced by the short is concentrated in the area of the short, which further degrades the insulation.

In the final stages of the short, assume that the resistance has dropped to 1 ohm. The concentrated heat has melted away nearly all of the insulation between the wires, and the wires are essentially conductor to conductor now. Presuming that the toaster has not yet been separated from the circuit due to melting, the equivalent resistance of the toaster and short circuit in parallel with each other is 0.952 ohms. This results in a current consumption of 126 amperes, which well exceeds the 15 ampere rating of the cord. Hopefully, this would cause the breaker for the branch circuit to trip. If not, it is likely that the conductor in the wire will melt away, as the theoretical power input is now 15.12 kilowatts.

If the breaker or fuse does not open the circuit, often the short will "follow" the conductor back to its source of power as far as it can while the circuit is intact. That is, the conductor will melt, spark, arc, and burn along its length back toward the source of its electrical power. The effect is somewhat like the older type dynamite fusee. When lighted, the burning follows the fusee cord back to the stick of dynamite. Of course, the short does not move away from the source of power, because the short itself has cut off any power in that portion of the conductors.

In the above example, several generalizations are of note.

- As the resistance in the short drops due to heat degradation, the overall resistance of the combined parallel circuit reduces.

- Until the short causes enough damage to sever the circuit, the operating appliance that is parallel to the short will continue to operate as it had, consuming its usual amount of power.
- As the resistance of the overall parallel circuit decreases, the amount of current consumed increases proportionally. This occurs despite the fact that the circuit has a constant voltage source. Heat choking of the current does not occur because of the parallel resistance arrangement.

It is apparent in the parallel resistance equations, that a parallel short will proceed and allow electrical current flow whether or not the appliance is operating. If the appliance is operating, the current is divided, with some of the current flowing through the appliance and some through the short. If the appliance is turned off, and "$R_L$" is therefore infinite, the circuit simply reverts to a simple, single resistance circuit, where the resistance of the short is the only resistance consuming power in the circuit.

For example, if the short circuit resistance was 10,000 ohms and the same toaster is considered, the equivalent resistance is 19.96 ohms. The current consumption for the circuit is 6.012 amperes: 6 amperes through the toaster and 0.012 amperes through the short circuit. The power consumption is 721.4 watts; 720 watts in the toaster and 1.44 watts in the short circuit.

With the toaster turned off, the resistance of the circuit is simply 10,000 ohms. This will result in a current consumption of 0.012 amperes through the short circuit resistance and a power consumption of 1.44 watts. Thus, when the electrical source supplies constant voltage to the circuit, it does not matter if the appliance is turned on or off. The short will proceed unaffected in either case.

This is one reason why so many cases of electrical shorting seem to occur at night or after work hours. If the work day is 8 hours long, then the nonwork day is 16 hours long. With the short proceeding whether or not the appliance is turned on or off, there is a 33% chance the short will cause a fire during work hours, and a 67% chance it will cause a fire during nonwork hours. Under these circumstances, the chances are much more favorable for a short circuit-caused fire to occur during nonworking hours.

Thus, to avoid a short occurring and causing a fire when no one is around, it is good practice to unplug appliances and extension cords when they will not be needed for extended periods. Most fire departments recommend this practice to homeowners to reduce the risk of fire when they are away on vacation.

## 8.4   Series Short Circuits

A series short circuit might occur, for example, in an appliance where there is an off-on switch in series with the appliance resistance. This is schematically shown in Figure 8.3.

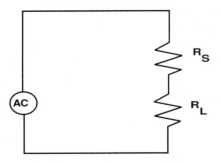

**Figure 8.3** Series short circuit.

The equivalent resistive load for resistances in series is given below:

$$R_{Eq} = [R_S] + [R_L]. \qquad \text{(xix)}$$

The current used in this circuit is then:

$$I = E/(R_S + R_L). \qquad \text{(xx)}$$

The voltage across the whole circuit is, of course, "E." However, the voltage across each of the resistances is given by the following:

$$E_S = I(R_S) \text{ and } E_L = I(R_L) \qquad \text{(xxi)}$$

where I = the current calculated from Equation (xix).

Thus, the power consumed by each resistance is given by the following:

$$P_S = I^2(R_S) \text{ and } P_L = I^2(R_L) \qquad \text{(xxii)}$$

$$P = P_S + P_L = I^2(R_S) + I^2(R_L) = E\,I$$

where P = total power consumed by the circuit, I = current calculated from Equation (xix), $P_S$ = power consumed by short circuit, and $P_L$ = power consumed by the load of the circuit.

Under normal circumstances, when the switch is in the closed position (the appliance is on), the value for "$R_S$" is typically very small, perhaps only a fraction of an ohm. Thus, the resistance associated with the switch is trivial, and is generally ignored in circuit calculations.

Consider the example given before, a toaster that has a 20 ohm-heating element with a 15 ampere-rated electric cord. The toaster is plugged into a wall outlet rated at 120 VAC. Assume that the resistance of the switch is 0.01

ohms. Thus, the total resistance of the toaster and switch is 20.01 ohms. When the toaster is operating normally, this results in a current and total power consumption as given below:

$$I = (120 \text{ v})/(20.01 \text{ ohms}) = 5.997 \text{ amperes.}$$

$$P = (120 \text{ v})(5.997 \text{ a}) = 719.6 \text{ watts.}$$

When the toaster is turned off and the circuit is open, the resistance of the switch is infinite. Thus, no current can flow, and no power is consumed.

However, if the switch is damaged, cracked, improperly assembled, deteriorated by the environment, or perhaps subjected to a high voltage surge such as a lightning stroke, the resistance can be reduced from infinity to some lower value. Assume that the resistance is reduced to 100,000 ohms.

When the resistance of the switch is 100,000 ohms, the total resistance of the circuit is then 100,020 ohms. This results in a current consumption of 0.0012 amperes, and a total consumption of 0.144 watts. Of the total amount of power, the switch is consuming 0.144 watts, and the toaster is consuming 0.0000288 watts. The power consumed by the toaster is trivial. This is because the voltage across the switch is 119.98 V, while the voltage across the toaster is 0.024 V. Thus, nearly all of the power being consumed is concentrated at the point of shorting in the switch. This expended power is converted to heat.

As the switch heats up, the resistance in the switch drops further due to heat-related deterioration. If it is assumed that the resistance drops to 1000 ohms, then the total resistance of the circuit is 1,020 ohms. This results in a current consumption of 0.118 amperes, and a total power consumption of 14.1 watts. Of the total amount of power, the short in the switch is consuming 13.84 watts, and the toaster is consuming 0.28 watts. The voltage across the switch is 118 V, and the voltage across the toaster is 2 V. As before, the power consumed in the switch is being converted to heat, which causes further deterioration of the switch.

When the resistance of the switch drops to 20 ohms, the total resistance is then 40 ohms. This results in a current consumption of 3 amperes, and a total power consumption of 360 watts. Of the total amount of power, the short in the switch is consuming 180 watts, and the toaster is consuming 180 watts. At this point, where the resistance of the switch and the resistance of the toaster are the same, the switch will be consuming the maximum power it can as long as the circuit remains intact.

In series shorting, as shown in the previous example, it is often the case that the current consumed does not exceed the rating of the cord, at least initially. However, if the heat buildup is sufficient to damage the circuit and sever the appliance load out of the series, then a second shorting phase may occur where the short is simply a dead short. A "dead" short is one where the short is the only load in the circuit. This situation most often occurs when the cord supplying electricity to the appliance has both the hot and cold conductors in the same casing, or if the hot conductor comes in contact with a ground, e.g., a grounded metal conduit.

With respect to series shorting, the following generalizations can be made.

- When it initiates, most of the power is consumed in the area of the short; very little power is consumed by the appliance or load.
- As long as the series circuit is intact, the current consumed by the short will normally not exceed the current normally consumed by the appliance or load. Thus, series shorts may rarely cause breakers to trip.
- When the resistance of the short equals that of the load, the power consumption of the short will be maximum, presuming that the series circuit remains intact. The power consumed by the short at this stage will generally be half that consumed by the appliance or load when operating.
- If the heat causes sufficient damage, it is possible for the series circuit to degenerate into a dead short circuit, i.e., a simple circuit in which there is only the short as the load.

Series shorting such as has been described in the example, occurs even if the switch to a particular appliance is in the "off" position. In fact, it *only* occurs when the switch is in the "off" position. This is because when the switch is in the "on" position, the resistance through the switch is too low for any significant heating effects to occur.

This is another reason why it is good practice to unplug appliances when people plan to be out of the house for an extended period. Even when the switches are turned off, an appliance can short circuit.

## 8.5   Beading

One of the more classic visual methods of detecting shorting after a fire has occurred is the observation of beading. Beading is simply melted conductor that has reformed into droplets or beads near or at the point of shorting. The heat of the shorting melts the metal conductor. Once melted, the surface tension of the material causes it to form drops or beads. When the bead of metal is out of the electric current loop, it cools and solidifies in the droplet or bead shape.

The two most commonly used metals for electrical conductors are aluminum and copper. Copper has a melting point of 1892°F and a boiling point of 4172°F. Aluminum has a melting point of 1220°F and a boiling point of 3733°F. Electrical solders, which are usually composed of various combinations of lead, tin, antimony, zinc, and silver typically melt and flow at temperatures well below 1000°F. For example, a standard solder composed of 48% tin and 52% lead will melt at 360°F. Most electrical solders have at least 40% tin for good electrical conductivity.

In typical building fires, the temperatures of burning wood, cloth, paper, and such are not ordinarily hot enough to melt copper conductors. Such fires will typically cause copper conductors to anneal and oxidize in varying degrees, but do not usually cause melting of the copper. However, electrical shorting can supply sufficiently concentrated energy to cause localized melting of the copper in the immediate area of the short. For this reason, the finding of beaded copper conductor in a typical building fire is a strong indication of electrical shorting.

It should be noted, however, that fires involving flammable liquids or gases, some types of combustible fuels, and some types of combustible metals can reach temperatures that can readily melt copper. Also, in confined areas where there is excess oxygen or air flow to fan the flames like a blacksmith's bellows, a fire that normally might not be able to melt copper, may be able to do so.* Thus, it is important to note if any of these fuels or unusual circumstances were present in the area where beading of the copper conductor was observed.

However, in the same typical building fire where there is burning wood, paper, etc. the temperatures are often sufficient to melt aluminum electrical conductors. For this reason, the finding of beaded aluminum conductors is not, in itself, a strong indicator of shorting, especially if the beading is

---

* It is even possible to melt iron with wood under the right circumstances. In Africa, it is known that iron ore was smelted into iron using large vertical, stone towers. These towers would be filled with layers of wood, iron ore, and limestone. The wood would be burned causing some of the wood to pyrolyze into charcoal. The large fire would also set up a high convective draft through the tower, which would supply excess oxygen to the process. The oxidation of the charcoal would provide sufficient temperatures to melt high-carbon-content iron, and the carbon monoxide from the charcoal helped to reduce the iron oxides. The limestone acted as a flux to remove the impurities from the iron, which would drop to the bottom in pig-like lumps.

A check of an iron-carbon diagram of iron shows why this process works. While steel does not melt until a temperature of about 3000°F is reached, an iron containing 4.3% of carbon, which is the eutectic point, will melt at a temperature as low as 2200°F. Thus, the trick in the process is to chemically reduce the iron oxide to iron carbide, and then control the amount of carbon. This is pretty slick chemistry when you consider that it was done more than 4000 years ago!

observed in an area corresponding to a hot fire area. It is possible that the fire alone can cause the aluminum conductors to bead. Thus, other indicators have to be considered to determine if the beading is due to shorting.

One of the other indicators used to determine whether or not beading is the result of shorting, is spreading of the strands in multistrand conductors. In large current-carrying conductors, a short circuit may cause a momentary, large overcurrent transient. As predicted by Maxwell's equations, this current transient then causes a correspondingly large magnetic field to momentarily develop in the area of the short. These magnetic fields can warp, dishevel, and distort the strands in the conductors. Often, a multistrand conductor will have its strands unwind and spread out in the area of the short due to these magnetic effects. A fire by itself will not cause this effect.

The amount of beading around a short can often be a crude indicator of how long the short was operating. This is because of the following.

- The heat input to the short is often a function of the fuse size, breaker rating, or wire size. The one with the lowest current-carrying ability will set the limit for the maximum sustainable current.
- A specific amount of heat is required to raise the temperature of a conductor to its melting point. Thus, given the mass of the beads or amount of missing conductor, the amount of electrical energy necessary to cause formation of the beads can be estimated.

For example, consider a number 14 copper conductor. This conductor, due to its diameter and material properties, has a resistance of 2.525 ohms per 1000 feet at room temperature, i.e., 68°F. Its weight is 12.43 pounds per 1000 feet. Copper has a heat capacity of 0.0931 BTU/lb °F, and a heat of fusion of 75.6 BTU/lb.

Given these facts, consider a two-wire copper conductor that is 60 feet long. It is fused at one end for 15 amperes, as per National Electrical Code standards, and has a dead short at the other end. Under these conditions the following occurs.

- The resistance of the conductor loop at room temperature is 0.303 ohms.
- The short will draw the maximum current of 15 amperes when the combined resistance of the short and conductors is 8 ohms. If the combined resistance drops below 8 ohms, the fuse will blow in accordance with the time-delay characteristics of the fuse.
- The amount of heat necessary to raise the temperature of the conductor loop to melting is 254 BTU, assuming no heat losses.
- Assuming a power rate of 120 VAC at 15 amperes, the electrical input to the short could be 1800 watts, or 6143 BTU/hr. To heat the con-

ductor to near melting would require about 150 seconds under those conditions (assuming no heat losses). Assuming a current rate of 7.5 amperes, it would take about 300 seconds or 5 minutes to heat the conductors up to the melting point.

- Assuming a current of 15 amperes, after the conductor reaches melting, the conductor would melt away and form beads at a rate of 13.2 in/sec along the length of both wires in a direction towards the source of power. The whole 60-foot-long section would melt away in 66 seconds.
- Assuming a current of 7.5 amperes, the conductor would melt away at a rate of 6.6 in/sec, and the whole 60-foot section would melt in 132 seconds.

In considering the above example, a number of items are noteworthy.

- If the fuse is found to be blown, then it could be concluded that the combined resistance of the short and conductors was less than 8 ohms.
- Even momentarily, the resistance of the short circuit would be no less than 0.303 ohms. Thus, the maximum, momentary current would be 396 amperes. This fact can be used to set a limit as to the response time of the fuse if the time-response function is known.
- If the fuse did not blow, and the wires were found melted all the way back to the fuse box, then it could be concluded that the combined resistance was 8 ohms or more.
- The minimum warm-up time of the whole 60-foot section of wiring to reach the melting temperature would be 150 seconds, or 2.5 minutes. This, of course, assumes that the short begins fully developed.
- The minimum time required for melting the wiring, assuming it was already at melting temperature would be 66 seconds. Thus, the minimum combined time for both warm up to melting, and then melting of the whole 60 feet is 216 seconds, or 3.6 minutes.

The above example also demonstrates an important point: it does not take long for a fully developed short to melt a lot of conductor material. In the above example, which assumed many ideal conditions, it required less than 4 minutes to heat up and melt 60 feet of two wire, #14 copper conductor with a dead short at one end. This is 1.49 pounds of molten copper, at 1892°F. Since wood and paper typically ignite at 500°F, this is more than enough to initiate a fire. In fact, many fires begin with the melting of just an inch or less of conductor.

An item of note with respect to beading is that in a two- or three-wire-insulated conductor that carries single-phase alternating current, generally it is the hot wire that will bead the most. Sometimes this fact is handy when the wires have not been properly color coded.

Beading also occurs in steel. In electrical systems, steel is often used in conduits, electrical boxes, and similar components. Occasionally, it is also used as a conductor. Shorting can occur between the hot copper or aluminum conductor and the conduit or metal box. Normally the conduit and metal boxes are grounded, which allows them to "complete the circuit." Typically, the shorting will produce a burn-through pattern in a metal box or conduit similar to that of a cutting electrode stick used by welders. When the arcing melts through the steel, the steel may form beads, which then solidify just below the penetrations.

Since typical residential and light commercial fires cannot melt steel unless the circumstances are extraordinary, and since only a few types of flammable materials can cause steel to melt in an open fire, the observation of steel beads in an electrical box or conduit in association with a "burn hole" is a very strong indicator of electrical shorting.

Further, even if there was fuel in the vicinity that could melt steel, it would be difficult to explain how the burning of the fuel could produce concentrated and specific points of melting in the conduit or electrical box. Only electrical shorting and welding-type effects can do this. The fire front developed by a burning fuel load is broad and general. It tends to affect large areas rather than specific points.

## 8.6   Fuses, Breakers, and Overcurrent Protection

It is a common, but mistaken, notion that fuses or circuit breakers will wholly protect a building against fire caused by electrical shorting. Fuses and common circuit breakers do not specifically protect against fires or shorting. They protect against overcurrent. They are not specifically short circuit detecting devices.

In other words, a fuse or circuit breaker simply opens an electrical circuit when that circuit uses more amperage than the current-carrying capacity or trip point of the fuse or breaker. If a short occurs and does not use more current than the trip point of the fuse or breaker, the short will not be suppressed. The short will continue to operate as long as its current usage is less than the trip point.

In fact, the National Electric Code states that the purpose of overcurrent protection, i.e., fuses and breakers, is to open the circuit if the current reaches a value that will cause an excessive or dangerous temperature in conductors or conductor insulation. In residences and light commercial buildings where there is no special-purpose equipment, fuses and breakers are sized to simply protect the particular branch circuit wiring from getting hot enough to damage its insulation. In sum, if the wiring isn't carrying enough current to get hot, the breaker or fuse will not trip.

It is for this reason that an oversized fuse or breaker is a safety hazard. While it allows a person to avoid the aggravation of changing fuses or resetting breakers in an overloaded system, it can allow a short circuit to proceed unchecked, or allow an overload to heat up the wiring until the insulation melts and a short develops.

A fuse is basically a small strip of metal, encased in a tube or housing, which is connected to terminals. The metal strip is made of a low melting temperature alloy, sometimes referred to as a fusible alloy. Fusible alloys generally have melting points in the range of 125°F to 500°F, and are usually made of some of the following materials: bismuth, lead, tin, cadmium, or indium. Usually the alloy's composition is at the eutectic, so that the melting point is well defined.

The electrical resistance of the fusible alloy metal strip is low, so that it normally acts as a conductor in series with the circuit. However, when the current in the circuit reaches the set point, the metal strip heats up due to "$I^2R$" effects. The metal strip then melts away, thus opening the circuit.

The rating or set point of a fuse is the amount of current it will allow to pass on a continuous basis without tripping. Under actual operating conditions, however, the trip point may vary slightly from the rating due to localized air temperature around the fuse. In general, as the air temperature in the area of the fuse increases, the trip point decreases. This, of course, is because of the temperature-sensitive nature of the metal strip that comprises the fuse. Less electrical energy is needed to melt a hot metal strip than a cold one.

For example, a Fusetron™ dual element fuse has a carrying capacity of 110% of its current rating at 32°F, 100% at about 86°F, and 90% at about 130°F. At 212°F, the carrying capacity will drop to about 63% of its current rating.

The above is very noteworthy. When heat from a nearby fire impinges on a fuse box, and causes the fuses to get hot, the fuses may trip even if there has been no short or unusual overcurrent in the circuit. A fire near the fuse box will often cause the fuses to trip (even when there is no short), which then cuts off the electricity to the circuit. This can prevent secondary shorts from occurring as the fire spreads to wiring and equipment.

Circuit breakers are similar in function to fuses. They open the circuit when the current exceeds the set point of the breaker. However, while fuses are not reusable, breakers can be reset after tripping open the circuit and reused. In general, a circuit breaker is simply a spring-loaded switch that pops to an open circuit position when the current is too high.

In smaller- and medium-sized circuit breakers, the unloading or tripping mechanism is likely a bimetallic strip through which the current flows.

When the strip heats up sufficiently by "$I^2R$" effects, the strip overcomes the spring load and opens the circuit. A "snap through" type mechanism is typically employed.

In medium- and larger-sized circuit breakers, the tripping mechanism may be a solenoid. The switch contactors on the solenoid will likely be held in place by a spring load. When the current is at the set point, the current flow through the solenoid produces enough magnetic field buildup in the solenoid to cause the contactors to overcome the spring load. Like the bimetallic version, a "snap through" type mechanism is typically employed to open the contactors.

In many circuit breakers, both bimetallic and solenoid current sensing elements are used. The bimetallic element is used to sense the lower overcurrents, typically in the "1 times rating" to "10 times rating" range, and the solenoid element senses the higher level overcurrents, typically in the range of more than ten times the rating. The latter often occurs when the short circuit or ground fault resistance is very low. Most circuit breakers used in residential and light commercial applications contain both bimetallic and solenoid sensing elements.

When circuit breakers contain bimetallic elements, they are sensitive to the ambient temperature conditions in the same way that fusible alloy fuses are. If a circuit breaker containing a bimetallic element is put into a hot environment, it may trip at lower overcurrent levels. For this reason, when circuit breakers must be placed in hot or varying temperature environments, dual solenoid type breakers are used. One solenoid is set for the low overcorrects, replacing the bimetallic, and the other for high overcorrects. This substitution is done because solenoid elements are less sensitive to ambient temperatures.

Both circuit breakers and fuses exhibit inverse time characteristics with respect to overcurrent protection. In other words, as the applied current above the set point increases, the time required for opening the circuit decreases. Table 8.2 lists a typical time current or minimum melt curve for a Bussman™, 200 ampere, 600 volt, low peak dual element fuse, at standard temperature.

**Table 8.2  Opening Time vs. Applied Current for Bussman™ Low-Peak LPS-RK 200 (RK1) Fuse**

| Applied Current (amperes) | Opening Time (seconds) |
|---|---|
| 2000 | 0.01 |
| 1500 | 0.20 |
| 1000 | 10.0 |
| 600 | 55 |
| 400 | 150 |
| 300 | 450 |

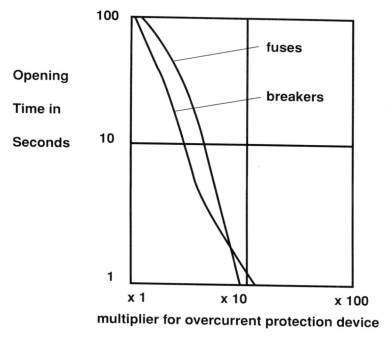

**Figure 8.4** Current-time plots.

The plot of applied current versus circuit opening time is called the "time current curve" for the particular fuse or circuit breaker. Usually the plot is done on a log–log graph. Figure 8.4 shows sample time-current plots for both fuses and breakers.

In large power applications, some circuit breakers are not only designed to trip when there is an overload current condition, but they may also trip when there is an underload condition, that is, when the current (or voltage in some instances) flow is too low. Underload type circuit breakers are typically employed when a low current condition would cause damage to a particular piece of machinery or equipment.

In other large power applications, automatic reclosing circuit breakers may also be used. These are circuit breakers that are designed to automatically reset themselves after tripping. They are often used to guard against unnecessary nuisance trips that are generally taken care of quickly by smaller circuit protection devices located downline or elsewhere in the system. If the overcurrent (or undercurrent) condition persists after two or three attempts at reclosing, the device will then lock itself in the open circuit position.

To protect against nuisance trips that occur during large motor start-ups, some breakers are equipped with time delay devices. These devices allow the circuit breaker to handle large, but temporary current surges without tripping. The time delay devices are usually calibrated to operate in the zone where electric motor current onrush normally occurs.

When there are very high voltages, high voltage circuit breakers are employed that either open in an oil emersion, or employ a blast of air. These techniques are used to break the electrical arc between the contactors as the contactors open.

With high voltages, even if the contactors are separated, it is possible for the electricity to arc from one contactor to the other and still maintain the circuit, especially if the air pathway has already been ionized by arcing at smaller separations. Once the air pathway is ionized, it is easy to increase the separation without breaking the arc because the air has lost much of its insulating property. (This is the same principle that allows the operation of florescent lights and neon signs.)

Since most oils are poor conductors, opening the electrical contacts while emersed in oil stops the arcing. Likewise, in the air blast type, an air blast is used to blow away the ionized air pathway. In essence, the air blast simply blows out the electrical arc.

The above brief discussion of circuit breakers and fuses only mentions the more common types. Manufacturers offer a wide range of fuses and circuit breakers that are tailor-made for specialty applications, or enhance or minimize various characteristics of the equipment. For this reason, the design of switch gear and protection equipment has become an engineering sub-specialty shared by both electrical and mechanical engineering disciplines.

In general, manufacturers have done a good job in producing overcurrent protection products that are reliable and longlasting. Most have in-house testing and quality controls to ensure that the products that leave the factory work as specified. When the equipment is installed properly and used appropriately, it will normally provide the required overcurrent protection.

However, as the complexity of the equipment increases, the opportunity for failure increases. Manufacturers and designers are not omniscient. Sometimes unforeseen conditions develop that were not anticipated by the designers. Damage sometimes occurs in transit or storage. Also, the equipment must be installed correctly and be appropriate for the specific application. If the application is inappropriate, or the installation is incorrectly done, even the best equipment will not do a good job. The more complex overcurrent protection equipment requires preventive maintenance and periodic inspection. The lack of this may allow premature failure of the equipment. And, of course, there are failures created by sabotage, vandalism, floods, fires, earthquakes, and other mayhem.

In short, for all kinds of sundry reasons, sometimes overcurrent protection equipment does not work right. Sometimes it may not open the circuit properly when an overcurrent condition occurs. It may "hang up" or only open one of the phases in a three-phase system. Sometimes it may wholly fail and not respond at all to the overcurrent condition. And, sometimes the overcurrent protection equipment itself may be the cause of an overcurrent problem. The equipment may develop an internal short circuit. Because of these possibilities, it is important to check the overcurrent protection that may have been present in the system when the cause of the fire is suspected of being electrical in nature.

## 8.7  Example Situation Involving Overcurrent Protection

Consider the following situation. In an older house, the branch wiring to a kitchen wall outlet was rated for 15 amperes. The fuse that protected that circuit was the older, screw-in type household fuse. To dispense with the aggravation of frequent fuse replacement, the homeowner had a 30 ampere fuse inserted in the socket, which was supposed to have only a 15 ampere fuse.*

One day the homeowner left an automatic coffeemaker unattended. Initially, the coffeepot was full. However, after a time all the coffee boiled off and the coffeemaker overheated. The overheating of the coffeemaker caused the plastic shell of the appliance to soften, collapse, smoke, and char. It also caused the appliance's electric cord to heat up. Where the cord entered the appliance, the plastic insulation broke down and the cord shorted. The short then followed the cord back into the wall outlet, and caused the branch circuit wiring to overheat near the outlet.

Because the branch circuit wiring had old-fashioned, flammable fiber insulation wrap, the insulation around the wiring caught on fire. The fire spread through the wall interior and then breached the wall. The resulting fire caused great damage to the house.

In the above situation, what was the primary cause of the fire?

- Was it negligence of the home owner who left the coffeemaker unattended?
- Was it a design shortcoming of the appliance maker who did not include a safety device to account for a dry pot, or overheat situation?
- Was it the 30 ampere fuse in a 15 ampere socket?
- Was it the old-fashioned wiring that was insulated with flammable fiber insulation?

Further, if the correct fuse had been used, would the fire have been prevented, or the damages at least mitigated? Can the manufacturer of the

---

\*    In older fuse blocks, 5, 10, 20, and 30 ampere-rated screw-in type fuses are interchangeable. In most current codes, this type of fusing is no longer acceptable in new installations.

coffeemaker correctly assert from an engineering safety point of view that proper branch circuit fusing obviates the need for any internal safety device to protect from a dry pot situation?

## 8.8   Ground Fault Circuit Interrupters

A ground fault circuit interrupter is a device that deenergizes a circuit when the current flow in a current-to-ground type short circuit exceeds some predetermined current set point. The current flowing to the ground that would cause a ground fault circuit interrupter to trip is often much less in magnitude than the current flow required to trip breakers or fuses on the supply side of the circuit.

In other words, if the current in a short circuit flows over to the safety ground, the GFCI will shut off the circuit. However, if the current from the short circuit flows over to the cold wire and not the safety ground, the GFCI will do nothing. In that case, it will be up to the overcurrent protection to take care of the short.

In single-phase, residential type applications, ground fault circuit interrupters monitor the current passing through the "third" wire or the safety ground wire of the system. The NEC requires that GFCI devices be installed in bathrooms, kitchens, garages, and other areas where people are exposed to electrocution hazards.

For example, one such hazard might be created when a person drops an electric razor into a sink full of water, and then reaches into the water to retrieve the razor. When the razor enters the water, the current from the hot line bleeds out into the water and is grounded through the chassis of the razor into the safety ground, i.e., the "third" wire ground. When this occurs, the GFCI senses the current flowing through the safety ground and opens up the circuit.

While GFCI devices are generally intended to reduce the risk of accidental electrocution, they sometimes can arrest shorting that would have otherwise resulted in fire. They can be very useful in detecting and arresting shorts that occur on the downline side of the load, from the cold wire to the safety ground. This is a type of series short, as discussed before in Section 8.4, except that the ground is now included in the circuit. In essence, the ground completes the circuit from the point of shorting, and substitutes for the cold conductor of the circuit.

Shorts of this type can be relatively slow in developing, and might not be detected by standard fuse or breaker overcurrent protection. This is because if the resistance at the point of shorting is relatively low, the current flow may not be noticeably higher than normal. In effect, the short would be equivalent to a wire-to-wire splice. However, if the placement of the GFCI in the circuit is in line with the current flow through the ground, it will detect that current is flowing through the ground instead of the cold wire, and open the circuit.

## 8.9 "Grandfathering" of GFCIs

In residences where the houses were built prior to the GFCI requirement of the NEC, usually it is not required that GFCIs be installed, although it is recommended. The original electrical installation is generally allowed to stay in place if it meets three conditions:

1. It continues to meet the standards current at the time it was installed.
2. There have been no undue changes made to the original installation.
3. Its continued operation creates no undue risk.

If significant modifications are made to the house, like extensive remodeling and rewiring, it is likely that local code enforcement officials will require that the remodeled portion be upgraded to comply with current code requirements, which may include the installation of GFCIs where appropriate. If the remodeling constitutes a basic reconstruction of the residence of more than a certain percentage, often more than 50%, the local code enforcement officials may require that that whole house be upgraded to meet current code requirements.

However, while such "grandfathering" is allowed for private residences, this is usually not the case for commercial buildings or public buildings. Generally, such places are required to adhere to the current code in effect with respect to GFCIs. This is to protect public safety.

As a reminder to the reader, the above references to code enforcement are simply generalizations. As with any such generalizations, they may not apply to your specific area or jurisdiction. To be sure what the requirements are for your area, check with your local code enforcement officials.

## 8.10 Other Devices

There are a variety of other electrical devices that are used to sense electrical problems in the lines. Lightning arrestors and surge arrestors are devices that sense high voltage or high current transients and shunt the transient to ground. In some such devices, when the voltage of the transient exceeds a set amount, sometimes called the clamping voltage, the line momentarily connects to ground to discharge the transient. In more sophisticated such devices, the device may monitor the increase per time of the transient. If either the current or voltage climbs faster than a set amount, the device shunts over to ground momentarily.

Some appliances that have electric motors are equipped with embedded bimetallic switches. These switches open up the circuit within the windings of the motor if the motor windings get too hot. The idea is to shut off the motor before it gets hot enough to short out and catch fire. Some of the bimetallic switches reset when the motor cools down. Other types may stay permanently in the open position once a problem has occurred.

In some appliances, notably some types of coffeemakers, there is a thermally sensitive resistor or diode. If the circuit gets too hot, the resistor or diode opens up the circuit. These devices do not reset. In many ways the device is like a fuse, except that it is not strictly an overcurrent device. Such devices are usually added to protect against "dry pot" situations.

Some types of electrical equipment "treat" the incoming power received from the utility. Undesirable harmonics, transients, and static are filtered from the line. This may be done by the use of sophisticated circuitry and solid-state devices, or more simply by the use of motor-generator sets. In the latter, the power from the utility is used to operate an electric motor that powers a generator. The idea is that the harmonics, static, and transients coming from the utility power lines are smoothed out by flywheel inertia effects within the motor and generator set. Also, there is no direct circuit link between the incoming power from the utility, and the outgoing power produced by the generator. Thus, the power coming from the generator is "cleaner," that is free of harmonics and spikes, than the power received by the motor from the utility.

Another type of treatment device is called a UPS device. The UPS acronym stands for *uninterruptable power service*. Such equipment not only filters out undesirable static, harmonics, and transients, but also switches over to batteries or alternate power generation equipment if the utility has a power failure, low-voltage condition, or similar.

In any case, it is worth noting that these special purpose devices that treat the incoming power often contain special-purpose electrical protection devices. In case of an overload or underload, they may not simply open the circuit and stop everything. They may contain relays that switch over circuits to new lines, or cause certain components to operate that were previously off-line. They may even shed certain circuits or operating equipment in an orderly sequence until the problem is isolated from the rest of the circuit. These special-purpose protection devices may, in addition to overcurrent or undercurrent, sense ground faults, voltage variance, power factors, or relative loading between phases.

## 8.11   Lightning Type Surges

While it is possible for lightning to cause electrical shorting problems that may result in fire, more frequently lightning causes circuit components to blow apart, which then opens the circuit. This is especially true in circuits that contain nonlinear components like inductors, capacitors, transistors, diodes, and IC chips. In inductive components, like motor windings, lightning or high voltage surges can cause shorting between the windings and the chassis ground. This occurs when the transient voltage is large enough to cause breakdown of the dielectric in the winding insulation.

Typically, a lightning type surge or transient will not cause fuses or breakers to directly trip. Since a lightning stroke will often occur in less than 50 microseconds, most fuses and breakers simply do not react fast enough to arrest it. A quick check of the generic current-time plot shown previously in Figure 8.4 shows that as the lapsed time decreases, the amount of overcurrent allowed to pass through the overcurrent protection device increases. Thus, generally, a lightning surge will not be stopped by a breaker or fuse.

In cases where the fuse or breaker provides significant impedance to the surge, the surge may simply arc across the fuse or breaker from terminal to terminal, or terminal to ground. It is not unusual to observe where a lightning magnitude surge has simply "jumped" around a fuse.

When a lightning magnitude surge passes through a fuse or breaker box, it may cause damage in a component well downline from the fuse or breaker box. For example, if the component is a motor, it may short out as previously discussed. Because of this shorting damage, eventually the current overprotection of that branch circuit will come into play. Thus, while lightning does not normally cause a fuse or breaker to trip directly, it may cause damages that eventually result in a fuse or breaker tripping.

## 8.12   Common Places Where Shorting Occurs

The following is a list of items or locations where electrical shorting leading to fires often occurs. The list is not intended to be exhaustive, but simply represents items that seem to occur regularly.

**Staples** — In residences and light commercial buildings it is common practice in some areas to simply staple Romax type conductors to wood

**Plate 8.4** Aluminum electrical cable. The cable heated from shorting and pulled apart forming "points" on the cable strands. Note white color of oxidized aluminum.

members. If the staple is applied too tightly, the insulation around the conductors can be crushed, cracked, or even torn. In some cases, the staple itself may bite or cut into the insulation. Over time, the damaged insulation may further degrade because of tears and penetrations in the protective sheathing. Current may leak from one conductor to another due to the damaged insulation and precipitate shorting.

**Corners** — When nominally straight conductors are bent at a sharp angle, the conductor cross-sectional area may be deformed. When the bend is sufficiently sharp, the conductors may develop a crease, or otherwise become distorted and lose some of their effective current-carrying cross-sectional area.

The resistance of a conductor with uniform cross-sectional area is given by the following:

$$R = (rl)/A \qquad\qquad (xxiii)$$

where r = the specific resistance of the conductor usually given in ohms per $cm^3$, l = length of the conductor, and A = cross-sectional area of conductor.

From inspection of Equation (xxiii) it is apparent that if the cross-sectional area is halved, the resistance is doubled. And if the resistance is doubled, then the amount of heat generated by the resistance at that point is doubled, as per the power equation given below.

$$P = I^2R \qquad \text{(xxiv)}$$

In addition to deformation of the conductor itself, sharp bending of a conductor may cause the insulation on the outside radius of the bend to become stretched. This will reduce its thickness, and may even cause small microholes or microtears to develop. When such microholes develop in plastic materials, it is often the case that the color of the plastic will "whiten." Clear plastic will often become milk white or opaque where the micro-holes are concentrated.

When the insulation wrap around the conductor stretches and becomes thin, or develops microholes due to excessive strain, the insulation will lose some of its dielectric properties. Because multiple conductors are often in common casings, the insulation between the hot and cold conductors may be sufficiently damaged for a current leak to develop between conductors.

**Skinning** — When wiring is installed, it is often pulled through conduit, weather heads, holes, or box openings. Sometimes as the conductors pass over edges, around corners, or over rough spots, the exterior insulation around the conductor will be "skinned," that is, some of the exterior insulation wrap will be abraded away. This causes the insulation to be thinner at that location. As such, the "skinned" area is an insulation "weak spot" in the conductor. If the thinning is sufficient, it is possible that current can leak across the weak spot.

Conductors not only have to handle the normal voltage supplied by the utility, but also the common voltage spikes and transients generated by switching activities within the utility system. If insulation is skinned, it may be able to withstand the nominal voltages, but may not be able to stand up to large spikes and transients. Breakdown of the skinned insulation may occur after repeated spikes, or after a single spike if it has sufficiently high voltage.

**Edges** — Wires and conductors are not immobile although they may appear to be wholly static. First of all, they expand and contract with temperature change. Also, they often jerk slightly in response to large current rushes, such as occurs when a motor is turned on. (The amount of jerk can be calculated by the application of Maxwell's equations.) Because the conductors are often fastened to various portions of a structure, portions of the conductor will move as the structure moves in response to wind, temperature, and loads. Thus, conductors in contact with a sharp edge may have their insulation cut after a period of time due to small relative motions between the conductors and the sharp edge.

Whenever conductors or other types of wiring are run through boxes, walls, or other items that have sharp edges, the NEC requires that smooth bushings be provided. (See NEC 410-30; 410-27(b); 370-22; 345-15; and 300-16.)

**Flexible cords** — Flexible electric cords typically have stranded conductors. The conductor is composed of many individual hair-like strands of conductor bundled together similar to a rope. Over time, the cord may be bent back and forth at a particular point in the cord causing fatigue damage to the individual strands.

As the strands break apart, the ability of the conductor to carry current diminishes. If enough of the strands break apart, the cord will develop a hot spot at that point for the same reasons as discussed under "corners" in the previous paragraphs: loss of current-carrying cross-sectional area.

From experience, the two most common points in a flexible cord where fatigue damage to the strands occurs is where the cord exits the appliance, and where the cord connects into the male plug. These are the two points where the cord will often be bent the most.

Some people have the habit of disengaging a plug from a wall outlet by pulling on the cord. When this is done, the force needed to disengage the male plug from the outlet is carried by the conductor, since the conductor is usually more rigid than the exterior insulation wrap. Not only can this cause breakage of the electrical connection between the conductor and the male plug, but if the cord were pulled while it was at an angle with the plug, individual stands may be broken by combined bending and tensile stresses in the strands. This is why it is recommended that a male plug be disengaged from a wall outlet by grasping the plug itself.

In addition to fatigue breakage, flexible cords often sustain physical damage to the cord due to mashing and folding. This may occur, for example,

**Plate 8.5** Beading at ends of individual strands of wire.

when an extension cord is run through a door jamb and the door is closed on it, when a knot in the cord is pulled tight, or when the cord is walked on often. I have observed where extension cords have been laid across the burners of cook stoves, laid across the rotating shaft of a bench grinder (take a guess on this one as to what happened), laid under rugs in busy hallways, and laid across garage floors where the car was regularly driven. I have also seen them nailed to walls with the nails driven between the conductors, stapled to walls and ceilings, tied around nails and hooks, and embedded into wall plaster to avoid having an unsightly cord hanging down the wall.

One of my favorites combines several types of abuses at once. A breeze box fan was suspended from the ceiling by its own cord, which had been tied with a square knot to a hot steam pipe. The fan had been suspended over the top of a commercial steam cooker, which was regularly opened and closed. This all goes to show, of course, that there is simply no end to the creativity a person can apply to the abuse of flexible cords.

Lastly, besides being physically damaged as noted above, flexible cords are often just electrically overloaded. Many light application extension cords are rated for 15 amperes. If a 20 or 25 ampere appliance is plugged into the cord, the appliance may draw more current than the extension cord can safely handle. The cord will then heat up due to "$I^2R$" heating effects. There will be some point along the cord where it will heat up a little more than elsewhere, and if left unchecked the cord may, in time, fail and short at that point.

It is not uncommon for a 15 ampere-rated extension cord to be equipped with three or more outlets. Thus, if three appliances, each drawing 10 amperes are plugged into the extension cord and are operated, the cord would then be carrying twice its rated load. Unfortunately, few homeowners are cognizant of the current ratings of extension cords. Most people simply use whatever extension cord is handy, has the appropriate length, or is the cheapest.

**Lugs and terminals** — Lugs and terminals are used to connect conductors together. The ability of the lug or terminal to carry current is a function of the contact area between the two materials, and the quality of the metal-to-metal contact between them. In most cases, the conductors, lugs, and terminals are cleaned prior to installation, and pressure is applied to the connection to ensure good contact.

If the contact area between the two conductors is reduced or the contact quality is degraded, the connection will lose its ability to carry current. The connection point can then become a hot spot due to "$I^2R$" heating effects. Two of the more common reasons for a lug or terminal to lose its ability to safely carry current are: looseness and corrosion.

Looseness of the connection can occur because of vibrations, impacts to the panel box, temperature effects, material creep, chemical attack, and a host of other less obvious causes. Looseness can cause a loss of current-carrying area because in most cases, the amount of contact area is partially a function of the compression between the two materials. When the compression is firm, the two materials are in intimate contact. When the compression is not firm, there may be an air space between the materials that acts as an insulator. A firm connection also tends to push through the light layer of oxide that usually forms on the conductors during storage and shipment.

It is not uncommon for some manufacturers of screw-type lugs to recommend that the lugs be periodically checked for tightness. Some manufacturers of industrial type bus bars have employed "break away" lug connections to ensure that screw-type connections do not back off and become loose. Other manufacturers have employed crushable threaded lugs to prevent lug back off. Most have specific tightening specifications to ensure a tight connection that will not back off.

Corrosion at a lug or terminal usually causes problems in two ways. First, the products of corrosion are often not good conductors of electrical current. While copper and aluminum are excellent conductors, copper oxide and aluminum oxide are not. A layer of corrosion products between the conductors typically increases the resistance of the connection, which then results in heating of the connection.

Secondly, corrosion may cause material damage to the connection. It may result in material loss, weakening of the material, or even dimensional distortion. In the latter case, the distortion may occur due to a change in material properties while being subject to the same loading or stress.

Corrosion of lugs and terminals is often caused by exposure to water, constant high humidity, or chemicals. A water drip may occur directly over an electrical box, or the humidity in the area may be constantly saturated. Chemical vapors from processes or materials storage may also contribute to and accelerate corrosion. It is important to note whether the electrical box and connections have been rated for the specific environment in which they are used. There are many types of electrical boxes that are variously rated for outside use, use in high humidity, use in explosive environments, etc.

The corrosion of lugs and terminals can also be caused by the electricity itself flowing through certain combinations of dissimilar conductor materials. This is called galvanic corrosion, and is basically the same phenomenon used to produce gold or silver plating in jewelry.

In galvanic corrosion, the passage of electrical current literally causes one of the materials to plate out onto the other. This results in material loss to the donor material, which can be substantial depending upon the circumstances. The material receiving the transferred material does not benefit by this plating action either. Typically, the transferred material quickly oxidizes and becomes a nonconducting crust of hoary fuzz or "crud" around the lug or terminal.

The formation of this crud further accelerates the galvanic process because the crud itself acts as a surface for gathering water moisture to the connection. The presence of moisture around the connection helps to promote galvanic corrosion because the moisture provides a medium, or a chemical solution if you will, in which the reaction can take place.

Thus, the primary factors influencing the rate of galvanic corrosion include.

- The type of materials in physical contact with one another.
- The amount of current flow.
- The voltage across the terminal connection itself (a relatively high-resistance connection will help promote the process).
- The humidity or moisture that may come into contact with the lugs or terminals.

For these reasons, it is not proper to directly connect aluminum conductors to copper conductors. Copper and iron is also a bad connection combination due to its tendency to quickly corrode. It is also not proper to connect aluminum conductors to lugs and terminals that are rated only for copper conductors, and vice versa.

If a mixed connection between a copper conductor and an aluminum conductor must be made, a dual-rated connection box should be used. Also, connections between copper and aluminum conductors can be safely done when an intermediary material is used that is compatible to both. Tin is often used for this purpose. Thus, copper conductors and aluminum conductors that will be connected together are often tin coated or "tinned" at the point where direct contact will be made. Other types of solder coatings or "tins" have been developed to be used as intermediaries.

However, these coatings must be used with great care. Over time the coating may crack, become abraded, corrosively degrade, etc. Breaching of the coating may then allow the two dissimilar metals to come into direct contact with another and set up a galvanic corrosion cell.

With respect to dissimilar conductors, the National Electric Code, Section 110-14, states in part that:

Conductors of dissimilar metals shall not be intermixed in a terminal or splicing connector where physical contact occurs between dissimilar conductors (such as copper and aluminum, copper and copper clad aluminum, or aluminum and copper clad aluminum) unless the device is suitable for the purpose and conditions of use.

When a lug or terminal connection is loose or corroded, often a "chattering" or buzzing noise can be heard emanating from the problem connection. This noise is generated by the 60 Hz alternating current arcing across or within the connection. Because there is not enough current-carrying contact area, current is literally arcing across air gaps in and around the connection.

When such arcing occurs, it typically causes pitting of the connection, and scatters tiny blobs of conductor material in the vicinity of the arcing. The area around the chattering is often blackened due to the formation of carbon residues from the air by the arcing. The local temperatures of the arcs themselves will range from 2500°F to as high as 10,000°F, which is the same range of temperatures found in arc welding or lightning.

If the chattering is allowed to continue, it usually results in overheating of the connection, electrical shorting and failure, and possibly fire. Sometimes after a fire, an occupant of the building may recall having heard a chattering or buzzing sound coming from the electrical box.

**Motor burn out** — When electric motors become worn out, or the rotating shaft becomes locked, perhaps due to a seized bearing, the windings in the motor can overheat and short out. This is especially true of motors that are not equipped with internal thermal switches. Such switches, also known as high-heat-limit switches, shut off the motor should its windings overheat. Some older motors have flammable insulation shellac around the windings that can ignite and then sustain fire.

**CB radio coaxial cables** — Because citizen band radios do not require licenses, the people who use them are often not trained very well in the basic principles of radio and radio transmission. Some CB hobbyists inappropriately (and illegally) use powerful linear amplifiers on low power rated equipment to boost their signal output. They may also mix and match various types of antennas with their transmitters without matching antenna impedance to transmitter impedance. Thus, one of the common points where CB radios overheat and cause fires is the antenna coaxial cable, leading from the transmitter or linear amplifier to the antenna. Usually the whole line will heat up, especially after the transmitter has been in use a long time during an extended gab session. When sufficiently hot, the antenna line can melt away its insulation sheath and ignite materials that come in contact with it.

## Further Information and References

*Fire Investigation Handbook*, U.S. Department of Commerce, National Bureau of Standards, NBS Handbook 134, Francis Brannigan, Ed., August 1980. For more detailed information please see Further Information and References in the back of the book.

*Manual for the Determination of Electrical Fire Causes*-1988 ed., NFPA 907M. For more detailed information please see Further Information and References in the back of the book.

*National Electric Code*, published by the National Fire Protection Association, Quincy, MA., various editions, published approximately every 3 years. For more detailed information please see Further Information and References in the back of the book.

*SPD — Electrical Protection Handbook*, published by Bussmann-Cooper Industries. For more detailed information please see Further Information and References in the back of the book.

# Explosions 9

For a charm of powerful trouble,
Like a hell-broth, boil and bubble.
Double, double toil and trouble;
Fire burn: and, cauldron bubble.

— **The Witches,** *MacBeth,* Act IV, Scene I
*William Shakespeare, 1564–1616*

## 9.1 General

An explosion is a sudden, violent release of energy. It is usually accompanied by a loud noise and an expanding pressure wave of gas. The pressure of the gas decreases with distance from the origin or epicenter. Explosions resulting from the ignition of flammable materials may also be accompanied by a high temperature fireball that can ignite combustibles in its path.

Explosions caused by the sudden release of chemical energy are classified into two main types: deflagrating explosions and detonating explosions.

A deflagrating explosion is characterized by a relatively slow, progressive burn rate of the explosive material. The progressive release and dispersion of energy through the explosive material in a deflagrating explosion is accomplished by normal heat transfer dependent upon external factors such as ambient pressure and temperature conditions.

Deflagrating explosion generally causes damage by pushing things around because of pressure differentials. This includes things like walls, ceilings, floors, large pieces of furniture, etc. Higher pressure gas emanating from the explosion epicenter impinges on a surface with a lower pressure gas on the other side. The pressure difference between the two sides results in a net force being applied to the surface that may be sufficient to move the item or tear it away from its anchor points.

Deflagrations generally have a low ability to cause fissile- or brisance-type damages. Small objects near the epicenter of the deflagration are often left undamaged as the pressure wave passes around them. The pressure differences on their external surfaces are often insufficient to cause breakage or disintegration.

A detonating explosion is characterized by a relatively high burn rate, high energy release rate, and a high peak explosion pressure. The progressive

**Plate 9.1** Building decimated by gas explosion.

release and dispersion of energy through the explosive material is accomplished by shock waves and their associated pressure forces and stresses. For this reason, transmission of energy through the detonating material is not dependent upon ambient conditions of pressure or temperature.

Detonating explosions have higher fissile abilities than deflagrations. It is this quality that makes them useful in blasting work. Objects near the epicenter of a detonating explosion are torn apart, often like so much smashed and shattered glass, due to the transmission of intense shock waves through their materials.

Some detonating materials have high enough fissile qualities that they are even used to cut large pieces of steel. A small amount of the material, which is in the form of a putty, is simply applied along the "cut line" of the steel beam. The resulting shock waves generated by detonation are sufficiently intense to cause the steel to break apart where the putty had been applied.

A general distinction between deflagrating explosions and detonating explosions is that the former have subsonic pressure propagation rates with the explosive material while the latter have supersonic pressure propagation rates.

In addition to deflagrations and detonations, there is a third category of explosion that involves the sudden expansion of high-pressure gases, as might occur from a ruptured high-pressure vessel or pipe. This category also includes the sudden expansion of pressurized liquids into gas, such as would occur when pressurized boiler feed water flashes into steam when the pressure is suddenly lowered. This category of explosion will be considered first.

## 9.2   High Pressure Gas Expansion Explosions

The third category of explosion mentioned in the previous section is sometimes classified as a polytropic expansion. This type of explosion does not involve the release of chemical energy via a chemical reaction. It simply involves the rapid expansion of pressurized gases to ambient conditions. In essence, it is the conversion of enthalpy energy to irreversible "P-V" work, with the final state of the gas at equilibrium with ambient conditions.

Polytropic expansions involve a change of state that is usually represented by the general expression:

$$PV^n = Constant \qquad (i)$$

where $P$ = pressure, $V$ = volume, and $n$ = polytropic gas constant.

When $n = 1$, the expansion occurs at isothermal conditions, and gas behavior conforms to the ideal gas law. Of course, when this occurs, the release is not sudden. In order to accomplish isothermal expansion, the release must be very slow in a quasi-reversible manner. This is not an explosion. At most, this might be a small, slow leak at a low escape velocity where the pressure difference is slight.

When $n = k$, where $k = [C_p/C_v]$, the expansion occurs at adiabatic conditions. The reader may recall that when a process occurs very fast, there is no time available for the transfer of heat to the surroundings. Hence, the process is adiabatic. This is the mathematical model corresponding to explosions.

When "$n$" is a value intermediate between "1" and "$k$," the corresponding thermodynamic process is also intermediate between an adiabatic and isothermal process, or perhaps some combination of the two. This is not a model for an explosion, but might model certain processes that occur less rapidly than an explosion. It could also perhaps model a two- or three-step process where one of the steps involves a quick, adiabatic expansion.

When a reservoir of gas under high pressure is suddenly released into a lower pressure environment through a hole or rupture, the velocity of the escaping gas is determined by equating the change in enthalphy from the high pressure state to the low pressure state at the hole to its kinetic energy equivalent. In essence, this is a simple conversation of energy equation, as shown below.

$$\Delta(enthalpy) = \Delta(kinetic\ energy) = \Delta(energy\ between\ states) \qquad (ii)$$

$$m(h_2 - h_1) = (1/2)mv^2 = \Delta E$$

$$v = [2(h_2 - h_1)]^{1/2}$$

where h = specific enthalphy of the gas, m = mass, v = velocity, and $\Delta E$ = total energy change.

In Equation (ii), it is assumed that the escaping velocity "v" is either equal to or less than the speed of sound for the gas. For various reasons, the escape velocity will not exceed Mach 1.

Because the shape of the hole at the point of rupture can affect the "efficiency" of the above process, a correction factor "C" is often added to the equation as shown below.

$$v = C[2(h_2 - h_1)]^{1/2} \qquad \text{(iii)}$$

where C = coefficient to account for hole shape and venturi constriction effects.

In general, values for "C" are usually between 0.95 and 0.99 when the escaping gases pass through a regular hole, that is, a hole that has been drilled or is reasonably smooth and symmetric by design. It is assumed that the escaping gas has a high Reynolds number, that is, a relatively high rate of flow. When the vessel simply comes apart, and there is no flow through a hole, then "C" is equal to 1.0.

With a little algebraic hocus-pocus, Equation (iii) can be converted into the following expression for gases, provided no condensation occurs in the gas during the escape.

$$v = C\{2[P_1V_1][k/(k - 1)][1 - (P_2/P_1)^{(k-1)/k}]\}^{1/2} \qquad \text{(iv)}$$

where v = velocity of escaping gas, C = coefficient for opening, $P_2$ = pressure outside vessel, $P_1$ = pressure inside vessel ($P_1 > P_2$), k = gas constant, $C_p/C_v$, and $V_1$ = specific volume of gas inside vessel at $P_1$.

Of course, the total expansion of the released gas can be calculated by simply assuming that the total mass is conserved and undergoes a change of state from conditions $(T_1, P_1, V_1)$ to $(T_2, P_2, V_2)$.

In general, the sudden release of compressed gas affects only the space into which the gas expands to ambient pressure. In many respects, the blast or explosion effects of a polytropic expansion are about the same as a deflagration explosion given equal initial pressures and temperatures. However, in a polytropic expansion explosion, there are generally no fireball or fire front effects to consider.

## 9.3 Deflagrations and Detonations

In the study of fires and explosions, by far the most common type of explosion encountered is the deflagration. Deflagrations often occur when flammable gases or dusts have accumulated to levels above their lower limits of flammability. Examples of deflagrating explosives include:

1. Explosive mixtures of natural gas and air at room conditions.
2. The decomposition of cellulose nitrate, an unstable compound often used in propellants.
3. Black powder.
4. Grain dust.

Detonations are encountered often in arson or sabotage cases. Occasionally, accidental detonations occur, usually in construction work, quarry work, or similar situations. Examples of detonating explosives include:

1. Dynamite.
2. Nitroglycerine.
3. Mercury fulminate.
4. Trinitrotoluene (TNT).
5. Ammonium nitrate fuel oil (ANFO).*

Under special conditions, a normally deflagrating explosive can be made to detonate. Such special conditions include the application of high pressures, strong sources of ignition, and long flame run-up distances. With the rare exception of the last condition, in uncontrolled fires and explosion, deflagrations generally remain deflagrations.

When an explosion occurs in an unconfined, open area, the pressure wave will harmlessly expand and expend itself until the pressure difference becomes insignificant. When an explosion occurs in a confined space, the pressure wave will push against the confining structure. This is why when a small amount of loosely piled gunpowder is ignited, it simply burns quickly with a moderate hiss. However, if the same amount of gunpowder is wrapped tightly in a paper container and ignited, it becomes an ear-splitting firecracker that bursts apart.

When an explosion occurs within a typical building, the building is generally damaged. While most building codes require that buildings or inhabited structures be able to withstand externally applied downward loads due to snow, rain, ice, and wind, the building codes do not require that the structures be able to withstand the outwardly directed loads generated by an explosion located within or adjacent to the structure.

---

* During the trial for the World Trade Center bombing, the type of explosive the conspirators were caught in the act of making at the time of arrest was a type of ANFO. The authorities described it as a "witches' brew." Hence, the quotation at the beginning of this chapter is an obscure reference to this heinous act of sabotage.

In buildings, many accidental explosions are typically caused by some of the following:

1. Ignition of natural gas leaks.
2. Ignition of vapors from improperly stored gasoline, cleaning solvents, copy machine chemicals, or other volatile flammable liquids.
3. Ignition of leaking liquid propane vapors (LP).
4. Ignition of grain dust, coal dust, flour dust, textile dust, and other types of dust from combustible materials.
5. Ignition of certain types of fine metal powders, such as aluminum and magnesium.
6. Ignition of atomized flammable liquids.

Casual inspection of the above list shows that most accident explosions occurring in buildings are of the deflagrating type.

The materials involved in deflagrating explosions can be ignited in a number of ways. The most common source of ignition is an electric spark. As was noted in Chapter 2, the smallest spark that can cause ignition of combustible vapors is called the minimum ignition energy (MIE). Usually the spark as measured in laboratory tests is supplied from a capacitor across an air gap to the fuel. The optimum air gap distance that can cause ignition is referred to as the minimum ignition quenching distance.

Table 9.1 lists the MIE values for the vapors of several fuels at stoichiometric conditions in air.

One of the reasons that extra precautions must be taken around areas where there is concentrated oxygen, such as in hospitals, is that the MIE in pure oxygen can be many times less than that of the MIE in air. Table 9.2 lists the MIE values for the same fuels at stoichiometric conditions in pure oxygen instead of air.

**Table 9.1  Minimum Ignition Energies for Combustible Gases in Air**

| Fuel | MIE |
| --- | --- |
| Natural gas | 0.00030 joules |
| Propane | 0.00026 joules |
| Ammonia | > 1.00 joules |
| Methanol | 0.00014 joules |
| Normal butane | 0.00026 joules |
| Trichloroethylene | 0.300 joules |

Source: U.S. Department of the Interior, Bureau of Mines, Bulletin 680, p. 33.

**Table 9.2   Minimum Ignition Energies
for Combustible Gases in Oxygen**

| Fuel | MIE |
|------|-----|
| Natural gas | 0.000003 joules |
| Propane | 0.000002 joules |
| Ammonia | Not Available |
| Methanol | Not Available |
| Normal butane | 0.000009 joules |
| Trichloroethylene | 0.018 joules |

Source: U.S. Department of the Interior, Bureau of
Mines, Bulletin 680, p. 33.

A comparison of Tables 9.1 and 9.2 finds that some of the combustible vapors will ignite in oxygen at MIE levels two orders of magnitude less than in air.

This extreme sensitivity of pure oxygen to even the smallest spark has been a bane to NASA. A small spark in an oxygen-rich atmosphere is what triggered the fire in Apollo 1 where three astronauts were killed. The breathing atmosphere inside the capsule utilized an oxygen-rich mixture. This mixture was changed in later missions to reduce the hazard.

Oxygen sensitivity to spark ignition was also the cause of the explosion that disabled Apollo 13 enroute to the moon. There were cracks in the insulation of electrical wires in an oxygen tank on the side of the command ship. When the tank was utilized and the oxygen contacted the wires, the ensuing explosion blew off an entire side panel. Only by very remarkable teamwork and some smart thinking on the part of many people, both on the ground and in the spacecraft, were the lives of the astronauts saved.

The following is a partial list of common sources of electrical sparks in homes and commercial businesses that can cause ignition of combustible vapors.

1. *Electric motors* — the commutator slip ring sliding contact. Sump pumps are famous for igniting propane vapors that have accumulated in basements.
2. *Loose electrical plugs in wall sockets* — when the appliance turns on, the plug prongs may arc to the contacts. This occurs in refrigerators and freezers where the appliance turns on and off intermittently.
3. *Static discharge due to frictional action between two electrically dissimilar materials* — the effect is called triboelectrification, and can occur between any combination of gas, liquid, and solid. A particularly hazardous situation is the transfer of flammable liquid from one container to another, especially when one is a plastic bucket.

4. *Open relays and switches* — both will spark when the contacts open and close.
5. *Lightning.*
6. *Electric bug killers* — the high voltage needed to zap a bug can also zap an explosion if there are flammable vapors around.
7. *Thermostats* — these also have open and closed contact points, although most operate at low voltages.
8. *Old-fashioned doorbells* — essentially, this is simply another version of a relay, except that the spark is more continuous while the bell is operating.

In addition to sparks, deflagrating explosions can also be set off by the following list of common ignition sources.

1. Pilot lights in hot water tanks, furnaces, heaters, ovens, stoves, and gas dryers.
2. Smoking materials such as cigarettes, matches, cigars, lighters, etc. In some cases, lighters have been known to fall into tight places, become wedged, and allow their butane to shoot out into another ignition source like a fan. This has occurred in vehicles where people often toss their lighters on the car dashboard, and the lighters fall through the vent openings into the ventilation system.
3. Sparks from the sliding contact of metals or abrasives. Examples include grinding wheels, gas-welding igniters, cigarette lighter flints, metal chisels, and hacksaws.
4. Stove or furnace electric igniters.
5. Hot surfaces, such as electric heater elements and cooking griddles.
6. Radiant energy. Some chemicals only need to be exposed to sunlight to explode. These fall into a special subclass of photoreactive chemicals.
7. Heat from chemical reactions, or between items being mixed.
8. Running automobiles or other internal combustion engines, especially in confined garages. Various exposed parts of the engine become very hot.

## 9.4  Some Basic Parameters

In both deflagrating and detonating explosions, the maximum pressure occurs when the explosion is wholly confined, which is a constant volume process, and the explosive mixture is close to stoichiometric concentrations. The maximum pressure for many hydrocarbon-based deflagrating explosion mixtures will range between seven and nine times the ambient pressure. The

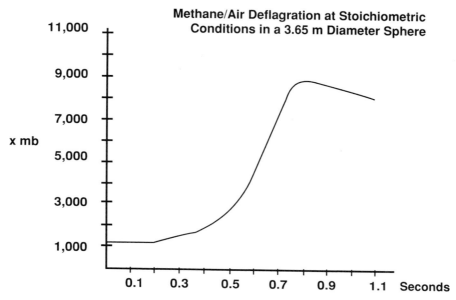

**Figure 9.1** Pressure versus time plot. (U.S. Department of the Interior, Bureau of Mines, Bulletin 680, p. 16.)

typical maximum pressure for a gaseous detonating explosion will be nearly double that of a deflagrating explosion under the same conditions.

Thus, if the ambient pressure is 1013.3 millibars, 1 standard atmosphere, the maximum pressure for a typical deflagrating explosion could range from 7100 mb to 9120 mb. A gaseous detonating explosion maximum pressure might then be 14,200 mb to 18,240 mb.

The graph in Figure 9.1 shows the pressure rise versus time as measured in absolute pressure of a methane and air deflagrating explosion at stoichiometric mixture, 1013.3 mb ambient pressure, and 25°C. The explosion was done in a laboratory vessel 3.65 meters in diameter and the explosion occurred under constant volume conditions. It is notable that the maximum pressure attained in the explosion, 8600 mb, occurred in less than one second.

In deflagrating explosions, there is normally a mixture range of fuel and air in which explosions are possible. For example, methane or natural gas will explode when the concentration is between 5% and 15% in air by volume. If the concentration is less than 5% or more than 15%, no explosion will occur. The ratio of air to fuel will be such that the flame propagation will be self-quenching.

While a methane and air deflagration explosion can reach pressures of 8600 millibars gage pressure when the mixture is optimized for maximum pressure, which usually occurs near or at the stoichiometric mixture, the

**Table 9.3  Explosive Limits of Some Common Gases**

| Fuel | Limits (% v/v) |
|---|---|
| Methane | 5.0–15.0 |
| Ethane | 3.0–12.4 |
| Propane | 2.1–9.5 |
| Acetone | 2.6–13.0 |
| Ammonia | 15.0–28.0 |
| Gasoline | 1.3–6.0 |
| CO | 12.5–74.0 |
| Methanol | 6.7–12.0 |

explosion pressure drops off quickly at mixture concentrations either less than or more than the optimum point.

For example, at a methane concentration of 5%, the lower limit for methane and air explosions, the explosion pressure will only reach about 3200 mb. At the upper limit of 15%, the explosion pressure will only reach about 3900 mb. Similarly, the flame temperature of the explosion, which is maximum at or very near stoichiometric conditions, drops significantly when the mixture is either less than or more than the optimum mixture.

Table 9.3 lists some common deflagrating fuels and their explosive limits.

For dusts such as flour and grain, the lower explosive limits are about 40 to 50 grams per cubic meter of air. However, some items, such as birch bark wood dust, have a lower explosive limit of 20 grams per cubic meter of air. Bituminous coal and lignite dusts have similar explosive lower limit numbers as agricultural dusts. The upper explosive limits for both types of such dusts are not clearly defined.

Table 9.4 lists the lower explosive limits for some metallic dusts. Like the grain dusts, the upper explosive limits for metallic dusts are not clearly defined.

**Table 9.4  Lower Explosive Limits for Metallic Dusts**

| Material | Lower Limit (grams per cubic meter) |
|---|---|
| Aluminum | 80 |
| Iron | 120 |
| Magnesium | 30 |
| Manganese | 120 |
| Sulfur | 35 |
| Uranium | 60 |
| Zinc | 480 |

## 9.5   Overpressure Front

Overpressure is the amount of pressure in excess of the usual ambient pressure. Engineers usually call this the gage pressure and distinguish it from absolute pressure, which is the pressure measured from a vacuum. In the English system, gage pressure is measured in pounds per square foot gage (psfg), and absolute pressure is measured in pounds per square foot absolute (psfa). When the abbreviation "psf" is used, gage pressure is often implied.

A reasonable model for calculating the pressure at the explosion front, the boundary where the explosion pressure is maximum, is the inverse cube rule. This is because the pressure envelope that expands outward from the epicenter is three dimensional, and the pressure of a gas is a function of its volume. The general equation is shown below.

$$P_r = K[I/r^3] \qquad\qquad (v)$$

where $P_r$ = pressure at explosion front located distance "r" from epicenter, K = arbitrary constant for conversion of units and efficiency of explosion, I = explosion intensity, amount of explosive, or amount of energy release, and r = distance from explosion.

When the pressure at one point is known, it is possible to use Equation (v) to calculate pressures at other locations.

$$P_1 = C, \text{ a known amount at distance r1.} \qquad\qquad (vi)$$

$$P_1[r_1^3] = KI$$

$$P_2[r_2^3] = KI$$

$$P_2 = P_1[r_1^3/r_2^3]$$

Damage markers can often help in the solution of Equation (vi). For example, it is known that glass windows break out when the overpressure exceeds 0.5 to 1.0 psig (34.46 mb to 68.9 mb). Thus, using the farthest distance at which windows are known to have blown, an estimate of the pressure at various points along the pathway of the explosion front can be made.

Other markers that can be used to estimate the pressure of a deflagrating explosion at a specific point are:

- Items that have been lifted whose weight is known or can be estimated, like roofs, ceilings, etc.

- Items that have been hurled from the explosion area. In such cases it is possible to calculate the velocity parameters associated with their trajectory, and from that estimate the initial kinetic energy imparted to the item by the explosion. Since the change in kinetic energy is equal to the work done on the item, a measure of the pressure that caused the item to be hurled can be estimated by equating "P-V" work done on the item to its kinetic energy as shown below:

$$\text{Work} = [P_2V_2 - P_1V_1]/[1 - k] = \Delta KE = (1/2)mv^2$$

- Items that have been pushed a specific distance or overturned such that the work needed to accomplish this can be determined and equated to "P-V" work.

In the detonation of explosive materials like ANFO or dynamite, the scatter of debris items can be roughly related to the explosive yield by the following relation:

$$W_E = [\ r^3]/K \qquad\qquad\qquad (vii)$$

where r = distance from epicenter to farthest scatter of debris, $W_E$ = amount of explosive yield in equivalent kilograms of TNT, and K = scaling factor, 91,000 $m^3$/kg.

It should be emphasized that the above relation is only to be used as a "first cut" estimate of the amount of explosive.

Explosive yield is a term used in association with detonations to indicate the amount of explosive effect produced. Usually, the explosion under consideration is equated to an equivalent amount of TNT that would cause the same level of damage or effect. TNT is often used as a standard against which other explosions are measured because there are so many different types and variations of explosives.

A gram of TNT, by the way, produces about 4680 joules of explosive energy. However, energy equivalence is not the whole story, and is only a rough indicator of explosive effect. Explosive effect equivalence must also take into account detonation velocity and fissile effects.

Thus, an explosion with a yield of 100 kg of TNT simply means that the explosion produced about the same explosive effect as 100 kg of TNT. For example, 100 kg of AMATOL is equivalent in effect to about 143 kg of TNT. ANFO, an inexpensive explosive often used for blasting surface rock, has 142% the explosive effect of TNT.

With respect to equivalent explosive yield, it is inappropriate to assign a deflagrating explosion with a TNT-equivalent explosive yield. It's simply an "apples and oranges" comparison. A review of the basic differences between detonations and deflagrations in the first part of this chapter will make this point clear.

Despite this comparative disparity, however, it is commonly observed that some investigators spend a significant amount of time calculating an equivalent explosive yield in TNT kilograms of natural gas explosions, ammonia explosions, and so on. In one prominent case involving a boiler failure in the Midwest, the explosion resulting from the flash expansion of hot, pressurized boiler feed water from ruptured header pipes was even equated to so many kilograms of TNT.

This equivalence calculation is usually done by simply equating the energy content of so many kilograms of TNT to the energy released by the deflagration of so many kilograms of natural gas, ammonia, etc. Of course, there obviously may be an energy equivalence on a joule-to-joule basis between the two, just as there can be a joule-to-joule equivalence of the electricity in a flashlight battery to some amount of gasoline. However, this certainly is not the basis for an equivalence of explosive effect.

Consider, for example, a cubic foot of air at 2000 psia and 20°C. This has an energy content of $1.93 \times 10^5$ joules. A one-pound-stick of dynamite has an energy content of about $2.09 \times 10^5$ joules. From just their energy content, it might be said that a cubic foot of air compressed to 2000 psia is about equivalent to a stick of dynamite.

However, the sudden release of a cubic foot of compressed air in a blast hole will have little fissile effect on the surrounding rock. It will likely "whoosh" out the top of the hole harmlessly. On the other hand, the detonation of a stick of dynamite in the same hole can break apart solid rock, causing it to fracture into many pieces. Alternately, it is nearly impossible to slow down the explosion of a stick of dynamite to power a gas turbine. A stick of dynamite shoved in a gas turbine will simply blow it apart. However, it is possible to slowly release the cubic foot of compressed air to make the gas turbine operate and extract work.

Thus, to examine the damage at a site where the explosion has been caused by a deflagration, and then to equate it calorically to so many sticks of dynamite or pounds of TNT is not a particularly valid comparison. In fact, except for determining the final amount of heat released, it provides no valid comparative information concerning the amount or severity of damage. Unfortunately, some lawyers, reporters, public officials, and investigators insist that this equivalence be done as a "wow" factor for the jury, newspaper readership, viewing audience, etc. despite its dubious scientific value.

The pressure wave or front that propagates from the explosion epicenter has a pressure distribution over time similar to that shown in Figure 9.2.

As is shown in the figure, the initial pressure front is a "wall" or "spike." The pressure on one side of the spike is the ambient pressure, and just at or

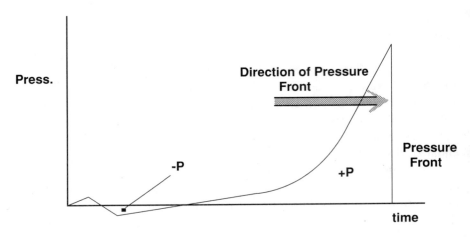

**Figure 9.2** Profile of the pressure wave or front.

slightly behind the spike, the pressure is maximum for that distance from the origin of the explosion.

As the front passes by a certain point, the pressure level drops. At some time after the front has passed, the pressure even drops below ambient and a slight low pressure front occurs. In motion pictures showing the passage of pressure fronts in actual explosions, this negative pressure area sometimes looks like a slight reverse surge, especially after having observed the main pressure front initially push through with great destructive force. After the negative pressure subsides, the pressure stabilizes back to ambient.

This is why so much attention is paid to the determination of the maximum pressure. The destructive power of an explosion is basically contained in the pressure front. If a structure can survive that, it is likely that it will be able to survive the lower level pressure zone following the front.

Of course, the above situation assumes that the explosion has taken place in the open. The presence of buildings, hills, valleys, and other features can complicate the profile of the pressure wave because of reflections, refractions, and shadowing. The pressure wave of an explosion can bounce off or be deflected by buildings and geographic features like waves in a pond bouncing off the side of a boat. There can even be destructive and constructive interference in certain situations.

## Further Information and References

"Air Blast Phenomena, Characteristics of the Blast Wave in Air," reprint of Chapter II from *The Effects of Nuclear Weapons*, Samuel Glasstone, Ed., April 1962, U.S. Atomic Energy Commission, pp. 102–148. For more detailed information please see Further Information and References in the back of the book.

*ATF Arson Investigation Guide*, published by the U.S. Department of the Treasury. For more detailed information please see Further Information and References in the back of the book.

*Chemical Thermodynamics*, by Frederick Wall, W. H. Freeman and Co., San Francisco, 1965. For more detailed information please see Further Information and References in the back of the book.

*Code of Federal Regulations 29, Parts 1900 to 1910, and Part 1926* (two volumes). For more detailed information please see Further Information and References in the back of the book.

*The Condensed Chemical Dictionary*, revised by Gessner Hawley, 9th ed., Van Nostrand Reinhold Co., 1977. For more detailed information please see Further Information and References in the back of the book.

*Explosive Shocks in Air*, Kinney and Graham, Second Edition, Springer-Verlag, 1985. For more detailed information please see Further Information and References in the back of the book.

*Fire and Explosion Investigations*-1992 ed., NFPA 921. For more detailed information please see Further Information and References in the back of the book.

*Fire Investigation Handbook*, U.S. Department of Commerce, National Bureau of Standards, NBS Handbook 134, Francis Brannigan, Ed., August 1980. For more detailed information please see Further Information and References in the back of the book.

*General Chemistry*, by Linus Pauling, Dover Publications, New York, 1988. For more detailed information please see Further Information and References in the back of the book.

*Handbook of Chemistry and Physics*, Robert Weast, Ph.D., Ed., 71st ed., CRC Press, Boca Raton, FL. For more detailed information please see Further Information and References in the back of the book.

*Hydrogen-Oxygen Explosions in Exhaust Ducting*, Technical Note 3935, by Paul Ordin, National Advisory Committee for Aeronautics, Lewis Flight Propulsion Laboratory, Cleveland, Ohio, April 1957. For more detailed information please see Further Information and References in the back of the book.

*Investigation of Fire and Explosion Accidents in the Chemical, Mining, and Fuel Related Industries — A Manual*, by Joseph Kuchta, U.S. Department of the Interior, Bureau of Mines, Bulletin 680, 1985. For more detailed information please see Further Information and References in the back of the book.

*NFPA 495: Explosive Materials Code*, National Fire Protection Association, 1990 ed. For more detailed information please see Further Information and References in the back of the book.

*Pressure Integrity Design Basis for New Off-Gas Systems*, by C.S. Parker and L.B. Nesbitt, General Electric, Atomic Power Equipment Department, May 1972. For more detailed information please see Further Information and References in the back of the book.

*Principles of Combustion*, by Kenneth Kuo, John Wiley & Sons, New York, 1986. For more detailed information please see Further Information and References in the back of the book.

"Sensitivity of Explosives," by Andrej Macek, U.S. Naval Ordnance Laboratory, *Chemical Reviews*, 1962, Vol. 41, pp. 41–62. For more detailed information please see Further Information and References in the back of the book.

*Thermodynamics*, by Joachim Lay, Merrill Publishing, 1963. For more detailed information please see Further Information and References in the back of the book.

# Determining the Point of Ignition of an Explosion

# 10

You can observe a lot by looking.

— **Yogi Berra**

Hey Boo-Boo.
Do you see what I see?

— **Yogi Bear**
*Hanna-Barbera Productions*

One must learn to see things with one's inner vision.

— **Yogi Maharishi**

## 10.1  General

The explosive limits of a gaseous fuel are useful in calculating the amount of fuel that may have been involved in an explosion. If the room or space in which the explosion occurred is known, its volume can be determined. If the type of gas that fueled the explosion is known or assumed, then the amount of fuel necessary to cause the explosion can be estimated from the lower explosive limits.

Consider the following example. Suppose an explosion occurred in a small bedroom, 12 ft × 10 ft × 8 ft, which contained a natural gas space heater. The explosion occurred when a light switch located halfway down the 8 ft high wall was thrown, after the room had been closed up for two hours. Singeing was noted on the ceiling and walls down to the light switch, and faded below it. How much gas had leaked, at what rate had it leaked, and what size leak would be necessary to cause the explosion?

Since natural gas is lighter than air, much of the natural gas would initially collect on the ceiling and fill the ceiling space downward, like an inverted bowl. The explosion could only occur if the concentration of gas at the light switch was at least 5% or more by volume. Assuming a cloud of 5% methane and air in the top portion of the room, then about 24 cubic feet of methane would have been required. Since the room had been closed for two hours, the leak rate was about 0.2 ft³/min.

191

Because most household natural gas systems operate at a pressure of about 0.5 psig, application of the Bernoulli equation finds that the escape velocity of the natural gas from the gas lines or heater would be 2.32 ft/sec. To have a leak rate of 0.2 ft³/min, a leak cross-sectional area of 0.21 in.² is required. Thus, this estimate of leak size gives a hint as to what size hole or opening to look for in the gas lines and components around the space heater.

## 10.2   Diffusion and Fick's Law

It is commonly thought that when a light, bouyant gas is introduced into a room full of air, the light gas floats to the top and forms a distinct layer. Initially, this is often true. However, over time this is not true. Given time, the light gas will diffuse into the air in the room until the concentration at the top of the room is not significantly different from the bottom of the room.

If gases were to layer out according to their relative specific gravities, then the earth's atmosphere would indeed be very different than it is now. Next to the ground would be the heaviest gases found in the air: argon and carbon dioxide. Just above that would be a relatively thick layer of oxygen. On top of the oxygen, well above most people's heads and the tops of buildings would be a layer of nitrogen. Finally, the topmost layer would be composed of water vapor, methane, and the lighter gases.

However, our firsthand experience indicates this is not the case. The relatively lighter gases do not rise to the top of the atmosphere, and the relatively heavy gases do not sink to the bottom and hug the ground. All the gases are well mixed in proportions that do not significantly change from sea level to the top of Mount Everest. In fact, the proportions do not significantly change from that at ground level until the upper areas of the stratosphere are reached, where other factors are at work to change the relative composition of the atmosphere.

Given time and provided that the gases do not chemically react with one another, all the gases in a confined mixture mingle until the composition at the top is the same as at the bottom. This is called Dalton's law of partial pressures. In essence, Dalton's law states that each gas tends to diffuse into a contained space as if the other gases were not present. This mixing tendency is also true of liquid solutions when the solvent and solute are miscible with one another.

At a given temperature, the rates at which different gases diffuse are inversely proportional to the square root of their molecular weights. This observation is consistent with the Kinetic theory of gases, where the pressure, or concentration of a gas, is proportional to its average kinetic energy.

$$P = C[(1/2)mv_{ave}^2] = (3/2)CkT \qquad\qquad (i)$$

where C = an arbitrary constant that also embeds the number of molecules per unit volume, $v_{ave}$ = average velocity of gas molecule, m = weight of gas molecule, k = Boltzman's constant, $1.38 \times 10^{-23}$ joules/°K, and T = absolute temperature, °K.

It follows that diffusion is a function of the average velocity of the gas as it wanders about, colliding with the various other gas molecules. Gases with a high average velocity diffuse faster than gases with a low average velocity. Thus, algebraic manipulation of Equation (i) yields:

$$m_1/m_2 = v_2^2/v_1^2 \quad\text{or}\quad v_2/v_1 = [m_1/m_2]^{1/2} \qquad\qquad (ii)$$

For example, if one mole of hydrogen gas were released into a room full of air, what would be its diffusion rate as compared to a mole of methane released in the same fashion?

A mole of hydrogen has a molecular weight of 2. A mole of methane has a molecular weight of 16. Thus, the hydrogen would diffuse into the room full of air at a rate 2.83 times faster than the methane.

In the case of a solution of gases where the concentration is not uniform, the diffusion of the solute gas into the solvent gas from a region of high concentration to an area of lower concentration is modeled by Fick's law, which is given below for the one-dimensional case.

$$dm/dt = -DA[dc/dx] \qquad\qquad (iii)$$

where dm/dt = mass of solute gas diffusing per unit time, −D = diffusion constant for the particular solvent gas at a given temperature, A = cross-sectional area of the diffusion flow, and dc/dx = concentration gradient in the direction "x" perpendicular to the cross-sectional area, "A."

By Fick's law, the diffusion of material across the boundary stops when dc/dx = 0, that is, when the concentration of the solute gas in the solvent gas is the same everywhere in the space.

For example, consider a room that is 3 meters wide, 4 meters long, and 3 meters high. At the bottom of the room is a layer of ethanol vapors. The layer is about 50 cm thick, with the concentration being zero at the air/ethanol interface, and maximum nearest the floor. It is known that 92 grams of ethanol (2 moles) have evaporated to form the layer. What would be the diffusion rate of ethanol into the rest of the room when the temperature is 40°C?

In this case:

A = (300 cm)(400 cm) = 120,000 cm²

D = 0.137 cm²/sec @ 40°C

dc/dx = [(2 moles/(300 cm × 400 cm × 50 cm)]/[50 cm] =
                   6.67 × 10⁻⁹ mole/cm⁴

dm/dt = –DA[dc/dx] = –[0.137 cm²/sec][120,000 cm²]
                   [6.67 × 10⁻⁹ mole/cm⁴]

dm/dt = 0.0001 moles/sec or 0.0046 grams/sec.

An interesting point about the form of Fick's law is that it is very similar to the one-dimensional equation for thermal conductivity. Note the comparisons below.

$$dm/dt = -DA[dc/dx] \quad \text{Fick's law} \qquad \qquad \text{(iv)}$$

$$dq/dt = -KA[dT/dx] \quad \text{Thermal conductivity}$$

The reasons for the simularity are rooted in the fundamental effects involved in both processes. Both processes involve the exchange of kinetic energy from molecule to molecule, and both processes depend on a simple potential gradient to make the process go. In the case of the conduction law, the potential gradient is the change in temperature per distance along the material. In the case of Fick's law, the potential gradient is the change in concentration per distance along the gas solvent.

Further, both processes involve very fundamental aspects of the property called entropy. While the concept of entropy has not been specifically addressed in this text, it is nonetheless interwoven throughout. It is left to the reader to research the connections between the conduction law and entropy, and Fick's law and entropy.

## 10.3  Flame Fronts and Fire Vectors

When an explosion occurs, especially in deflagrating explosions, it is often the case that some of the fuel will be pushed along with or ahead of the pressure front creating a fireball or flame front effect. Fuel will often still be burning in the gaseous mixture as the explosion is in progress. The flame front will eventually quench itself when either the temperature drops below the ignition point or the fuel is exhausted.

As the pressure wave moves outward and expands, the burning gas will cool. Since the pressure diminishes as the cube of the distance, the temperature of the flame front also drops quickly. Usually the flame front will die out well before the pressure front has harmlessly expended itself.

Thus, in many deflagrations, there is a large, general area of pressure damage within which is a smaller area that contains singeing or fire damage.

As a fireball passes through an area, it will impinge on the side perpendicular to its travel, but it will not impinge on the backside of the object. The effect is similar to that of a surface wave on pier supports. As the wave approached the vertical support, the wave will crash on one side, but will only slightly affect the backside, or "shadow" side.

In the fire-affected area of an explosion, there will be objects that exhibit heat damage or singeing on one side, the side that directly faced the explosion. These objects act as directional vectors, and indicate the direction from which the fireball came. Mapping of these various vectors and the intensity of burn or singeing associated with each one provides a ready map of the pressure front of the explosion, and its temperature intensity. Taken together, these vectors will plot out the expansion of the pressure front, and will lead back to the point of ignition of the explosion.

Due to the very short time available for heat transfer from the flame front to the object, objects with higher thermal inertia will be less affected by the momentary flame front.

## 10.4   Pressure Vectors

When the pressure front moves through an area causing damage, a rough estimate of the pressure front at that location can be correlated to the type of damage or injury that occurred there. Table 10.1 lists some representative pressure values and their usual associated damages.

**Table 10.1   Pressure Intensity vs. Type of Damage or Injury**

| Pressure Level | Type of Damage or Injury |
| --- | --- |
| 0.5–1.0 psig | Breakage of glass windows |
| > 1.0 psig | Knock people down |
| 1.0–2.0 psig | Damage to corrugated panels or wood siding |
| 2.0–3.0 psig | Collapse of nonreinforced cinder block walls |
| 5.0–6.0 psig | Push over of wooden telephone poles |
| > 5.0 psig | Rupture ear drums |
| > 15.0 psig | Lung damage |
| > 35 psig | Threshold for fatal injuries |
| > 50 psig | About 50% fatality rate |
| > 65 psig | About 99% fatality rate |

Like a flame front, the pressure front will impinge most strongly on surfaces normal to its path of travel. Thus, the movement or shifting of items from their normal positions by the passage of the pressure wave can be used to trace the path of the pressure front back to its epicenter.

For example, if an explosion occurred in a certain room interior to the house, it would be expected that the room might have the following.

- All four walls might be pushed outwards, away from the room.
- The floor might be pushed downward, perhaps even collapsing into the basement.
- The ceiling might be pushed upward, perhaps even lifting the roof or upper floor upward.
- Metal cabinets or ductwork might be dented inward on the side facing the epicenter.

On the other hand, if the explosion occurred in the basement, the following effects might be observed.

- If the basement walls are block, the walls might have moved outward creating a "gap" between the wall and the surrounding dirt.
- The floor above the basement might have lifted upward, and then collapsed inward.
- Ductwork might be dented inward on the side facing the epicenter.
- Free-standing shelves next to walls may have been pushed into the walls and rebounded, falling into the floor.

## 10.5   The Epicenter

When a gas leak occurs, the gas will initially form an irregularly shaped cloud of fuel that diffuses away from the leak. If the gas is methane and the air is more or less calm, the gas will slowly rise at first to the ceiling and accumulate in high areas. If the gas is propane, it will sink to the floor and accumulate in low areas.

As the fuel cloud diffuses and moves away from the point of leakage, its boundary may come into contact with an energy source capable of igniting the fuel, provided the fuel cloud is above the lower limit of flammability in air. At first contact, the fuel cloud likely will not have sufficient concentration for ignition. However, given time and the right conditions, the concentration of the cloud near the energy source may increase sufficiently for ignition to occur.

In cases where the source of ignition is steady, like a pilot light, the fuel cloud will be ignited as soon as its concentration exceeds the lower limit of flammability. If the energy source is intermittent, like a relay or switch, the cloud may ignite after surrounding the ignition source.

As noted in previous sections, the point of ignition can be determined from the fire and pressure damage vectors. The plot of these directional vectors on a fire scene sketch will collectively point to the epicenter of the explosion, the origin of the fireball and pressure fronts. This is the primary reason why the documentation of debris scatter after an explosion is important.

However, walls, hallways, and other features within a building can direct and alter the path of the fireball and pressure fronts. Thus, the investigator should not presume that all the directional vectors will point radially inwards toward an epicenter. Some will indicate that the pressure front followed nonradial pathways from the epicenter, having been guided by strong walls or passageways. Thus, the damage vectors may form a trail. The backtracking of several trails that converge at the same location is usually indicative of successful location of the epicenter.

## 10.6   Energy Considerations

The amount of energy contained in an explosion is a direct function of the type of fuel and the amount of fuel involved in the reaction. In general, the energy of an explosion is dissipated in the following forms.

1. *Acoustical* — the blast sound and pressure wave.
2. *Kinetic* — the displacement of objects away form the point of origin of the explosion.
3. *Heat and expansion energy* — lost to the surroundings.
4. *Fissile* — the propagation of intense shock waves through nearby materials resulting in their fracturing.

Items 1 through 3 above usually apply to deflagrations, because little energy is dissipated in fissile effects in a deflagration. Items 1 through 4 all apply to detonations of high explosives. In detonations, fissile effects can and usually are significant.

A quick reconnoiter of the explosion scene near the epicenter can usually provide evidence of whether there were significant fissile effects. If there are, it is advisable to take samples of materials that the blast front would have impinged upon near the epicenter. This is because high explosives will often embed small particles of the explosive material into nearby surfaces, like small embedded bullets. These particles can be detected by laboratory tests, and the explosive material clearly identified. In some cases, because there are identifying trace elements added to many types of high explosives, the place of manufacture of the explosive can also be determined. This is a significant starting point in the tracking of the explosive material from initial producer to user.

Because the pressure waves dissipate as the cube of the distance from the explosion volume envelope, if the pressure needed to lift a ceiling, push over a wall, or break glass in a window across the street can be determined, then the pressure of the explosion at the envelope boundaries of the explosion can be calculated or estimated.

Knowing the room or space where the explosion originated, it may be possible to narrow the list of possible fuels. In short, it is possible to gauge the energy content of an explosion and deduce something about the nature of the fuel that caused the explosion.

For example, it might be deduced that the explosion, which originated in a small bathroom, had a pressure of about 4 psig when the ceiling of the bathroom lifted. The bathroom was noted to have a natural gas space heater, and an open bottle of nail polish remover (acetone), which had evaporated. The source of ignition energy for the explosion apparently came from a running wall clock (motor commutator ring). The gas lines to the space heater were pressure checked, and a small leak was found. Interviews with the homeowner found that the 16-ounce bottle of nail polish remover was only half full when it was left open. How could a person tell which item caused the explosion?

First, from interviews it could be established how long the room had been left unattended. This would establish the amount of time available for either leaking or evaporation, since both items are smelly and would have been noticed. The evaporation and diffusion rates of acetone could then be checked to see if enough vapors would occur in that amount of time for an explosive mixture to have formed in the air. Similarly, the leak rate of the heating could be extrapolated to determine if enough fuel could be put into the room to cause an explosion.

If the heater leak had caused the explosion, there might be a singe pattern that would be most intense around the heater. Similarly, if the nail polish remover were the culprit, the singe pattern might center about it. Acetone is not initially bouyant in air, while methane is. A singe pattern high in the ceiling may indicate the fuel cloud was mostly methane that had not fully diffused into the room. Alternately, a singe pattern low in the room, from the floor upward, might indicate the fuel cloud was heavier than air.

Of course, it should not be overlooked in our hypothetical example that it is also possible that both fuel sources contributed to the explosion.

## Further Information and References

*The Engineering Handbook*, Richard Dorf, Ed., CRC Press, Boca Raton, FL, 1995. For more detailed information please see Further Information and References in the back of the book.

*General Chemistry*, by Linus Pauling, Dover Publications, New York, 1988. For more detailed information please see Further Information and References in the back of the book.

*Thermodynamics*, by Ernest Schmidt, Dover Publications, New York, 1966. For more detailed information please see Further Information and References in the back of the book.

# Arson and Incendiary Fires 11

The price one pays for pursuing any profession or calling is an intimate knowledge of its ugly side.

— **James Baldwin,** in *Nobody Knows My Name, 1961*

Pass the methyl, Ethyl.
Pour on the gas, Cass.
Turn on the fan, Stan.
Give it a spark, Clark.
Get the hell away, Ray.
Keep your alibi straight, Nate.
Look real sad, Tad.
Call the adjuster, Buster.
Sign the form, Norm,
and count the money, Honey.

— **Anonymous**

## 11.1 General

The specific legal definition of arson varies from state to state. However, a working definition of arson is: the malicious burning of homes, residences, buildings, or other types of real property. Malicious burning in this context is intended to also include incendiary explosions as well as fire.

In the Model Penal Code, Section 220.1(1), the definition of arson includes the starting of a fire or explosion with the purpose of destroying a building or damaging property to collect insurance money. The building or property may belong to another, or to the person who commits the arson.

It is worth remembering that if a person burns down his own home or building, it is not necessarily an arson. Barring local fire and public safety laws and ordinances, a person may usually do as he wishes with his own property, including burn it down. It is only when the building is burned down to fraudulently collect on insurance, perhaps deprive a bank or divorced spouse of their property rights, or avoid other types of obligations, does burning one's own property become an arson.

In some states, the crime of arson is divided into three degrees.

**Plate 11.1** Burn through of bedroom floor where accelerant was used.

- First-degree arson is usually the burning of an inhabited house or building at night.
- Second-degree arson is the burning at night of an uninhabited building. "Uninhabited" means no humans inside the building.
- Third-degree arson is the burning of any building or property with intent to defraud or injure a third party.

First- and second-degree arsons are usually considered felonies. As such, it is necessary to prove that the person accused of the arson intentionally caused the fire or explosion. Some third-degree arsons may be considered misdemeanors if the fire was small, and are legally considered similar to malicious mischief or vandalism.

If a person dies as a result of an arson, in some states the death is considered murder. The death may have occurred during the fire because the victim was fatally burned, or the death may have occurred sometime after the fire due to grievous injuries resulting from the fire. This may include injuries directly caused by the fire, such as burns and smoke inhalation, and also injuries indirectly resulting from the fire, such as when people leap from a building to escape being burned to death.

A fire set to conceal another crime is considered an arson in some jurisdictions. Sometimes criminals will set fires in the hopes that the fire will obliterate or at least obscure the evidence of a second crime, perhaps a burglary or murder. Perhaps, the criminal has shot a person, and then set fire to that person's house hoping that the investigators will believe the victim died as a result of the fire, rather than the shooting. Or, perhaps a burglar will set fire to the house he has just robbed hoping that the items taken will not be noticed as missing from the fire debris.

In some jurisdictions, properly authorized city- or state-employed arson investigators have the same investigation and arresting authority as police officers or marshals. Some may be allowed to carry weapons and police type equipment. They also have the same investigation and arresting responsibilities. This means that they have to follow the same rules that police officers and marshals follow, such as "Mirandizing" a suspect during interviews.

Of course, in nearly all jurisdictions, it is unlawful for a citizen to disobey the lawful orders of a fire marshal or fire chief during the extinguishment of a fire, or similar fire safety activity. This is to protect public safety and to ensure orderly suppression of fires and fire hazards.

Incendiary fire is also a term used to describe a deliberately set fire. Alternately, an incendiary fire may be defined as one that is not accidental, or that is not the result of natural processes. As opposed to the term "arson," the term "incendiary" also includes fires that may not qualify by legal definition as arsons. These are fires where there may not be malicious intent, or where there may not be any attempt to defraud for monetary gain. It is normally presumed, however, that the person responsible for an incendiary fire is aware that the fire should not have been set for legal, ethical, or safety-related reasons.

The term "suspicious fire" is typically used to designate a fire that has some of the characteristics of an arson or incendiary fire. Usually the term implies that findings are tentative or preliminary, and that further evidence gathering is required to properly confirm a finding of either arson or incendiary.

## 11.2   Arsonist Profile

The average person arrested for arson is male. Women arsonists constitute only about 11% of the total arrests. Most arsonists are over 18 years of age. Of the men arrested for arson, only 42% are under 18 years of age. Of the women arrested for arson, only 30% are under 18 years of age.

On the average, there are about 18,000 arrests per year for arson. This is about 30% more than the total arrests for embezzlement, and is just slightly less than the total arrests for murder and nonnegligent manslaughter. Arson appears to be a "growth" industry. Total momentary losses in the past decade in terms of constant dollars has increased.

It has been posited by some that there are seven fundamental motives for arson or incendiary fires. Whether this seven-item list is truly exhaustive is perhaps open to question, but it does contain most motives.

1. **Revenge:** To get even with someone for real or imagined slights. If the slight is very imaginary, see item 6 below.
2. **Personal gain:** This usually occurs when the arsonist needs money, and the property to be burned is expendable under the circumstances.
3. **Vandalism:** This is the "just plain orneriness" category. This is very popular in the adolescent age group, or where there are gangs and transients. In some cities there are actually traditions of vandalism on certain days. Detroit, for example, is famous, or rather infamous, for the incendiary fires annually set to mark "Devil's Night" or "Witching Night," the night before Halloween. The same event used to be called "Picket Night" in St. Louis, but thankfully the tradition there seems to have largely died out.
4. **To conceal another crime:** Burglaries and murders are often masked by fires. It is hoped that the missing items will be presumed to be burned up, or the shot person will be so badly burned that the bullet holes will be overlooked.
5. **Rioting:** Pillaging, looting, and setting fires have a long tradition in civil riots and unrest. "Burn, baby, burn," is not just a media-generated slogan. The riots associated with the Rodney King matter is a prime example. Other notable examples include the Watts Riots in 1968, the Civil War Draft Riots, and the French Revolution.
6. **Abnormal psychology:** The usual euphemism for being crazy or mad. This includes firebugs, persons who want to be heroes at fires, people wishing to purify the world, self-appointed angels of justice, etc.
7. **To deny use of property by another person:** An example of this would be neighborhood vigilantes burning down a vacant house being used by undesirable drug dealers. The military use of fire, to deny the enemy the use of a building or structure, also fits into this category.

## 11.3   Basic Problems of Committing an Arson for Profit

In order to be a successful arsonist, the arsonist must burn enough of his house, property, or inventory for it to be considered a total loss. In that way, he will collect the policy limits for the fire, instead of having the insurance company rebuild or replace the property. When the latter occurs, he usually does not make enough money. After all, it was the policy limit that was the temptation in the first place, and not the allure of a redecorated bathroom.

Because of this, when a fire is promptly extinguished by the fire department, or the fire has simply not spread very well and damages are minor, a second fire will sometimes "mysteriously" break out in the same building a day or so later. Usually the second fire is bigger and better. Lessons about fire will have been learned from the first attempt; "practice makes perfect."

A rational person might think that after the first unsuccessful attempt, no one would be so foolish as to try again, and especially in the same building. However, sometimes the motives that compelled a person to commit arson the first time are so powerful that normal prudence is abandoned. Sometimes it is replaced with unrealistic rationales. Some arsonists are imbued with the notion that they are too smart to get caught. Others who have more desperate motives, may believe that it doesn't matter. To them, jail may be less repugnant than not having the insurance money and having to face bankruptcy or foreclosure.

Thus, the fundamental problem facing the arsonist is how to set a fire or explosion such that it will have enough time to burn and consume the building or property before being spotted and put out by the fire department. The secondary problem is how to do this deed without there being enough evidence found later to be blamed for it.

With regards to the problem of blame, usually one of two strategies is involved. The first is to try to disguise the arson so that it looks like a "natural" fire. For example, the fire might be set up to appear as if it were accidentally caused by an overheated kerosene space heater that leaked fuel.

The second strategy is to make no real effort to disguise that the fire is an arson, but to make sure that the person has an alibi or some other reason to not be specifically blamed for the fire. For example, the fire might be determined to be incendiary, but is blamed on vandalism. Or, perhaps the arsonist is able to delay the start of the fire such that he can be at a social event to provide an "ironclad" alibi. Thus, while the fire might be determined to be an arson, it can't be connected directly to the arsonist.

In cities where the response time of the fire department is just a few minutes, the arsonist will likely need an accelerant to speed the fire along. This also means he will have to set fire in several places more or less simultaneously. He may also need to deliberately set fires in strategic areas to enhance the fire spread. Such areas might include wooden stairways or storage areas for flammable materials.

Another problem confronting the arsonist is that he must set the fire and remain unseen by witnesses. And, of course, he must do so with as little danger to himself as possible. With respect to this last point, it is not uncommon for less-than-astute arsonists to burn or blast themselves to pieces as they attempt to ignite liquid gasoline on the floor of a room full of explosive gasoline vapors.

In some cases, the arsonist will be part of a team. The person who will actually collect the insurance money, usually the owner of the property, will hire a second person to set the fire and assume the risk of being caught. The owner will, of course, create an alibi for himself by being in a public place when the fire occurs. Presumably, both will share in the insurance money.

## 11.4   The Prisoner's Dilemma

The obvious problem with the "team" arson strategy, is that for the rest of their lives, both persons share a nefarious secret. In a pinch, will both partners keep silent, or will one betray the other for more favorable treatment?

Interestingly enough, this problem has been well studied by mathematicians who specialize in game theory, and is called the "prisoner's dilemma." The basic problem, as considered in terms of game theory, was first discussed by the Princeton mathematician, Albert Tucker, in 1950. Essentially, the problem is this: assuming that both persons are arrested for arson and are not allowed to communicate with one another during interrogation, the available options for the two arsonists are as follows:

1.   They both keep mum in hopes of either beating a conviction, or receiving a lighter sentence due to a lack of corroborative evidence.
2.   One arsonist tells on the other to obtain a lighter and perhaps commuted sentence, while the other one stays mum.
3.   They both tell on each other and both receive maximum sentences.

If the object is to minimize the total prison time of both culprits, option 1 is likely the best choice. However, this requires that each prisoner trust the loyalty of the other. While option 1 does minimize the total prison time of both culprits, it does not minimize the individual prison time of either prisoner.

If one person wants to get off as lightly as possible and has no abiding loyalty to his partner, option 2 is the best option. He simply "rats out" his partner in return for the lightest possible sentence. This occurs, of course, while the partner is keeping mum, thinking that his coconspirator has also maintained loyalty.

However, there is the risk that if both prisoners think that option 2 is the best deal, then they may both end up getting option 3, which is the worst possible outcome.

Skillful questioning by law investigative authorities usually attempts to convince each suspect that option 2 is the best bet. That is why suspects are usually kept separated when questioning is done. When they are kept separate, they cannot collude and reinforce each other's testimony as their stories are told. When an alibi story has been fabricated by the suspects to create a plausible lie, it is difficult for both parties to separately invent all the tiny but obvious details that a person would know and remember who had actually been there. The two testimonies are then compared for discrepancies, and these discrepancies can be used to confront the witnesses.

## 11.5 Typical Characteristics of an Arson or Incendiary Fire

An arsonist has several problems to overcome in order to accomplish his goal. The ways in which an arsonist solves these problems are the same ways in which identification of an arson or incendiary fire is provided. The following is a short list of the more common characteristics of an incendiary fire.

1. Multiple origins of fire, especially several points of fire origin that are unconnected to each other. There are no fire pathways between the various points of origin to account for the simultaneous break out of fire in multiple locations.
2. The point of origin is in an area where there is no rational ignition potential. For example, a breakout of fire in the middle of a fireproof carpet with no obvious ignition source available.
3. Use of accelerants such as gasoline, kerosene, turpentine, etc. Often these agents are detectable by their lingering odor, the "pour" patterns they produce on floors when ignited, and by chemical analysis. Many times two points of fire origin will be connected by an accelerant pour pattern, called a trailer. Recently, specially trained dogs have been used to sniff out accelerants at fire scenes in the same way that they have been previously used to sniff out drugs. This is a very timely technique. Also, dogs are a little smaller and more agile than humans and may pick up scents in an area that might otherwise be overlooked by humans.
4. The presence of trailers. These are fire or burn pathways that exhibit flammable liquid pour patterns. Trailers are used by the arsonist to accelerate the spread of fire to more areas of the building. This reduces the burn time of the building so that more of the building will have burned up by the time the fire department arrives.
5. The finding of deliberately arranged fire load. This is where everyday items typically found in the building are rearranged to enhance the fire. Examples include closets stuffed with crumpled newspapers, flammable clothing piled in heaps on floor around stoves and heaters, mattresses laid over space heaters, blankets laid over torpedo type heaters, kerosene heaters placed under clothes hung on a clothesline, open gas cans placed near high wattage light bulbs, etc. In essence, solid flammable materials are substituted for liquid type accelerants.

6.  Buildings or residences that are missing personal items that would normally be present. This might include family photograph albums, special collections or instruments, trophies, prized dresses, tools, etc. These are items that the arsonist has a personal interest in.

7.  Buildings or residences with extra items not normally present, or which appear out of context. This is done by the arsonist to beef up the amount of contents lost in the fire. Useless junk items are carried into the building, which after the fire are claimed to be valuable furniture.

8.  An unusually fast consuming fire for the time involved, and a very high burning temperature in areas where the fire load is, in all respects, very ordinary.

9.  Tampering with fire protection and alarm systems. This is done to give the fire a longer time to burn.

10. Unnatural fire pattern. A fire pattern that does not follow the rules and has burned in an unusual or unnatural sequence. Natural fires always follow the physical laws of heat transfer and chemical combustion in a logical progression through the fire load. Human intervention usually subverts the normal logical progression of the fire, making the fire progression appear out of order.

11. The finding of timers and incendiary devices. Timers are devices used to delay the start of the fire so that the arsonist can get away safely, and perhaps have an alibi. A timer may be as simple as a candle placed so that the fire will not start until the candle has burned halfway down, or it may be as complicated as a device placed in the telephone that will cause ignition when the telephone rings. Hand in hand with a timer is usually an incendiary device. This is typically some kind of accelerant or highly flammable material that is set to catch on fire by the timer. For example, a tall taper is set in the floor of a closet surrounded by newspapers. On the other side of the closet is a rubber balloon filled with an accelerant. The taper takes an hour to burn down to the point where the flame can ignite the crumpled newspapers laying low on the floor. The ensuing newspaper fire spreads across the floor of the closet, and the flames leap up to the balloon, causing it to break open. The balloon then spills its load of accelerant into the newspaper fire. In this case, the taper is used as a timer, and the accelerant-laden balloon is the incendiary device.

12. Tampering with heating and air conditioning equipment to enhance fire spread. Moving air will help spread a fire through a building faster. Thus, many arsonists make sure that the blowers are on during a fire. In the winter time, this may mean that the thermostat is set as high as possible so that the furnace will run constantly. In the summertime, it may mean that the air conditioning system is set as low as possible

for the same reason. If the building is equipped with an attic type blower, the blower may have been operating under unusual circumstances, like in the dead of winter, or outside the time settings normally used on the timer for the blower's operation.

13. Tampering with utility systems. Sometimes the electrical wall outlets are rigged to short and catch fire. The arsonist hopes to create a "V" pattern emanating from the wall outlet, which the fire investigator will blame on electrical shorting in the building wiring. Unfortunately, because many fire investigators are unfamiliar with electrical equipment and codes, there is a tendency of some to label all fires associated with electrical equipment as being caused by shorting. This tampering ploy seems to be more prevalent in rural areas, where electrical codes are not strictly enforced and the quality of electrical workmanship is at the level of "a handyman's special." In such cases, it can be difficult to tell if the outlet or wiring was rigged by an expert, or wired by a layman. Similarly, sometimes a gas pipe may be loosened so that during a fire, the leaking gas will cause the house to explode or cause the fire to burn more fiercely. Fresh tool marks on pipes that heretofore have not been leaking, often allow easy spotting of this ploy.

## 11.6   Daisy Chains and Other Arson Precursors

A daisy chain is when a building is sold within a clique in order to jack up its value prior to an arson or incendiary fire crime being committed. It works like this: Joe buys a dilapidated warehouse building in an old section of town for $50,000. Joe sells the building two months later to his brother, Frank, for $60,000. In another four months or so, Frank sells the same building to his sister-in-law, Zelda, for $75,000. In another three months, Zelda sells the building to George, her cousin, for $100,000.

The purpose of the daisy chain is to artificially inflate the value of the building on paper. In each transaction, the building is insured against fire loss and the purchase price of the building is used to define the limits of the insurance policy.

At some point, the building burns down and the last owner collects the insurance money, which is based on the inflated purchase price of the last transaction. Often, no money actually changed hands during the various paper transactions, and the interim owners listed on the documents were simply "fronts" or "straw men" for the real owner. Some of the owners may not have even realized that they were temporary owners of the property. Their signatures may have been forged, or the property was purchased on a power of attorney basis.

Thus, when obviously dilapidated buildings suddenly have an unusual number of buying and selling transactions, and the paper value of the buildings has significantly increased without any corresponding improvement in the property, it may indicate that an arson is in the making. Similarly, after a fire, if such a daisy chain can be traced, it may be used to help establish that the fire was planned well in advance. Some cities have actually been able to predict fires in specific buildings by tracking real estate transaction activity in high fire areas.

There are also less sophisticated indications that an arson may be in the making. For example some arsonists may call in a number of false alarms or set small fires in the neighborhood near where they will eventually commit the arson. They do this to measure how long it takes for a fire to be spotted by neighbors and how long it takes the local fire department to respond. In this way, the arsonist determines the approximate time in which he has to do his dirty work.

This technique is also sometimes used to set up a plausible reason for the fire at the main target of the arsonist. "There have been a lot of small fires in this area lately. I guess they set this one, too. Damn delinquents!"

The arsonist hopes that the main fire, the arson, will be classified as just one more of these vexing fires being set in the neighborhood by bad people out of sheer orneriness. This ploy is actually somewhat interesting, because it is being conceded up front that the fire is incendiary. The fire can even be set in a very amateurish way, possibly giving further credence to the "damn delinquents" theory.

The conclusion that the fire was incendiary or created by an arson may be freely accepted by all parties. The rub is proving who committed it. Even if it is widely concluded by everyone involved that the fire in the building was an arson, as long as the person or corporation that owns the building did not do the deed, the insurance company will still pay out for the damages as per the provisions of the policy.

When the primary motive for an arson is to make a lot of money, following the money trail will often provide important insight. The insurance settlement from a fire will often take care of a lot of money problems for the arsonist that would otherwise not be resolvable without bankruptcy, foreclosure, or loss of control over property. Often a fire set for profit will be equivalent to winning the lottery for the arsonist; it is an instant financial relief. In accidental fires where there has been no arson, usually there is no unsolvable money problem prior to the fire.

The investigator should be well aware that money trail information should be considered very, very carefully. Not all fires that occur to people in debt are incendiary, and occasionally perfectly honest folks may reap needed windfalls from accidental fires. Thus, the mere fact that a fire has been propitious for a debt-ridden person is not evidence, per se, that there is something sinister going on. However, it doesn't hurt to look a little harder when this occurs to make sure.

The investigation of personal affairs, like financial information, is normally left up to the law enforcement agencies and private investigators in the employ of the insurance companies. In general, forensic engineers are primarily concerned with the evaluation of physical evidence. However, sometimes such information is discovered in the course of an engineering evaluation of physical evidence, so it pays to keep one's eyes and ears open.

For example, in one case, a fire in a bungalow had been spotted early by a neighbor and the fire department was able to extinguish the fire before extensive damage had been done. An initial survey of the premises found many characteristic indicators of a set fire: smoke detectors tampered with, gas pipes loosened with fresh tool marks on the pipe, furnace set to 110°F, and a closet stuffed with newspapers. The point of origin of the fire was the closet.

However, the most tantalizing piece of evidence found at the scene was a sheet of legal pad paper laying on the kitchen counter next to a copy of the insurance policy. On that sheet of paper was a budget schedule, listing all the debts of the owner. All of those debts were being subtracted from a single sum, which was equal to maximum value of the insurance policy on the bungalow. Apparently, the owner of the place had worked out a budget of how the insurance money would be spent, and had absentmindedly left it on the kitchen counter, possibly thinking that the fire might consume it.

## 11.7   Arson Reporting Immunity Laws

Many states now have laws that provide immunity for the reporting of arsons or incendiary fires by forensic engineers, adjusters, and other investigators. When evidence is discovered that a crime or arson has been committed, these laws allow the evidence to be confidentially disclosed to appropriate officials without the investigator or employer of the investigator (e.g., an insurance company) being subject to civil liability.

This is a valuable law. Without it, an engineer or investigator could be personally sued for reporting that there was an incendiary fire or arson, especially if criminal prosecution of the party was either unsuccessful or not pursued.

Some states require that evidence be properly disclosed when a crime is discovered. Some states indicate that disclosure is to be done when requested by proper authority. Still other states indicate that evidence may be disclosed to proper authorities. Since states vary in this matter, it is best to check which rule applies in the state a person is working.

Most states, however, have provisions in the immunity law that requires that the information be obtained and released in good faith. If it can be shown that there was maliciousness, falsification, or bad faith in the reporting of the crime, the immunity can be withdrawn.

Also, there may be penalties for not adhering to the immunity law. For example, in a state that requires that disclosure be made, not disclosing evidence to the proper authorities may be a punishable offense itself. There may be certain procedures to be followed in making such a disclosure. Not following these procedures may be considered "bad faith" or be otherwise punishable.

## 11.8   Liquid Accelerant Pour Patterns

Gasoline, lighter fluid, barbecue fluid, kerosene, and other light distillate hydrocarbon compounds are usually the accelerants of choice, especially among the do-it-yourself incendiary crowd. They are easy to obtain, easy to conceal, cheap to buy, a person doesn't have to be 21 years old to have them, and a little bit goes a long way. The more common method for applying accelerant is what is shown on television crime shows and in the movies: just pour it over what you wish to burn, light it, and quickly run away.

An interesting aside to this involves gasoline and similar highly volatile hydrocarbon liquids. Since gasoline has a very low flashpoint, about −32°F, it quickly makes copious vapors even in cold weather. If an arsonist splashes gasoline in a closed area like a house or building, and takes too much time to do it, he may accidentally blow himself up when he strikes the match. This is because while he was busy spreading gasoline about the place, the gasoline already on the floor formed explosive vapors. It is not uncommon for the arsonist himself to sustain burns and related injuries as a result of the fire he set.

All of the flammable liquids previously noted burn at high temperatures. Usually, they burn at much higher temperatures and release heat faster than the flammable material upon which they have been applied, as in the case, of gasoline on a wood floor. Because of this, the surface where the flammable material has been applied will usually have a recognizably deeper and more severe char pattern than the rest of the surface that burned normally.

Also, for all practical purposes, the entire, contiguous pool of accelerant ignites at the same time. Thus, instead of a point of origin, there is an area of fire origin: the continuous surface area wetted by the accelerant. The resulting fire will generally burn the accelerant quickly, sometimes exhausting it before the normal fire has had much of a chance to spread.

For these reasons when accelerant is applied to surfaces and ignited, it often leaves what is called a pour pattern. The surface area upon which the accelerant pool has laid, called the pour area, will often be clearly and distinctly outlined by the unusually deep or severe charring pattern resulting from intimate contact with the burning accelerant. The area outside the pour pattern will usually be burned less severely. There will likely be indications that the fire spread more slowly, and burned cooler than it did within the pour area.

In the case of painted sheet metal, the painted areas in intimate contact with the burning accelerant will generally have lost their outer coating of paint, undercoats of primer, and any layers of protective galvanization. (Zinc melts at 786°F and will simultaneously oxidize in that type of situation). The high burning temperature of the accelerant will typically leave a very distinguishable pattern on the metal. If a day or two has passed such that the bare metal has rusted due to exposure, the pour pattern on steel sheeting will often appear as a bright rust pattern.

As long as the flooring or surface upon which the accelerant was applied has not shifted position because of the fire, the pour pattern will generally follow the drainage gradient. Thus, an investigator could recreate a pour pattern by applying a liquid similar in terms of wetness to the same location and observing if it follows the same general spread pattern. The purpose of such a demonstration would be to show that the observed deep burn pattern was caused by a burning liquid following the natural gradient rather than some other effect.

(Of course, it is assumed that the above experiment would not be done unless all sampling from the pour pattern was complete so that the area would not be inadvertently contaminated by the test.)

Unfortunately, given sufficient time, the burning liquid will usually create shallow, concave depressions in a wood floor or other flammable material. In concrete or ceramic type materials, it will create spalls. Both types of damage tend to mask the original drainage pattern of the surface.

Pour patterns are not always certain indicators of incendiary intentions. During a fire, it is possible that containers holding flammable liquids might spill their contents, which are then ignited by the spreading fire. The resulting damage may be a pour pattern on the floor. Thus, sometimes pour patterns are simply the result of fall down effects during a fire.

Such secondary pour patterns can be distinguished from primary pour patterns in a way similar to that used to distinguish secondary electrical shorting from primary shorting. Thus, a primary pour pattern has the following general characteristics.

- It occurs at or is the point or area of origin of the fire. There are indications of fire spread away from the area, and the area often exhibits the severest burn damage.

- In consideration of the whole body of evidence, it is the area of burn that must have occurred first in the timeline of the fire.

Similarly, a secondary pour pattern has the following general characteristics.

- It occurs in locations away from the point of origin of the fire. There are indications of fire spread to the area from other locations as opposed to the other way around.
- Little relative movement or travel of the fire from this area has occurred. The fire appears to have been active for only a short time as evident from the amount of fire load consumed, and may have been engulfed itself by simultaneous fire spread from other areas.
- There is nearby fall down which explains ignition and spread of the flammable liquid. The fall down was caused by the approach of an external fire front.
- In consideration of the whole body of evidence, it is an area that may have burned at any time during the course of the fire, and has had little or modest effect on the overall destruction of the property.

Fire patterns that appear like pour patterns can also occur when solid burning fall down materials make close contact with a floor area. Depending upon fire load, material burning temperature, and access to oxygen, some materials can fall on a floor and hug it closely while burning with a relatively hot flame. This can result in a burn pattern that somewhat resembles a pour pattern. However, it will generally be distinguishable from a pour pattern made by a flammable liquid because the burn and char damages will be uneven and more diffuse. Whereas a liquid spreads out evenly in a film along the gradient, solid materials fall in random, lumpy patterns that do not closely follow the drainage gradient.

However, there is an important exception of note to the above: powders or granular materials. Under certain circumstances, spilled powders can spread out over a surface like a liquid. If the powder is flammable, the resulting burn pattern may very closely resemble a liquid pour pattern.

In sum, it is worth remembering that while most liquid accelerants will produce pour patterns, not all pour patterns indicate an incendiary fire.

## 11.9  Spalling

If an accelerant has been poured on a concrete floor and ignited, it sometimes causes spalling and other temperature-related damage to the concrete. Con-

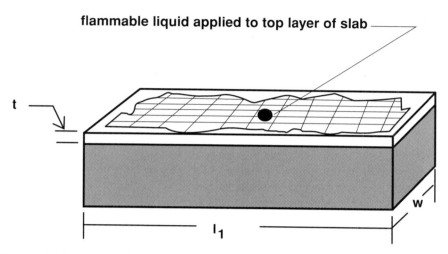

**Figure 11.1** Flammable liquid on concrete slab.

crete itself is not a quick conductor of heat. However, it is a good heat sink, which means it stores thermal energy well.

If the top surface of a concrete slab becomes relatively hot in a short amount of time, the heat does not quickly disperse into the slab. It stays concentrated near the surface. Thus, the topmost portion of the slab will expand as it heats up while the lower portion stays relatively cool and does not expand. This creates a shear stress between the expanding top layer and the lower layer. When the shear stress between the hot portion and the cold portion is sufficient, the top layer separates or spalls.

For example, high strength concrete has a compressive strength of 4000 p.s.i., and a tensile strength of 350 p.s.i. In direct shear, the same concrete has a shear stress strength of 600 p.s.i. to 800 p.s.i. The coefficient of expansion for concrete is $4.5 \times 10^{-6}$ per degree F, and Young's modulus for concrete is about $29 \times 10^6$ p.s.i.

Using the above information, consider the following simple model of a 6-inch concrete slab with burning accelerant on its top surface. It is assumed that the top 1/4 inch of the concrete slab evenly warms up due to the burning effects of the liquid accelerant, but the lower 5 3/4 inches of the slab remains at 55°F, since it is in contact with the earth and is massive enough to absorb most of the heat conducted into it. This situation is depicted below in Figure 11.1. The question is how hot would the upper 1/4 inch have to be to shear away from the lower portion?

Equation (i) is the usual equation for expansion of a material due to heating.

$$\Delta l = (l_1)(\alpha)(\Delta T) \tag{i}$$

where $l_1$ = original length of section, $\alpha$ = coefficient of expansion, and $\Delta T$ = change in temperature.

If the upper 1/4 inch is firmly constrained at both ends, the force generated at the ends of the 1/4-inch layer of the slab by the expansion of the material is given by the following:

$$F_1 = \sigma A_1 = \varepsilon E(w)(t) = (\alpha)(\Delta T)E(w)(t) \tag{ii}$$

where $\sigma$ = equivalent stress required to resist thermal expansion of the upper section of slab, E = Young's modulus, $A_1$ = area of face at end, i.e., (t)(w), t = thickness of top layer, w = width of section of top layer, and $\varepsilon = (\Delta l)/(l_1)$ = strain induced by thermal expansion.

Now, instead of the upper 1/4 inch layer being constrained at the ends by an unnamed force, consider that it is constrained by the shear between it and the lower, cooler layer of the slab. The force that resists the expansion of the upper layer is provided by the shear force between the two layers. This shear force is given by the following:

$$F_2 = \tau A_2 = \tau(w)(l_1) \tag{iii}$$

Thus, Equations (ii) and (iii) can be used to calculate when the force due to expansion will overcome the shear force the material can provide. By setting up an inequality and then solving for the temperature difference, the minimum temperature difference between the top and lower portions of the slab that will cause spalling can be calculated. This is done in the following inequality.

$$F_1 > F_2 \tag{iv}$$

$$(\alpha)(\Delta T)E(w)(t) > t(w)(l_1)$$

$$(\Delta T) > t(w)(l_1)/(\alpha)E(w)(t)$$

In this case assuming that "$l_1$" is the length of a one-foot-long section of slab, then:

$$(\Delta T) > (800 \text{ lb/sq in})(12 \text{ in})/(4.5 \times 10^{-6}/°F)(29 \times 10^6 \text{ lb/sq in})(1/4 \text{ in})$$

$$(\Delta T) > [6.13] [l_1/t] °F$$

$$(\Delta T) > 294°F.$$

Thus, an increase of 294°F in a one-foot section of concrete slab may cause the top 1/4-inch layer to shear off or spall from the slab due to expansion.

It can also be seen in Equation (iv) that the ratio of "$l_1$" to "t" determines how great the temperature difference must be to spall a certain size piece of material. In other words, with respect to the same thickness piece of material, large pieces spall when the temperature difference is great, and small pieces spall when the temperature difference is small. Thus, the size of the spall plate can be a crude indication of how hot the average temperature of the top layer of the concrete was during the fire.

It needs to be emphasized that the size of the spall plate is an indication of how hot the top layer of concrete became during the fire, not necessarily of how hot the burning material on top of the concrete was. Since heat transfer between the burning material on top of the floor and the floor slab itself depends upon lapsed time, contact area, the interface temperature, and the total heat capacity of the slab section, that is, the amount of "heat sink" present, the size of the spall plate is not a direct indication of the temperature of the material that burned on top of the concrete.

In fact, if heat transfer between the burning flammable liquid and the concrete slab is poor, it is likely that no spalling will occur. A smooth concrete floor with a surface coating of paint or sealant may provide sufficient interface insulation to prevent sufficient heat transfer to the slab for spalling to occur. Thus, it is quite possible for a flammable accelerant to have been applied over a concrete slab floor and ignited, but no spalling to have occurred.

Spalling can also be caused by nonincendiary effects. For example, during a fire it is possible that bottles of flammable liquids may fall on the floor, break open, and then be ignited by fire already present in the area. Thus, like a pour pattern, it is important to determine if the spalling is primary or secondary.

Freeze-thaw damages, deterioration by chemical attack such as salt, floor wear, repeated Hertzian-type compressions by wheeled vehicles and objects, point loads, and other effects not related to fire that may have preceded the fire may also produce effects similar to a spall pattern. Exposure to fire and heat may then exacerbate these damages.

As noted before in the previous section on pour patterns, sometimes burning solid materials can fall down and "hug" a floor area. If the fall down materials have sufficient heat intensity, they can also cause spalling. However, as noted with pour patterns, the resulting spalling will often be more diffuse than had it been caused by a flammable liquid. This is due to the evenness of a liquid film as opposed to scattered solid materials of varying sizes and shapes.

Concrete that has not been properly cured is more sensitive to heat than well-cured concrete. Uncured concrete has a relatively high moisture content. Thus, if the temperature in the upper layer of the slab exceeds that which is necessary to create steam, the phase change of liquid moisture to steam in the concrete will cause a type of spalling. However, this type of spalling often causes the concrete to crack in small pieces, and the pieces are often crumbly. In the slab itself, it will more resemble large-scale pitting.

Lastly, spalling can occasionally be caused by fire extinguishment operations themselves. If the fire has been long in duration, and the slab floor has been exposed to heat for a long time, it is possible for the whole slab to heat up. If cold water is then poured on the floor, the top surface will try to contract. The situation will be similar to that modeled in Equations (i) through (iv). However, instead of expansion, the top layer will be subject to contraction; the direction of shear will be opposite, and the temperature difference will represent how much cooler the top portion of the slab is than the lower portion.

In sum, while spalling can result from the application and ignition of a flammable liquid on a concrete floor, the presence of spalling is not an automatic indicator that the fire was incendiary.

## 11.10   Detecting Accelerants after a Fire

Contrary to popular myth, accelerants do not completely burn up with the fire. Often, small amounts of the accelerant will be absorbed by the material to which it was applied. This includes wood, concrete, tiles, textiles, and other common construction materials that have some porosity. Sometimes, sufficient amounts will be absorbed in the material, which then outgases after the fire. This sometimes creates a recognizable odor that is noticeable directly after the fire has been extinguished. Most firefighters are instructed to be alert for such odors during overhaul after the fire.

Various types of portable electronic "sniffers," as was discussed in Chapter 3, can be used at the scene directly after a fire to detect residual vapors of light distillates. Some of these "sniffers" are very sensitive, and are useful in the detection of accelerants.

Most recently, the B.A.T.F. and other federal and state agencies involved with fires and explosions have started using trained dogs to sniff out accelerants at fire scenes. The dogs are trained like the dogs used to sniff out illegal drugs. When the dog smells an accelerant, it barks and alerts its trainer as to the location of the accelerant residue. The use of dogs for this type of work is very efficient, since the dog can sniff out a relatively large area in a short time.

In pour pattern areas, some of the accelerant may have soaked into the material, especially if the material has some porosity. Surprisingly, even fine-grained concrete has the ability to absorb accelerant sufficient to be detectable after a fire. With appropriate sampling and analysis, sometimes the accelerant can be identified by chemical analysis. However, to ensure that the evidence will stand up in court later, the sampling and analysis must be done properly.

Often the presence of a hydrocarbon accelerant in standing water will be signaled by the presence of light and dark banding, or rainbow colors on the surface of the water. An example of this can be observed by adding a small amount of gasoline or light oil to water. Light directed onto the surface and then reflected away from the surface will be diffracted as it passes through the hydrocarbon film on top. This produces the familiar rainbow color effect.

Also, the light reflecting off the top surface of the hydrocarbon film may interfere with light that reflects off the top surface of the underlying water. This interference effect depends on the thickness of the film and the degree of absorption of the incident light by the film.

If the incident light that passes into the film is mostly absorbed, there is no significant second light beam to provide interference with the light beam bouncing off the top of the film. However, when absorptivity in the film layer is low, light passing through the film can bounce off the upper surface of the water and return through the film to emerge at the surface. This then provides a second beam of light, which may produce interference with the first. The observer sees light and dark banding on the surface, especially at the edges of the film. It is possible for both the rainbow effect, and light and dark banding to occur at the same time.

When standing water is thought to contain accelerant residues, samples can be collected simply by using a clean, uncontaminated eyedropper or syringe. If there is only a small amount of liquid, the liquid or moisture can be blotted up using sterile cotton balls, diatomaceous earth powder, common flour, or other moisture-absorbing materials that will not react with or chemically mask the presence of the accelerant.

Of course, the container in which the sample is placed must be clean and free of contaminants. It must not react with the sample. It should be airtight, and it should not mask the presence of the accelerant in subsequent chemical testing. For this reason, clean glass jars are often used, especially those that do not have glued cap liners or seals. The glue often contains materials that may contaminate the sample. Some types of rigid plastic containers are also suitable.

One possible mistake in sampling solid materials for accelerants, for example, is the use of some types of plastic sandwich bags for the preservation and storage of samples. Such bags are commonly used by police and law enforcement agencies to store many other types of evidence with no problem. They are also commonly used by fire investigators to preserve and store small items found at a fire scene for follow-up examination and inspection.

The common use of this type of bag to legitimately store other kinds of evidence might lead a person to assume that it is also acceptable for storage of samples possibly containing accelerants. It may also be rationalized that since the bags can be sealed air tight, that they would help hold in any light distillate materials that might otherwise evaporate before chemical analysis could be done, which is certainly true.

Unfortunately, the problem with some types of plastic sandwich bags is that they themselves outgas light distillate hydrocarbon gases. Thus, the sample suspected of containing trace amounts of accelerant may absorb outgassed hydrocarbon vapors from the sandwich bag during storage and become contaminated by them. Problems occur if this outgassed vapor is then analyzed and identified as a possible accelerant in the sample.

While it is true that the outgassed hydrocarbon from the sandwich bag can be specifically identified and subtracted from the test results of the sample, this is an unwanted complication that may make the chemical analysis suspect during trial, possibly even inadmissible as evidence. The situation is even more serious if the hydrocarbon accelerant found by chemical analysis is very similar to the hydrocarbon material outgassed by the bag. In such situations, it is likely that the laboratory test results will be inadmissible as evidence.

However, in situations where timely collection is important, and there are no better alternatives, plastic bags may be used as long as the suspected accelerant is clearly distinct from the type of vapors outgassed by the bags. It would also be a good idea to keep an empty, sealed bag from the same lot. This bag would then serve as the comparison sample. The analysis of vapor from the empty bag would provide the basis for identifying the contaminant vapors in the sample material, and subtracting or segregating them from the rest of the analysis.

Special plastic bags are available for the collection of evidence that may contain accelerants. These bags do not have the outgassing problem that may occur in some types of common sandwich plastic bags. However, these bags must be special ordered from law enforcement specialty supply houses.

One good way to preserve and store solid material samples that may contain accelerants is to use clean metal cans with lids that can be hammered down to make an airtight seal, such as unused paint cans. These containers will also preserve any volatile vapors that may come off the sample during storage. By using head space analysis techniques, everything that was put into the can, even the vapors, can be checked for the presence of possible accelerants.

In the head space analysis method, the metal can itself is gently heated, and any accelerants are vaporized into the head space of the can. By their nature, accelerants are usually light distillates that vaporize easily at warm temperatures. A small hole is made in the can, and the gases in the can are run through a gas chromatograph, infrared spectrophotometer, or similar apparatus for analysis.

A less desirable method of chemical analysis is the solvent method. In this method, the sample is first washed with a solvent, often an alcohol. The idea is that if there is an accelerant present, the solvent will absorb some of it during the wash. The solvent is then analyzed in a gas chromatograph, infrared spectrophotometer, or similar apparatus.

The problem with the above method is that if the accelerant is chemically similar to the solvent, the accelerant may be confused with the solvent wash or the solvent wash may mask its presence. The analysis would then produce a false negative result, that is, it would fail to detect the presence of accelerant even though it was actually present.

The chemical analysis for accelerants should also be open to "creative" accelerants. While gasoline, kerosene, alcohol, lighter fluid, and other light distillates are popular, other items can also be used. For example, red or white phosphorous have been used as well as finely milled aluminum powder. Even steel wool can be used as an accelerant. A forensic chemist who is directed to look only for one type of accelerant, e.g., gasoline or diesel fuel, may not notice that a metal or inorganic compound was the accelerant of choice.

Readers who have been boy or girl scouts and have had to build a fire from scratch, probably already understand the usefulness of a ball of fine steel wool. Many a scout will have some securely tucked away for use on a rainy day when his regular tender has absorbed the dampness. Steel wool will ignite from a flint spark very well and will even burn when it has been damp. In fact, it is possible for very fine steel wool to spontaneously combust when exposed to moisture laden air.

It is important to obtain samples as soon as possible after a fire. Delays in obtaining samples greatly diminish the chances of detecting accelerant in the sample if it was present to begin with. After a week, it is doubtful that any of the typical light distillate hydrocarbon accelerants can be meaningfully detected by chemical analysis. Excessive hosing of an area during fire extinguishment or during overhaul may also diminish the chances of detecting accelerant.

There are standards in the literature describing the proper way to secure and test samples for the presence of accelerants. A commonly cited standard is contained in ASTM E1387, *Standard Test Method for Flammable or Combustible Liquid Residue in Extracts from Samples of Fire Debris by Gas Chromatography.*

## Further Information and References

ASTM E1387, *Standard Test Method for Flammable or Combustible Liquid Residue in Extracts from Samples of Fire Debris by Gas Chromatography.* For more detailed information please see Further Information and References in the back of the book.

*General Chemistry,* by Linus Pauling, Dover, New York, 1970. For more detailed information please see Further Information and References in the back of the book.

*Van Nostrand's Scientific Encyclopedia,* 5th ed., Van Nostrand Reinhold Co., New York, 1976. For more detailed information please see Further Information and References in the back of the book.

# Simple Skids

# 12

In 1895, the gas engine automobile was invented by Daimler, Mayback, and Benz. One century later, there is nearly one car, bus, truck or motocycle for every human in the U.S., and there are about 43,000 deaths per year resulting from their use.

— The author

## 12.1 General

When a vehicle moves, it possesses a quantity of kinetic energy in proportion to the magnitude of its velocity and mass. In order for the vehicle to come to a stop, its kinetic energy must be dissipated. Skidding is one common way to dissipate vehicle kinetic energy.

In a panic situation, it is common for the driver to apply his brakes sufficiently to lock the wheels. In this situation, the wheels of the vehicle do not roll, they simply slide over the roadway surface. As the locked wheels slide, the tires will abrade on the roadway surface leaving visible marks.

Skidding is a simple type of work. If no other significant dissipative effect is at work, then the kinetic energy of the vehicle can be dissipated by the work done in skidding. It follows then, that if the amount of skidding is known, the amount of kinetic energy can be determined, and the initial speed of the vehicle can be calculated.

## 12.2 Basic Equations

The basic equation of engineering mechanics is Newton's second law:

$$F = m\,a \qquad (i)$$

where F = the applied force, m = the mass of the item, and a = the acceleration.

Since the acceleration, "a," is equal to the derivative of the velocity, then Equation (i) can be rewritten as follows:

$$F = (d/dt)(mv) \qquad (ii)$$

where v = velocity.

The "mv" term in Equation (ii) is also refered to as the momentum.
By rearranging terms in Equation (ii), the following equation is generated:

$$F(dt) = m(dv) \tag{iii}$$

The integration of Equation (iii) is called the impulse, and is equal to the change in momentum.

$$I = \int F(dt) = mv_2 - mv_1 \tag{iv}$$

Equation (iv) is properly a vector equation, where the terms in the equation have both magnitude and direction. More will be done with this particular equation in later chapters.

By multiplying both sides of Equation (iii) by "v," velocity, the following equation is formed:

$$Fv(dt) = mv(dv) \tag{v}$$

Integrating both sides of Equation (v) produces the following:

$$\int_{t_1}^{t_2}[Fv(dt)] = \int_{v_1}^{v_2}[mv(dv)] \tag{vi}$$

$$Fv[t_2 - t_1] = (1/2)m[v_2^2 - v_1^2] \tag{vii}$$

By substituting distance = velocity x time, the left term in Equation (vii) becomes the standard term for work, force x distance. The right term in Equation (vii) is the standard form that expresses the change in kinetic energy. Thus, the change in kinetic energy of a system is equal to the work done by or to the system.

$$F[d_2 - d_1] = (1/2)m[v_2^2 - v_1^2] \tag{viii}$$

## 12.3  Simple Skids

A simple skid is one where the vehicle starts from some initial velocity and comes to a full stop without significant impact by skidding.

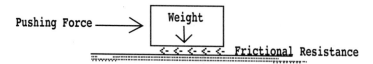

**Figure 12.1** Sliding frictional forces.

The work done by skidding is given by the following:

$$U = Wfd \qquad \text{(ix)}$$

where U = the work done by skidding, W = the weight of the vehicle, f = the coefficient of friction between the two contacting surfaces, and d = the distance skidded.

The term "Wf" is equal to the force necessary to push the item parallel to the ground (Figure 12.1). "W," the weight, is the force normal to the ground while "f" is a coefficient that when multiplied by "W" gives the lateral force needed to scoot the item along the ground.

Since the initial velocity of the vehicle is some unknown "v," and the final velocity of the vehicle in this case would be zero, then Equation (viii) can be rearranged as follows:

$$(1/2)mv^2 = Wfd. \qquad \text{(x)}$$

Solving for "v," the initial velocity before skidding began, then Equation (x) becomes:

$$v = [2(W/m)fd]^{1/2}. \qquad \text{(xi)}$$

Since the weight of an object divided by its mass is equal to the gravitational acceleration, Equation (xi) is further simplified to the following:

$$v = [2gfd]^{1/2} \qquad \text{(xii)}$$

where g = 32.17 ft/sec² or 9.81 m/s².

Equation (xii) is the common "skid formula" taught to police officers. Often, a nomograph is provided for easy solution of "v."

It is interesting to note that the weight or mass of the vehicle cancels out of the equation in this simple case. Calculation of the initial velocity of the vehicle depends solely upon knowing the length of the skid and the coefficient of friction.

Table 12.1 lists some typical friction coefficients for calculating forward motion skidding work.

**Table 12.1   Some Common Frictional Coefficients for Car Tires on Roadways**

| Surface Type | f-Value |
|---|---|
| Gravel and dirt road | 0.35 |
| Wet, grassy field | 0.20 |
| Dry asphalt | 0.65 |
| Wet asphalt | 0.50 |
| Dry concrete | 0.75 |
| Wet concrete | 0.60 |
| Snow covered | 0.20–0.25 |
| Ice | 0.10–0.15 |
| Loose moist dirt that allows tire to sink about 2" (5 cm) | 0.60–0.65 |

## 12.4   Tire Friction

The figures provided in Table 12.1 are reasonably accurate for relatively short skids. However, it is known that the coefficient of friction is not wholly constant during a skid. As the skid progresses, the coefficient of friction will decrease as a function of the forward velocity. This decrease in the coefficient of friction can be approximated by the following linear equation:

$$f = f_1 - cv \qquad (xiii)$$

where f = coefficient of friction, $f_1$ = initial coefficient of friction term (static friction), c = constant determined by experimentation, and v = forward velocity.

Table 12.2 lists some general values for solving Equation (xiii) for automobile tires.

A comparison of the constant values for "f" given in Table 12.1 with the linear equation function for "f" finds that they are equal to one another when the velocity of the car is 40 mph on dry pavement. For wet concrete pavement, the values are the same when the velocity of the car is 20 mph.

**Table 12.2   Typical Values for Solution of "f"**

| Conditions | $f_1$ | c |
|---|---|---|
| Dry concrete | 0.85 | 0.0017 |
| Wet concrete | 0.70 | 0.0034 |

Note:  The above values are to be used when "v," velocity, is given in ft/sec units. Use an appropriate scalar for use with metric units.

In addition to varying with forward velocity, in trucks, the road-to-tire coefficient of friction will vary with truck loading. As the loading on the tire is increased, the coefficient of friction decreases. The following equation approximates that effect:

$$f = f_1 - 0.22(L/N) \qquad \text{(xiv)}$$

where $f_1$ = initial coefficient of friction term (static friction), L = load carried by the tire, and N = rated load of the tire at the proper inflation pressure.

While the factor "$f_1$" is essentially the same as used before from Table 12.2, it is worth noting that truck tires typically have a coefficient of friction about 10% greater than passenger car tires. This is due to a different rubber blend used in truck tires.

The two factors, load and velocity, can be combined into a single equation for trucks as follows:

$$f = f_1 - cv - 0.22(L/D). \qquad \text{(xv)}$$

Because the road-to-tire friction can be affected by excess dirt, chemical films, surface polishing by traffic wear, moisture, the brand and size of tires, and other miscellaneous factors, it is important to recognize that the "f" values may exhibit significant variation even along the same area of roadway. The frictional coefficient of the inside lane may even vary some from the outside lane in a four-lane highway.

When very accurate values of "f" are required, it may be necessary to obtain the values experimentally. This can be done by the following.

- Renting a car similar to the one in question, deliberately skidding it at various initial speeds, measuring the skid lengths, and calculating "f." For obvious reasons, this method is not generally recommended for amateur drivers.
- Pulling a weighted sled with a fixed wheel and tire, and measuring the draw of the sled. The ratio of the draw to the sled weight gives a value of "f."
- Towing a similar car with the rear wheels locked and measuring the draw. Because the front end is lifted, the "f" value is the ratio of draw to the weight on the rear tires (or the weight of the car minus the lifted weight at the tow hoist).

## 12.5   Multiple Surfaces

In some instances, a vehicle may skid over several different surfaces in succession. This could occur, for example, if the vehicle began skidding on

the roadway, skidded on the shoulder, and then skidded on the dirt along
the roadway.

In such cases, the skid trajectory is simply divided into segments according to the surface type. Each segment is a work term, and the sum of the
work terms is set equal to the initial kinetic energy of the vehicle.

With multiple surfaces, Equation (x) is modified as follows:

$$(1/2)mv^2 = (W)\sum_{i}^{n}(f_i d_i).\qquad\text{(xvi)}$$

Solving for "v" in Equation (xvi), gives the following:

$$v = [2g(f_1 d_1 + f_2 d_2 + \ldots + f_n d_n)]^{1/2}.\qquad\text{(xvii)}$$

In multiple surfaces skids, it is sometimes useful to set each skid segment
work term equal to the change in kinetic energy of the vehicle during that
skid segment. In this fashion, the various individual velocities during the
overall skid trajectory can be calculated.

For example, assume that a vehicle skidded first on pavement and then
on soil. The skid trajectory would have two segments, and correspondingly,
two skid work terms. Let the initial velocity of the vehicle and the start point
of the pavement skid be designated by subscript 1. Let the end point of the
pavement skid and the beginning of the soil skid be designated by subscript
2. Then let the end point of the soil skid, where the car finally stopped, be
designated by the subscript 3. Thus, the following equations would describe
the situation:

$$v_1 = [2g(f_p d_p + f_s d_s)]^{1/2}$$

$$v_2 = [v_1^2 - 2gf_p d_p]^{1/2} = [2gf_s d_s]^{1/2}$$

$$v_3 = 0.$$

With all three velocity terms solved, a plot of "v" versus distance can be
constructed. From such a plot, intermediate velocities at any point along the
skid trajectory can be picked off. Such plots are very useful in pedestrian
accidents. If the location of the impact with the pedestrian is known from
pavement marks or other indicators, its location can be measured from one
end of the skid. From this measurement, the velocity of the vehicle at the
point of impact with the pedestrian can be determined.

In accidents involving pedestrians and cars or heavier vehicles, it is assumed that the impact with the pedestrian has no significant effect upon the trajectory of the vehicle. This may not be true in a motorcycle impact, where the weight ratios are closer.

## 12.6   Calculation of Skid Deceleration

Occassionally, it is useful to know the deceleration (i.e., negative acceleration) for a given skid. In this respect, the deceleration of the vehicle by skidding is simply the frictional force opposing travel divided by the mass of the vehicle.

The opposing friction force is given by:

$$F_f = Wf \qquad \text{(xviii)}$$

where $F_f$ = friction force opposing travel, W = weight of the vehicle, and f = coefficient of friction.

Dividing both sides of Equation (xviii) by the mass of the vehicle, the resulting term will be the deceleration of the skid.

$$a_f = g\,f. \qquad \text{(xix)}$$

## 12.7   Speed Reduction by Skidding

Sometimes skidding will not occur long enough to bring the vehicle to a stop. In such cases, the skidding simply reduces the initial velocity of the vehicle to some lower value. If either the initial velocity just prior to skidding, or the velocity just after skidding is known, the other can be calculated.

$$(1/2)m[v_2^2 - v_1^2] = Wfd \qquad \text{(xx)}$$

$$v_2 = [2gfd + v_1^2]^{1/2}$$

$$v_1 = [v_2^2 - 2gfd]^{1/2}$$

## 12.8   Some Considerations of Data Error

In reviewing Equation (xii), it is noted that only two terms are subject to possible error: "f" and "d." Since the length of the skid mark is usually measured with reasonable precision with a tape measure, the greatest apparent source of error is then the estimation of "f."

However, due to the particular construction of the skid equation, that is Equation (xii), the error is actually smaller than one might suppose. This is due to the square root effect of decreasing the error range.

To illustrate this point, suppose a vehicle skidded 100 ft and the estimated coefficient of friction is 0.70. The calculated initial velocity is then 67.1 ft/sec. Assuming that the skid measurement is accurate to the nearest foot, if the "f" value is increased 14% to 0.80, then the initial velocity calculates to be 71.7 ft/sec, an increase of 7% from the previous figure. Similarly, if the "f" value is decreased 14% to 0.60, the calculated initial velocity is then 62.1 ft/sec, a decrease of 7%.

Thus, the amount of error caused by either an imprecise estimation of "f" or measurement of "d," does not cause the same amount of error in the calculation of the initial velocity. The amount of error in the calculation of initial velocity is a function of the square root of the combined error of the "f" and "d" terms.

For example, if the "df" term is in error by +21%, the resulting initial velocity calculation will only be in error by 10%. Similarly, if the "df" term is in error by −19%, the resulting initial velocity calculation will only be in error by 10%. Thus, the final answer can be more accurate than the data upon which it relies.

## 12.9   Curved Skids

Work and energy are both scalar quantities. They have magnitude but not direction. In vector algebra, work is defined as the dot product of force and direction.

$$F \cdot d = |Fd(\cos a)| = U \qquad \qquad \text{(xxi)}$$

where U = work (a scalar), F = force (a vector), d = distance (a vector), and a = the acute angle between the force and the distance.

If a skid curves or does not otherwise follow a straight line, the skid should be measured along the curved line. If a vehicle spins while it skids, the length of the skid should be measured along the curves and loops of the skid to obtain the total work of the skid.

This is because energy is dissipated by the tire sliding on the pavement in the direction of the slide. It takes work to push forward a vehicle with locked wheels. It also takes work to spin a vehicle with locked wheels even if the vehicle does not end up moving forward any distance.

Thus, to be precise in the amount of energy disspiated by skidding, the curvilinear length of the skid should be used rather than the start to finish straight line distance. If is often the case that the lengths of skid marks noted in police reports are given as the straight line distance from where the skid initiated to where it stopped. When necessary, allowances should be made to account for any nonlinear skid mark distances.

## 12.10   Brake Failures

Occasionally the brakes on a vehicle will fail prior to usage, or just when skidding occurs (and the internal hydraulic pressure is at a maximum). In such cases, only two of the four wheels will lock up and skid. This is because most automobile brake systems have dual hydraulic systems or loops. Thus, when a leak in the hydraulic brake system occurs, only half of the system becomes inoperative.

In such cases where there has been a failure in the hydraulic brake system, the vehicle will be decelerated by only two of the four wheels. Thus, the weight distribution of the vehicle becomes important.

If the weight in a four-wheeled vehicle was evenly distributed, each wheel would support one-fourth of the weight of the vehicle. The sum of the work done by each wheel in decelerating the vehicle would then be as follows:

$$U = (1/4)W\sum_{i=1}^{n=4}(f_id_i) \qquad\qquad (xxii)$$

where the subscripts 1, ..., 4 indicate each of the wheels on the vehicle.

If all the distances skidded by the individual wheels are the same, and all the coefficients of friction are the same, Equation (xxii) reduces to the standard skid formula as before.

However, most cars and trucks do not have an even distribution of weight among the wheels. In cars with an engine up front, it is common for the front two wheels to carry 65% of the total curb weight of the vehicle. (Curb weight is the amount the car weighs with no one in it, and no freight or cargo.) In such a car, Equation (xxii) would become:

$$U = Wf[0.325d_1 + 0.325d_2 + 0.175d_3 + 0.175d_4] \qquad (xxiii)$$

where the subscripts 1 and 2 are for the front wheels, and 3 and 4 are for the rear wheels. The above also assumes that the term "f" is the same for all the tires.

If one or more of the tires were on a different surface, then it might also be necessary to assign a specific "f" value for each tire.

If the hydraulic brake system is one where the front right wheel is on the same hydraulic circuit as the left rear wheel, and vice versa, then the work done by skidding will be done by the two wheels, front and back, of the remaining operable hydraulic circuit. Applying Equation (xxiii), then the work done by the partially disabled brake system would be as follows:

$$U = Wf[0.325d_1 + 0.175d_3]. \qquad \text{(xxiv)}$$

If $d_1 = d_3$, then Equation (xxiv) reduces to the following:

$$U = (1/2)Wfd. \qquad \text{(xxv)}$$

Similarly, if the brake system were designed such that both front brakes and both rear brakes were on independent hydraulic circuits, the work done by a skid by either pair of brakes would be as follows:

$$U = (0.65)Wdf \qquad \text{For front wheels.} \qquad \text{(xxvi)}$$

$$U = (0.35)Wdf \qquad \text{For rear wheels.} \qquad \text{(xxvii)}$$

Equation (xxvii) is also applicable if the emergency brakes are used. It is customary for the emergency brake to engage the two rear wheel brakes. The emergency brake will normally be actuated by a direct mechanical cable linkage to the two rear wheel brakes.

For pickup trucks that are unloaded, and for other kinds of commercial vehicles, the weight distribution may favor the front end even more. Certain vehicles, like the Volkswagon "bug" or commercial vehicles that are fully loaded, may be heavier in the rear than in the front. The weight distribution of each vehicle must be individually assessed.

In general, most vehicles will have their GAWR (gross axle weight ratings) posted for both the front and rear axles on the certification label, which is required by law. This label is usually found on the driver's door post of passenger cars. The GAWR-R (front axle) and the GAWR-R (rear axle) should equal the overall vehicle GVWR (gross vehicle weight rating).

When the appropriate work term is determined, the value for "U" is then set equal to the kinetic energy term, and the initial velocity of the vehicle is solved as before.

The preceding analysis has assumed that the load distribution between the front and rear wheels follows the static load distribution. For many purposes, this assumption is reasonably accurate. However, during hard braking there is a dynamic effect that modifies this assumption. Section 12.12 discusses this.

## 12.11   Changes in Elevation

In all of the previous analyses, it has been assumed that there was no change in elevation from where the skidding started to where it ended. In cases where there is a significant change in elevation, a potential energy term must be added to the energy balance. Intuitively, it is reasonable to expect that a vehicle will skid farther in a downhill direction than uphill.

In previous equations, it was generally assumed that the kinetic energy of the vehicle was wholly converted into work done by skidding, that is:

$$\Delta KE = U \qquad\qquad (xxviii)$$

where $KE$ = the kinetic energy of the vehicle and $U$ = the work done by skidding.

To account for changes in elevation during skidding, a potention energy term is added to Equation (xxviii).

$$\Delta KE + \Delta PE = U \qquad\qquad (xxix)$$

Potential energy is often described as stored energy, such as, for example, when a spring is compressed. Work is done to compress the spring, and the work can be recovered as the spring is released. During the time that the spring is compressed, the work is stored in the spring as potential energy.

When a stone is dropped, it is accelerated by gravity. Ignoring air friction, the velocity of the stone after being dropped is given by:

$$v = gt \qquad\qquad (xxx)$$

where $v$ = the velocity after time "$t$" has elapsed and $g$ = the gravitational acceleration.

By substituting "$d(x)/dt$" for "$v$," and integrating, the following equation relating the distance the stone has fallen to the elapsed time is obtained.

$$h = (1/2)gt^2 \qquad\qquad (xxxi)$$

where $h$ = the height fallen by the stone.

Substituting "$v = gt$" into Equation (xxxi) and multiplying both sides of the equation by "$g$," Equation (xxxi) is transformed into the following:

$$gh = (1/2)v^2. \qquad\qquad (\text{xxxii})$$

By further multiplying both sides of Equation (xxxii) by the mass "m," the following equation is obtained:

$$mgh = (1/2)mv^2. \qquad\qquad (\text{xxxiii})$$

The left term of Equation (xxxiii) is the expression for the changes in potential energy due to a change in elevation. The right term is the standard term for kinetic energy. Thus, when a stone is dropped, its potential energy is converted to kinetic energy. Conversely, kinetic energy is consumed when a stone is elevated and its potential energy is increased.

If a vehicle skids downhill, the change of potention energy is converted into additional kinetic energy that must be dissipated by the skid: the skid will be longer. Similarly, if a vehicle skids uphill, kinetic energy is converted into potential energy: the skid will be shorter. The following equation describes this effect.

$$(1/2)m[v_2^2 - v_1^2] + mg[h_2 - h_1] = Wf[d_2 - d_1]. \qquad (\text{xxxiv})$$

If it is assumed that the vehicle comes to a stop, i.e., $v_2 = 0$, Equation (xxxv) simplifies to the following:

$$v = [2gdf - (h_2 - h_1)]^{1/2}. \qquad\qquad (\text{xxxv})$$

A quick way to determine the relative effect of an elevation change on the calculation of initial velocity is to apply the following:

$$D = 1.00 - \{[fd - h_2 + h_1]/fd\}^{1/2}. \qquad\qquad (\text{xxxvi})$$

Obviously, if the change in elevation is negligible with respect to the "fd" term, then the elevation terms can be dropped and simpler equations can be applied with no significant loss of accuracy.

Most police reports do not document changes in roadway elevation unless the vehicle has driven over a cliff or careened into a deep ditch. Generally, an assessment of elevation change will have to be done in an on-site inspection of the accident area.

Occassionally, roadway topographical drawings may be available from the local unit of government responsible for the maintenance of the roadway in question. However, sometimes these drawings are out-of-date and may not document recent repairs or modifications that may affect an accurate assessment of elevation in the accident area.

## 12.12   Load Shift

When a vehicle skids, the actual load carried by the tires can change. This is due to the different forces present during braking as opposed to when the vehicle is at rest. The change in load distribution is observable as "brake squat," where the front end of the car tilts down and the rear end tilts up as the car brakes. (A similar but reversed "squat" effect occurs during acceleration where the front end raises up and the rear end drops down.)

When a car is at rest, the usual static equation applies to the loads carried by the front and rear tires.

$$W = Q + R \qquad \text{(xxxvii)}$$

where W = weight of the car, Q = load carried by the rear tires, and R = load carried by the front tires.

Let "b" be the distance from the front axle to the rear axle and "a" be the distance from the center of gravity of the car to the front wheel axle. The moment equation about the front axle is then as follows:

$$Q(b) - W(a) = 0. \qquad \text{(xxxviii)}$$

In the above moment equation, the specific point of rotation is considered to be where the front wheels contact the road. By combining the above two equations, the front and rear load distributions can be determined.

$$Q = W[a/b] \qquad \text{(xxxix)}$$

$$R = W[1 - (a/b)]$$

Of course, these same two equations can be used in a reverse sense to calculate the center of gravity of a vehicle when the rear and front loads, and the weight of the vehicle are known.

In the above analysis, it was assumed that all the forces are pointing up or down. However, when the vehicle is being braked, there is a forward directed force "F." This force is the mass of the vehicle times the braking acceleration (or deceleration), and is directed through the center of gravity of the vehicle parallel to the ground (i.e., horizontal). This force is given by the following expression:

$$F = ma = mgf = Wf \qquad \text{(xl)}$$

where f = coefficient of friction.

The inclusion of "F" in the moment Equation (xxxviii) modifies it to the following:

$$Q(b) - W(a) + F(h) = 0 \qquad \text{(xli)}$$

$$Q(b) - W(a) + Wfh = 0$$

where h = the height of the car's center of gravity from the ground.

The loads carried by the front and rear wheels then changes to the following:

$$Q = W[(a/b) - f(h/b)] \qquad \text{(xlii)}$$

$$R = W[1 - (a/b) + (fh/b)]$$

As the equations show, the load on the rear end lightens and the load on the front end increases while the car is skidding and decelerating. Assuming that the car has a load distribution of 60% in front and 40% in the rear while at rest, then "a/b" in the above equation is 0.40. Further, assuming that f = 0.70 and h/b = 0.20, then skidding will cause the load in the rear to lighten by 14% and the load in the front to increase by 14%.

In general, the load shift caused by skidding does not affect skid calculations. However, it may become important when partial brake system failures occur, or when the emergency brakes are used in lieu of normal braking. When either condition occurs, the term "Wf(h/b)" has to be modified to agree with the reduced braking force.

For example, when only the rear brakes are applied, the following equations indicate the load shift during skidding:

$$Q = W(a/b)[1 - (fh/b)] \qquad \text{(xliii)}$$

$$R = W[1 - (a/b) + (a/b)f(h/b)]$$

Using the same assumptions as before (f = 0.70, a/b = 0.40, h/b = 0.20), it is found that skidding will cause a load shift of only 5.6% when only the emergency brakes are used. This is because of the reduced braking capacity of the emergency brakes as compared to the full system.

## 12.13  Antilock Brake Systems (ABS)

It might be presumed that the use of antilock brakes would eliminate skidding and skid marks. This is not quite the case. While it is true that antilocking brakes do not leave a heavy black skid mark like brake systems that can lock up the rotation of the wheels, a light skipping or alternating pattern on the pavement can still be discerned in most cases.

Antilock brake systems are generally designed to release the wheels to allow them to roll and not fully lock up. While this reduces the sliding or slip between the tire and the road, it does not fully eliminate it. In fact, many antilock braking systems are calibrated to release when slip just exceeds 30%. Slip is defined as follows:

$$\text{Slip} = [1 - v_t/v] \qquad\qquad (\text{xliv})$$

where $v$ = forward velocity of the vehicle, and $v_t$ = tangential velocity of the wheel where it contacts the road.

When slip is 100%, the vehicle is skidding; when slip is 0%, the vehicle has full traction between the tire and the roadway. Between 15% and 30% slip on dry pavement, the maximum braking effect is achieved with minimal sideslip. On wet pavement, maximum braking effect with minimum sideslip occurs when slip is between 30% and 60%.

Thus, even when antilock brake systems are employed, there is still some slippage between the tire and the road surface. It often occurs in a skipping pattern because the antilock system will alternately open and release the brakes as the slip set point is exceeded. Because the skid pattern of antilock brakes is different in appearance than conventional braking skid marks, it is often overlooked at the accident scene. In some cases, once it was learned that a car involved in the accident had ABS brakes, it was automatically assumed that there were no skid marks; so no search for marks was conducted.

In general, a full skid with locked wheels will bring a car to a stop in the shortest distance. However, during a full skid, the driver has little control over the vehicle. With ABS brakes, the actual braking distance may be slightly longer, but steering control can still be maintained.

On dry pavement, the same calculation methods for locked brake skids are applicable to ABS type skidding. On wet pavement, the same calculation methods can be used also, but the coefficient of friction should be considered to be about 10% greater.

## Further Information and References

"The Amateur Scientist" by Jearl Walker, *Scientific American*, February 1989, pp. 104–107. For more detailed information please see Further Information and References in the back of the book.

"The Amateur Scientist" by Jearl Walker, *Scientific American*, August 1989, pp. 98–101. For more detailed information please see Further Information and References in the back of the book.

"The Physics of Traffic Accidents" by Peter Knight, *Physics Education*, 10(1), January 1975, pp. 30–35.

*Bicycle Accident Reconstruction: A Guide for the Attorney and Forensic Engineer*, by James Green, P.E., 2nd ed., Lawyers and Judges Publishing Co., Tucson, AZ, 1992. For more detailed information please see Further Information and References in the back of the book.

*Consumer Reports*. For more detailed information please see Further Information and References in the back of the book.

*Engineering in History*, by Kirby, Withington, Darling, and Kilgour, published by Dover Publications, 1990. For more detailed information please see Further Information and References in the back of the book.

*Insurance Institute for Highway Safety/Highway Loss Data Institute*. For more detailed information please see Further Information and References in the back of the book.

*Manual on Uniform Traffic Control Devices for Streets and Highways*, National Joint Committee on Uniform Traffic Control Devices, U.S. Department of Commerce, Bureau of Public Roads, June 1961. For more detailed information please see Further Information and References in the back of the book.

*MathCad*, Version 8.0, Addison-Wesley, 1999. For more detailed information please see Further Information and References in the back of the book.

*Physics*, by Arnold Reimann, Barnes and Noble, New York, 1971. For more detailed information please see Further Information and References in the back of the book.

*Research Dynamics of Vehicle Tires*, Vol. 4, by Andrew White, Research Center of Motor Vehicle Research of New Hampshire, Lee, New Hampshire, 1965. For more detailed information please see Further Information and References in the back of the book.

*Statistical Abstract of the United States*, U.S. Printing Office. For more detailed information please see Further Information and References in the back of the book.

*The Traffic Accident Investigation Manual*, by Baker and Fricke, 9th ed., 1986, Northwestern University Traffic Institute. For more detailed information please see Further Information and References in the back of the book.

*The Way Things Work*, by David Macaulay, 1988, Houghton Mifflin, New York. For more detailed information please see Further Information and References in the back of the book.

*The Way Things Work: An Illustrated Encyclopedia of Technology*, 1967, Simon and Schuster, New York. For more detailed information please see Further Information and References in the back of the book.

*Work Zone Traffic Control, Standards and Guildlines*, U.S.D.O.T., Federal Highway Administration, 1985, also A.N.S.I. D6.1-1978. For more detailed information please see Further Information and References in the back of the book.

# Simple Vehicular Falls

# 13

I don't care to be involved in the crash-landing unless I can be in on the take-off.

— **Harold Stassen,** *perennial presidential candidate*

## 13.1  General

For various reasons, vehicles will sometimes leave the roadway and go over a cliff, drop into a ditch, or otherwise free fall from a precipice. In the reconstruction of such an accident, it is often important to determine the velocity of the vehicle just prior to the fall. In general, if the height of the fall is known and the horizontal travel is known, the velocity just prior to fall can be calculated directly from that information.

## 13.2  Basic Equations

Unless the vehicle meets another object on the way down to change its trajectory, its motion during a simple fall consists of two independent motions in a plane: one in the vertical or "y" direction, and the other in the horizontal or "x" direction.

When an object falls, some of its potential energy due to its elevation in a constant gravitational field converts to kinetic energy. The change in potential energy, that is, the change in elevation from the start of the fall to the point where the fall stops, will equal the change in kinetic energy.

$$PE_2 - PE_1 + KE_2 - KE_1 = 0 \tag{i}$$

Substituting the usual expressions for kinetic energy and potential energy into Equation (i), the equation can be restated as follows:

$$W[h_2 - h_1] = (1/2)m[v_2^2 - v_1^2] \tag{ii}$$

where W = weight of the object, m = mass of the object, $h_2$ = height of the point of impact after the fall, $h_1$ = height of the point where fall began, $v_2$ = velocity at point of impact after the fall, and $v_1$ = velocity when fall began.

Since most vehicles do not initially have any significant vertical velocity at the point just prior to fall, "$v_1$" is often zero. Also, since "$h_2$" will be the low point, it is usually considered to be zero also. Equation (ii) can thus be modified and reduced to the following:

$$v = [2gh]^{1/2} \qquad \text{(iii)}$$

where v = the vertical velocity at impact and h = the vertical distance the vehicle fell.

Because the acceleration of gravity is constant for such situations, the vertical velocity at impact can also be calculated by the following:

$$v = gt \qquad \text{(iv)}$$

where t = the time of fall.

By substituting "$v = dy/dt$" and rearranging terms, the following is obtained from Equation (iv):

$$dy = gt(dt). \qquad \text{(v)}$$

Integrating Equation (v) with respect to time and letting $y_2 - y_1 = h$, the following relation is derived:

$$h = (1/2)gt^2 \qquad \text{(vi)}$$

where t = time lapsed during the fall.

Of course, while the vehicle was falling in the vertical direction, the vehicle was still moving horizontally. In fact, neglecting air resistance, the vehicle would continue to move horizontally at the same velocity it had prior to the fall, and would continue to do so until impact with the ground occurs. The amount of time that the vehicle will continue to travel horizontally will be the same "t" found in Equation (vi) above. Solving for "t" in Equation (vi) gives:

$$t = [2h/g]^{1/2}. \qquad \text{(vii)}$$

Let "x" be the horizontal distance traveled by the vehicle from the point of take-off to the point of impact. Then the velocity of the vehicle just prior to the fall is given by the following:

$$u = x/t = x[2h/g]^{-1/2} = x[g/2h]^{1/2} \tag{viii}$$

where u = the horizontal velocity prior to the fall.

The velocity at impact, which included both the horizontal and vertical velocity components, is given by the following vector equation:

$$V = (u)i + (v)j \tag{ix}$$

where "i" and "j" are the horizontal and vertical unit vectors.

In absolute terms, the total velocity is given by:

$$V = [u^2 + v^2]^{1/2}. \tag{x}$$

The angle of impact is then:

$$a = \arctan(v/u) \tag{xi}$$

where a is measured from the impact point and upward.

A vehicle with no forward horizontal motion would have an angle of impact of 90° with the ground, assuming it is flat, and the vehicle would be found directly below the edge of the cliff or precipice.

## 13.3  Ramp Effects

If the take-off point is not level, an additional upward or downward velocity component must be taken into consideration. Let "f" be the acute angle that the vehicle's take-off ramp makes with the horizontal. Then, the velocity of the vehicle can be separated into orthogonal components as follows:

$$v = V\sin(f) \tag{xii}$$

$$u = V\cos(f) \tag{xiii}$$

where V = the velocity of the vehicle at take-off in the direction of travel.

Assuming that "v" is upward and positive, then the vertical motion equation is as follows:

$$v = V\sin(f) - gt. \tag{xiii}$$

The highest elevation the vehicle will attain during its trajectory is calculated as follows:

$$t_a = V(\sin f)/g \tag{xiv}$$

$$y_a = (1/2)gt_a{}^2 = (1/2)V^2(\sin^2 f)/g$$

where $t_a$ = the time it takes to get to the top of the arc in the trajectory and $y_a$ = vertical distance from the take-off point to the top of the arc.

The total fall of the vehicle will then be the height of the ramp from where the vehicle impacted plus the height it attained during its trajectory above the end of the ramp. In terms of time of fall, the total time of fall will include the time it takes to reach the apex of the arc plus the fall time from that same point. Let the total height fall, from the apex of the arc to the point of impact, then be as follows:

$$H = h + y_a \tag{xv}$$

$$H = h + V^2(\sin^2 f)/(2g) = h + (1/2)gt_a{}^2 .$$

The time needed to fall from the apex of the trajectory arc to the point of impact is given by the following:

$$t_b = [2H/g]^{1/2}. \tag{xvi}$$

The total time of the fall then, from the take-off point, through the upward trajectory arc, and down to the point of impact is as follows:

$$t = t_a + t_b = V(\sin f)/g + [2H/g]^{1/2} \tag{xvii}$$

$$t = V(\sin f)/g + [2h/g + V^2(\sin^2 f)/g^2]^{1/2}.$$

As before, the horizontal speed of the vehicle, "u," can be calculated by the following:

$$u = V(\cos f) = x/t. \tag{xviii}$$

If Equation (xviii) is substituted into Equation (xvii), the following is derived:

$$x = V^2(\cos f)(\sin f)/g + V(\cos f)\{2h/g + V^2(\sin^2 f)/g^2\}^{1/2}. \tag{xix}$$

Equation (xix) gives the horizontal distance "x" traveled by the vehicle in terms of the initial velocity, the angle of the ramp, and the height from the ramp to the point of impact. As would be expected, when the angle "f" is zero, Equation (xix) reduces to the following:

$$x = V[2h/g]^{1/2} \quad \text{or} \qquad \text{(xx)}$$

$$u = x[2h/g]^{-1/2}.$$

Equation (xx) is, of course, a restatement of Equation (viii), which is used when the precipice is flat, that is, there is no ramp angle.

When there is no drop-off, and the vehicle simply is catapulted by driving up and over a ramp, Equation (xix) reduces to the following:

$$x = 2V^2(\cos f)(\sin f)/g, \qquad \text{(xxi)}$$

which is the distance traveled from the end of the ramp to the point of impact, which occurs at the same elevation as the end of the ramp. The elevation above the end of the ramp attained during such a trajectory is given by Equation (xiv).

In general, Equation (xix) is solved for "V" by an iterative process. An initial guess for "V" is selected, and the equation is solved for "x." This "x" is then compared to the measured value for "x," the horizontal distance traveled during the fall by the vehicle. A new value of "V" is then selected and the process is repeated until the calculated "x" matches the measured "x."

In cases where the ramp is inclined at a negative angle, that is, an angle that dips below the horizontal, a slightly different derivation is required. With the ramp pointing downward:

$$u = V(\cos(-f)) = V(\cos f) \qquad \text{(xxii)}$$

$$v = V(\sin(-f)) = -V(\sin f)$$

It follows then that:

$$h = -V(\sin f)t + (1/2)gt^2. \qquad \text{(xxiii)}$$

The solution of the time of fall, "t," in Equation (xxiii) involves a quadratic equation in which there are two possible values of "t."

$$t = -V(\sin f)/g \pm [V^2(\sin^2 f)/g^2 + 2h/g]^{1/2}. \qquad \text{(xxiv)}$$

Except for unusual cases, the positive sign in front of the second term in Equation (xxiv) is used.

As before, the horizontal velocity is determined by dividing the horizontal distance by the lapsed time, "t."

$$u = V(\cos f) = x/t \tag{xxv}$$

where in this case, "t" is obtained from the solution of Equation (xxiv).

The method of solution for this situation is as follows.

- Set the "t" in Equation (xxiv) equal to "x/V(cos f)."
- Choose a good guess for "V" and solve for "x."
- Appropriately alter "V" iteratively until the calculated "x" matches the actual "x."

It is observed that when the above procedure is followed, the result produces a solution for "x" similar to Equation (xix) except that the first term is negative, while in Equation (xix) it is positive. In fact, Equation (xix) is appropriate for both situations, an uphill ramp and a downhill ramp, if the appropriate positive and negative sign conventions are observed with respect to the ramp angle.

## 13.4  Air Resistance

In the previous derivations, it was assumed that the air resistance was negligible. This is generally true when fall distances are small. In such cases, the omission of air resistance factors produces a conservative solution of the vehicle's velocity, that is, the calculated velocity will be slightly less than the actual velocity. But, from time to time it may be useful to estimate the air resistance in a vehicular fall, especially if the trajectory is long.

The drag due to air resistance is given by the following:

$$f_D = C\rho AV^2 \tag{xxvi}$$

where $f_D$ = air resistance force, $V$ = velocity through the air, $\rho$ = density of the air, $A$ = projected area of the vehicle normal to the air speed, and $C$ = a constant that includes units conversion and the "slickness" of the vehicle.

For dry air at sea level and 0°F, its density is 1.293 g/l or 0.0807 lbs/ft³. For variations in pressure or temperature, the application of the perfect gas law will suffice for four-place accuracy.

$$P/\rho = nRT \tag{xxvii}$$

$$P/(\rho T) = nR = \text{constant}$$

$$P_1/(\rho_1 T_1) = P_2/(\rho_2 T_2)$$

where P = pressure, T = temperature in absolute degrees, $\rho$ = density of the air, n = number of moles of air, and R = gas constant.

Dry air at sea level has a density 11% greater than sea level dry air at 30°C. Corrections for temperature and pressure are sometimes significant, and should be considered. The relative humidity can also cause variations in the air density.

Inspection of Equation (xxvi) finds that the drag varies with the square of the velocity directed into the air. Thus, at slow speeds the drag caused by air resistance is negligible. However, as the velocity increases, the drag quickly becomes more important.

One way to measure the frontal directed drag for a specific vehicle is to drive the vehicle along a level roadway when the air is still and note the loss of speed over time as the vehicle coasts to a stop with the engine disengaged. Table 13.1 lists actual data obtained experimentally from a 1983 Mazda GLC when allowed to slow down on a level concrete roadway where there was no measurable wind.

Frictional forces in a free-rolling vehicle can be modeled by the following equation:

$$f = kW + C\rho AV^2 \tag{xxviii}$$

where k = rolling resistance of the vehicle, and W = weight of vehicle.

In general, the rolling resistance of a vehicle is basically constant in typical road speed ranges for most vehicles.

**Table 13.1  Slow Down Data —**
**1983 Mazda GLC[a]**

| Speed (mph) | Lapsed Time (sec) |
|:-----------:|:-----------------:|
| 70 | 0 |
| 65 | 7.0 |
| 60 | 14.3 |
| 55 | 21.8 |
| 50 | 9.8 |
| 45 | 37.9 |

[a] Particulars: T = 100°F; radial tires; curb weight of vehicle and contents = 2240 lbs; frontal area = 3000 in².

If Equation (xxviii) is divided by the mass, "m," of the vehicle, the retarding acceleration caused by the air resistance is obtained.

$$a = kg + C\rho AV^2/m \qquad\qquad (xxix)$$

Using data from Table 13.1, the retarding acceleration can be calculated for several points using Equation (xxix). For example, the acceleration between 70 mph and 65 mph is $-1.052$ ft/sec$^2$. The acceleration between 55 mph and 50 mph is $-0.913$ ft/sec$^2$. Substituting these values back into Equation (xxix) and solving two simultaneous equations gives the following:

$$kg = 0.7 \text{ ft/sec}^2$$

$$C\rho AV^2/m = 0.0000359 \text{ ft}^{-1}.$$

Given that g = 32.17 ft/sec$^2$, then k = 0.0218.
Given that A = 20.833 ft$^2$, $\rho$ = 0.0663 lbt/ft$^3$, and W = 2240 lbf, then C = 0.00181 sec$^2$/ft.

The solved equation for the acceleration retardation by air resistance and rolling resistance is therefore:

$$a = -0.7 \text{ ft/sec}^2 - (0.0000359 \text{ ft}^{-1})V^2.$$

If the Mazda were traveling at 100 ft/sec or 68 mph, the horizontal acceleration due only to air resistance would be $-0.36$ ft/sec$^2$. Therefore, over the velocity ranges normally encountered in falls, neglecting air resistance introduces little error.

## Further Information and References

*Mathematics in Action*, by O.G. Sutton, Dover Publications, 1984. For more detailed information please see Further Information and References in the back of the book.

# Vehicle Performance

# 14

Of all inventions, the alphabet and the printing press alone excepted, those inventions which abridge distance have done most for civilization.

**— on the door to the Transportation Building at the World's Columbian Exposition,** *Chicago 1893*

... at the end of a year crammed with work, he has a little spare leisure, his restless curiosity goes with him traveling up and down the vast territories of the United States. Thus, he will travel five hundred miles in a few days as a distraction from his happiness. ... At first sight there is something astonishing in this spectacle of so many lucky men restless in the midst of abundance. But it is a spectacle as old as the world; all that is new is to see a whole people performing in it.

**— a description of Americans by Alexis De Tocqueville,** *from* Democracy in America, *published 1835–1840, translated by George Lawrence.*

## 14.1 General

This chapter examines some of the basics concerning vehicular acceleration and deceleration. One type of deceleration, skidding, has already been discussed in some detail in a previous chapter. This chapter is broader in scope, and discusses the theoretical underpinnings that affect vehicle response and performance.

## 14.2 Engine Limitations

Each motor vehicle has an engine with a set maximum amount of power it can deliver. Usually this is its horsepower rating. In the English units system, a horsepower is defined as 550 ft-lbf/sec. In the metric or International System of Units, power is usually measured in kilowatts, where 1 kw = 1.341 hp.

Power is defined as the amount of energy delivered per unit time.

$$P = d(E)/dt \qquad (i)$$

where P = power, E = energy, and t = time.

In an engine, the amount of power that can be delivered is normally a function of the engine's rpm (revolutions per minute), and the torque exerted by the engine on the drive shaft prior to the clutch or transmission.

$$P = 2\pi(\text{rpm})T \tag{ii}$$

where $T$ = shaft torque of the engine and rpm = revolutions per minute, a measurement of angular velocity.

As Equation (ii) shows, for a given power output, the torque increases as the engine rpm's decrease, and vice versa. Since the velocity of the vehicle is directly related to the engine rpm's, then it can also be said that the vehicle's speed increases as the torque decreases.

As was noted in a previous chapter, the two main ways that energy is dissipated while a vehicle is traveling are tire friction with the road and air resistance. In Chapter 13, Equation (xxviii) shows the frictional forces associated with rolling resistance and air resistance, and Equation (xxix) shows these same effects as a deceleration.

In general, rolling resistance is taken to include not only the specific tire and road rolling friction, but also the friction of bearings and other moving parts when the car is free-wheeling, i.e., when the engine is disengaged from the power train. For manual transmission cars, this is when the clutch is fully depressed. These friction factors are often all lumped together into a single term designated as the rolling friction of the vehicle.

When the available engine power exceeds the energy dissipated via rolling resistance, air resistance, and any changes in elevation, the vehicle can accelerate. When the available power equals the dissipating effects, the vehicle will have reached its maximum speed (at those conditions). If any of the dissipating factors cause the dissipation rate to exceed the engine's available power, the vehicle will slow down until equilibrium is reestablished.

$$F_R = KW + CA\rho V^2 \tag{iii}$$

where $F_R$ = total frictional forces, $K$ = rolling resistance constant, $W$ = vehicle weight, $\rho$ = density of air, $C$ = shape and units conversion constant (see Chapter 3), $A$ = front area of the vehicle, and $V$ = vehicle velocity.

It is presumed in the above that the air velocity with respect to the ground is zero, i.e., the air is calm. If the air is not calm, the wind speed along the velocity direction of the vehicle would have to be vectorially added or subtracted accordingly.

Work is defined as the force times the dot product of the distance over which it is applied. With respect to the energy dissipation factors, the work done by frictional resistance is as follows:

$$U = KW(x) + CA\rho V^2(x) \tag{iv}$$

where $x$ = the distance through which the force is applied.

If there is a change of elevation, then this must be taken into account by the factor "mgh," where "h" is the change in elevation. If the change in elevation is more or less linear, that is, $h = s(x)$ where "s" is the change in elevation per horizontal distance, Equation (iv) becomes the following:

$$U = KW(x) + CA\rho V_2(x) + mg[s(x)] \tag{v}$$

where $m$ = mass of the vehicle.

The dissipation of energy per unit time is therefore given by the following:

$$dU/dt = KWV + CA\rho V^3 + mgsV \tag{vi}$$

where V has been substituted for $dx/dt$.

Setting the rated power of an engine equal to the right side of Equation (vi) above, the maximum theoretical forward velocity of the vehicle can be determined.

$$P = V(KW + mgs) + CA\rho V^3 \tag{vii}$$

As was already noted in Equation (ii), the engine power is a function of engine shaft torque and angular velocity. If this is considered equivalent to a force applied to the vehicle at a particular velocity, then $P = FV$ where "F" is an equivalent force applied to the vehicle in order to move it, and "V" is the velocity it is traveling forward. Thus, the force applied to the vehicle diminishes as its velocity increases and vice versa.

The above, of course, is known by experience to most drivers. More power is needed to pull out from a dead stop or go over a hill than is needed to maintain a cruising speed.

Using $P = FV$ with Equation (vii), gives the following expression for the force being applied to the car versus the dissipative fractional factors.

$$F = KW + CA\rho V^2 + mgs \tag{viii}$$

The above equation allows calculation of the equivalent force being applied to the vehicle under the conditions at a steady velocity. It is similar to Equation (xxviii) in Chapter 13.

Given the horsepower rating of an engine, Equation (vii) can be used to determine the maximum forward velocity of the vehicle provided the other factors are known. The maximum forward velocity can then be substituted for "V" in Equation (viii), and the force that the engine can apply to the vehicle at that velocity can be determined.

Since P = dE/dt and "dE" is the change in kinetic energy of the vehicle, then the following holds:

$$P[\Delta t] = (1/2)mV^2 \qquad \text{(ix)}$$

where P = engine horsepower rating, $\Delta t$ = time interval, V = velocity, and m = mass of vehicle.

The above equation assumes that the initial velocity was zero when the power was applied.

Because there is a maximum "V" as already discussed, and assuming that the power output of the engine is constant over the time interval, then the amount of time it takes for the vehicle to reach the maximum velocity is as follows:

$$t = (1/2)mV_{max}^2/P \qquad \text{(x)}$$

The overall or average acceleration from zero to maximum speed for the vehicle is then the following:

$$a = V_{max}/t \qquad \text{(xi)}$$

where t is from the solution of Equation (x).

Generalizing, if it is assumed that the vehicle starts from a dead stop, then the velocity of the vehicle after time "t" is as follows:

$$V = [2Pt/m]^{1/2}. \qquad \text{(xii)}$$

The above equation assumes that "t" does not exceed the amount of time needed to reach the maximum velocity. A plot of "V" versus "t" finds that initially, "V" is zero, but climbs fast. "V" tapers off somewhat asymptotically as "t" increases until the curve is essentially flat at the vehicle's maximum velocity.

Similarly, the vehicle's acceleration can be calculated as follows:

$$a = [2P/mt]^{1/2}. \qquad \text{(xiii)}$$

Again, it is assumed that "t" does not exceed the "t" for maximum velocity found in Equation (x). A plot of "a" versus "t" finds that "a" starts out asymptotically high, and then curves downward. As "t" increases, the curve asymptotically approaches the "t" axis. At the time when "$V_{max}$" is attained, "a" is essentially zero.

Equations (xii) and (xiii) neglect frictional factors, power loss through the transmission, gear changes, and a number of other practical effects. However, the two equations do provide a sense of the way velocity and acceleration are expected to vary with time, engine power, and vehicle mass.

To more accurately model "V" and "t," the frictional effects that would slow up the vehicle have to be accounted for. This is done as follows:

$$[P(V)]t = [(1/2)mV^2 + (1/2)KWVt + (1/2)mgsVt + (1/6)CA\rho V^3 t] \quad (xiv)$$

power output = kinetic energy + rolling friction work +
elevation change work + air friction work

where $P(V)$ = the engine power output as a function of vehicle velocity (or gear position).

Equation (xiv) assumes that the initial velocity is zero, and that velocity increases continuously and more or less linearly.

## 14.3   Deviations from Theoretical Model

In the preceding section, it was presumed that all the power from the engine was transmitted to the car with 100% efficiency. This, of course, does not actually happen. There are two significant factors that reduce the engine efficiency in transmitting power to move the vehicle: wheel slip and transmission efficiency.

A brief definition of wheel slip was previously provided in Section 12.13 of Chapter 12. When there is full traction between the tire and the road, there is no slip. This "no slip" condition occurs when the tangential velocity of the tires, where they contact the road, equals the forward speed of the vehicle. When the tangential speed of the tire differs from the forward speed, slip occurs.

As before, slip is defined as:

$$\text{slip} = S = 1 - [v_t/v] \quad (xv)$$

where $v_t$ = tangential velocity of the tire at the point of contact with the road and $v$ = forward velocity of the vehicle.

To account for slip during acceleration, the term "(1-S)" should be used to modify the velocity and acceleration calculations. For hard accelerations without "peeling," a slip factor of 0.15 to 0.20 is appropriate.

**Table 14.1   Common Automatic Transmission Efficiencies**

| Gear Position | Efficiency "e" |
|---|---|
| 1st gear | 0.1–0.3 at the high end |
| 2nd gear | 0.3–0.5 at the high end |
| 3rd gear | 0.5–0.8 at the high end |
| 4th gear | 0.8–0.95 at the high end |
| (if equipped) | |

The second, and perhaps more significant factor that affects the mechanical efficiency of transmitting power from the engine is the transmission efficiency. In an automatic transmission, is it not uncommon for the mechanical efficiency between power input and power output to be as found in Table 14.1.

This means that the available power "P" to move the vehicle has to be multiplied by an appropriate transmission mechanical efficiency factor, "e," when calculating values for "V" or "a" using the equations derived in Section 14.2.

## 14.4   Example Vehicle Analysis

A 1988 Ford Taurus has a curb weight of 3155 pounds and a net engine power of 140 hp. If the vehicle accelerates to 30 mph with an automatic transmission, it will be in first gear up to 15 mph, and then will switch to second gear.

Using the equation that neglects frictional forces, the time it would take to go from 0 to 30 mph would be as follows:

$$t = (1/2)mV^2/[(1 - S)Pe] \quad \text{Note: "e" and "S" terms added to Eq. (x).}$$

$$m = 3155 \text{ lbf}/32.17 \text{ ft/sec}^2 = 98 \text{ lbf-sec}^2/\text{ft}$$

$$V = 30 \text{ mph} = 44 \text{ ft/sec}$$

$$P = 77,000 \text{ lbf-ft/sec}$$

$$S = 0.10 \text{ (slip factor)}$$

$$e = 0.35 \text{ (mechanical efficiency)}$$

Substituting and then solving,

$$t = 3.9 \text{ sec} \quad \text{and} \quad a = 11.3 \text{ ft/sec}^2.$$

Similarly, for a speed of 60 mph, a slip of 10%, an average transmission efficiency of 60%, then the time and average acceleration are calculated to be:

$$t = 9.1 \text{ sec} \qquad a = 9.7 \text{ ft/sec}^2.$$

From actual test data published by *Consumer Reports* (June 1988, p. 399), the time required for acceleration to 30 mph was reported to be 4.1 seconds, and the time for acceleration to 60 mph was 10.3 seconds. The respective accelerations were 10.7 and 8.5 ft/sec². 

Not surprisingly, the approximation formula did well for the 30 mph regime, where wind resistance is negligible, but was less accurate in the regime above 30 mph, where wind resistance is more significant. Still, the correlation of the predicted values with the actual test values is good despite neglecting friction. This is a useful first order approximation.

The maximum speed of the Taurus is estimated as follows:

Let  $A = 21.0 \text{ ft}^2$

$K = 0.03$

$C = 0.002 \text{ sec}^2/\text{ft}$

$\rho = 0.075 \text{ lbf/ft}^3$

$S = 0.05$

$e = 0.80$

$s = 0$ (no elevation change).

Solving with the above values, finds that "$V_{max}$" for the Taurus is 227 ft/sec or 155 mph.

The air drag at 155 mph is calculated to be 1.65 ft/sec², and the rolling friction is calculated to be 0.97 ft/sec². The total calculated frictional deceleration is then 2.62 ft/sec² at maximum speed. It is estimated that maximum speed would be achieved in about 43 seconds in about 5000 ft. Overall acceleration from start to maximum would be 5.3 ft/sec².

## 14.5  Braking

Most vehicles have published data concerning their ability to brake. For example, the 1988 Taurus already discussed in the last section was found to be able to brake without locking wheels from 60 mph to zero in 180 ft.

Applying some algebra to the simply dynamic equations yields

$$a = V^2/(2d) \qquad \text{(xvi)}$$

where a = acceleration, V = the initial speed prior to braking, and d = the braking distance.

In the Taurus example, the braking deceleration is calculated to be 21.51 ft/sec$^2$ or 0.668g. Assuming that the driver was professionally trained, the brakes were likely kept at a slip of about 20%. Therefore, the coefficient of friction between the test pavement and the tires would have been as follows:

$$f = (a/g)/(1 - S) \qquad \text{(xvii)}$$

$$f = 0.67/0.80 = 0.83$$

In Chapter 12, Section 12.4, Table 12.2, a frictional coefficient of 0.85 was listed for dry concrete with little or no motion (static coefficient). Inserting the speed of 88 ft/sec into Equation (xiii) in Chapter 12, a coefficient of 0.70 is calculated (dynamic coefficient). Thus, the difference in the dynamic frictional coefficient between a locked wheel skid (0.70) and a controlled hard braking with maximum effect (0.67) is very small. In this case, it is a difference of only 4.6%. Under most experimental conditions, the difference is less than the data scatter that would be expected over several trials.

## 14.6  Stuck Accelerators

The previous sections have demonstrated some facts that have relevance with respect to "stuck accelerator" or "sudden acceleration incident (SAI)" cases. These cases usually involve people who claim that their cars or trucks suddenly took off on their own due to stuck accelerators. In some of these cases, people will report that they tried to stop the cars by pushing the brakes with all their strength, but that the cars just kept on going.

With the 1988 Taurus, it was shown that the initial acceleration was 10.7 ft/sec$^2$. This, of course, was done by a professional driver who knew how to milk all the speed possible from a car. It was also shown that locked-up brakes would generally cause deceleration at about 0.7g to 0.8g, i.e., 22.5 to 25.7 ft/sec$^2$. Comparing these two figures shows that no matter how hard the engine might try to run away with the car, the brakes exert over twice the deceleration needed to counter the runaway engine.

In short, the only way a stuck accelerator can cause a car to run away while someone is standing on the brake is for both the accelerator and the brake system to be dysfunctional at the same time.

Typically, a person who experiences an SAI indicates that one or more of the following occurs.

1. The car takes off by itself from a stopped position, usually when the driver shifts from park to drive or reverse.
2. The car takes off by itself despite the driver's efforts to apply the brakes.
3. The car's cruise control appears to engage itself, causing the car to take off.
4. The driver reports that the car does not slow down when the accelerator is released.

The news media often give more attention to reported Sudden Acceleration Incidents (SAIs) than other vehicular accidents of the same severity. SAIs that are reported in the newspapers are often couched in language suggesting that the event was some sort of a personal betrayal or act of rebellion on the part of the car. Often, SAIs are described with phrases such as, "it fought the driver's best efforts to control it" or "it acted as if it had a mind of its own."

Perhaps this is due to the special relationship between people and their automobiles. More than any other machine in our modern society, automobiles are depended upon for people's livelihood, recreation, socializing, and sense of personal freedom. Interestingly, several novels and motion pictures have explored this fear of car rebellion, most notably *Christine* by Stephen King.

## 14.7   Brakes vs. the Engine

A car that can go from 0 to 60 mph in 8 seconds is considered to be a sporty vehicle. This computes to an average acceleration of 11 ft/sec$^2$. Some of us who drive more economical vehicles can expect 0 to 60 mph times of 10 to 14 seconds. These times compute to average accelerations of 8.8 ft/sec$^2$ and 6.3 ft/sec$^2$, respectively.

In published acceleration tests from which these types of figures are obtained, the test-driver literally floors the accelerator pedal and makes the vehicle go as fast as possible. Typically, the tests are done with factory supplied, showroom-quality new cars that are tuned and adjusted for peak performance. Such test figures represent the upper limits of acceleration of the vehicles using standard factory supplied equipment.

While the engine supplies the power to accelerate the car, the brakes consume power to decelerate it. Unless there is some mechanical deficiency in the brake system, most cars have a braking system capable of decelerating the vehicle about 22 ft/sec² or more. In a tug-of-war contest between the brakes and the engine, the brakes win. The brakes can decelerate the car two to four times more than the engine can accelerate it. Even if only the emergency brakes are available (usually the two rear brakes), the deceleration effect of the emergency brakes alone is approximately equal to the acceleration the engine can produce.

Consider the following example. According to the May 1989 issue of *Consumer Reports* (p. 344), a 1989 Eagle Summit LX was tested and found to accelerate 0 to 30 mph in 4 seconds, and 0 to 60 mph in 13.1 seconds. In braking, the same vehicle was found to stop in 150 feet from a speed of 60 mph without locking up the wheels.

The acceleration tests of 0 to 30 mph and 0 to 60 mph compute to an average acceleration rate of 11 ft/sec² and 6.71 ft/sec², respectively. (As shown previously, the initial acceleration is higher than acceleration initiated at higher velocities. As the vehicle speeds up, there is less available power from the engine to accelerate the vehicle.) The braking deceleration of the 1989 Eagle Summit LX without locking up the wheels computes to 26.0 ft/sec².

If a car is at rest and the brake pedal and accelerator pedal are mashed down simultaneously while the engine is in gear, the engine will stall out, assuming the clutch doesn't burn out first. Similarly, if the emergency brake is fully engaged and the engine is in gear when the accelerator is floored, the car will either not move and stall out, or it will lurch forward slowly, dragging along the locked rear wheels.

If the car is able to lurch forward with the emergency brake engaged, it is usually because the car has front-wheel drive. Because the front of a car is typically heavier than the rear, deceleration by the locked rear tires alone may be slightly less than the acceleration transmitted through the unlocked front wheels in a powerful car. Therefore, some forward motion may occur. However, such motion would hardly be of a magnitude for consideration as an SAI, and would be easily controlled by the driver.

Thus, if a driver claims that the car accelerated into a "runaway" despite the driver's best efforts to apply the brakes, there are only a limited number of possibilities as to what occurred. These possibilities are listed as follows.

1. The brakes failed.
2. A pedal other than the brake was mashed down by the driver.
3. The driver "incorrectly" described the event.

If the car is still driveable after the accident, a simple drive around the block can verify point 1 above.

If sufficient information exists to quantify the acceleration that occurred in the SAI, such as a "peel out" mark or similar evidence, the acceleration magnitude of the runaway can be compared to the published 0 to 30 mph acceleration figures for the car model. If they are reasonably comparable, and the brakes are in working order, then it can be concluded that the driver simply mashed down on the accelerator instead of the brake as reported.

## 14.8   Power Brakes

In a car equipped with power brakes, the driver's foot effort in depressing the pedal is assisted by an auxiliary source of power. Usually this auxiliary power comes from one of the following.

- A pressurized hydraulic system powered by a pump operated off of the engine.
- A vacuum diaphragm mechanism that utilizes low-pressure air sources from within the engine or from a vacuum pump run-off of the electrical system.

In either case, the auxiliary source of power allows the driver to use much less effort to engage the brakes than would otherwise be required. Power brakes are often used in large vehicles such as trucks, vans, recreational vehicles, buses, and luxury cars where the driver's effort alone might not be enough to operate the brakes effectively.

If the auxiliary power system fails, the brakes will still function, but it will require much more effort for the driver to operate them. If a driver has become accustomed to operating power brakes and has never operated them without the power assist, the brakes may appear not to work. The brakes will be much harder to depress, and the resulting stopping effect may be reduced. If the driver is frail or weak, it may not be possible for the driver to operate the brakes.

In such cases, failure of the brake power assist could allow a driver to depress the brake pedal hard without noticeable braking response. The type of malfunction does mimic an SAI. If the power brakes malfunction as described, the engine drag can slow the car when the driver releases the accelerator pedal, and the emergency brake can be engaged for further stopping effect.

The loss of the brake power assist does not cause the engine throttle to open up and the car to suddenly accelerate. However, this type of malfunction may surprise or panic the driver when he discovers the lack of familiar response from the brakes. In such a panic, the driver may inadvertently step on the accelerator causing an SAI.

## 14.9   Linkage Problems

When a driver depresses the accelerator pedal, there is a mechanical linkage that transmits the pedal motion to the engine throttle. Sometimes the linkage is constructed of wire cable, sometimes a series of connected rods, and sometimes a combination of the two. Other methods of transmitting the pedal motion to the throttle are possible (e.g., electromechanical, pneumatic, hydraulic, etc.), but a mechanical linkage provides the most reliable operation.

Most U.S. built cars, and cars built to U.S. standards, have a spring load on the engine throttle such that if the linkage were to break or become disconnected, the throttle lever will return to the idle position. It does not return to an open or full throttle position.

In order to increase engine power output, the throttle must be opened. In order to do this, the throttle lever must be pulled against the opposing spring load. Whenever the force pulling on the throttle lever is relaxed, the throttle spring will return the throttle lever back to the idle position.

It is possible for the throttle spring to break or fail. In such cases, the accelerator might "hang" at the point where it was depressed furthest and not return to the idle position. If this occurs, when the driver lets up on the accelerator, the car would continue to go at the same speed as before.

Similarly, a linkage can sometimes become bound or develop an unusual amount of friction that could retard the return action of the spring. A bent rod, a "kinked" cable, or a wet cable that has frozen over can cause such a problem.

## 14.10   Cruise Control

Various types of control and sensing strategies are used in cruise control systems. Most cruise control systems utilize engine vacuum or vacuum created by a small DC electric pump to control the throttle. Speed is sensed by a magnetic pickup on the drive shaft, by factory installed sensors in the transmission, or by piggybacking off the signals of a digital speedometer, if the vehicle is so equipped.

Most cruise control systems have the following characteristics in common.

1. They have a set point for the speed that is to be maintained. The set point is usually engaged when the vehicle is traveling at the desired speed.
2. The set point is not absolute, but has a tolerance above and below it, before the throttle is incrementally opened or closed.
3. There is a minimum speed below which the cruise control will not operate. Most cannot be engaged when the car is in reverse or below 35 mph.

4. The set point can be increased or decreased without having to release the cruise control. The cruise control can therefore act like a hand throttle.
5. The cruise control is disengaged when the foot brake or clutch is tapped.
6. The cruise control can be reengaged at the set point if the control lever is tapped, or the resume control is operated (same thing).

Cruise control SAIs are often related to some of the following situations.

1. The driver uses the accelerator to drive faster than the cruise control set point and forgets that the set point is still engaged. When the accelerator is released, the car does not fully slow down, but slows down to the set point. The car appears not to respond to the released accelerator.
2. The driver engages the resume feature when traveling at a speed slower than the set point. Thus, the car appears to momentarily take off on its own.
3. The driver intends to tap the brake to disengage the cruise control but taps the accelerator instead. The car appears to surge forward.

In all of the above situations, the cruise control has not malfunctioned.

Because most cruise control systems utilize vacuum (actually an air pressure slightly lower than the ambient) for control of the throttle, the cruise control normally fails in a fail-safe mode. That is, when the cruise control fails and the vacuum is broken, the system ceases to function. A leak in any of the vacuum lines causes the system to stop, and a malfunction in any of the parts causes the system to stop. Also, when the brake pedal is depressed, this breaks the vacuum.

Cruise controls also utilize an integrated circuit "chip" to handle input and output signal information. While it is theoretically possible that such a chip could be affected by a nearby large electric or magnetic field, this has not been a practical concern. Induced false signal processing is thwarted by the fact that the internal redundancies in most cruise control systems require two or more simultaneous circuit failures to occur. In practice, this does not occur.

## 14.11 Transmission Problems

In automatic transmissions, the linkage between the gear selector and the transmission can sometimes get out of adjustment such that the indicated gear on the selector is not the gear being used. Such misalignments can cause "half step" shifts in gear position. For example, the neutral gear might not be at the "N" position, but it might lie halfway between the "R" and "N" positions.

When such adjustment problems occur, it can cause a driver who sloppily shifts the gear selector to believe he is in one gear setting when he is in the adjacent gear setting. This can cause unexpected movement of the vehicle and result in an SAI complaint.

Some misalignments occur slowly over time. Other times, it may be because some recent mechanical work on the car required temporary disconnection of the gear cable or linkage, and the reconnection work caused a shift in adjustment. Still, other cases might involve worn detent pins or other such parts that allow slippage or "stop" in the gear position. In such reported SAIs, the gear selection adjustment problem will also be apparent after the accident; such problems do not spontaneously "heal."

Some late model cars are equipped with automatic shift locks. This device requires that the driver put pressure on the brake pedal before the gear selector can be shifted out of "park." It is notable that since the introduction of this device, cars so equipped have had fewer reported SAIs. For example, during the period September 1986 through November 1988, Audis equipped with automatic shift locks had 60% fewer reported SAIs than identical model vehicles not so equipped. This result further corroborates the point that many reported SAIs are simply due to pedal misapplication.

## 14.12   Miscellaneous Problems

Some SAIs have resulted from floor mats, shoes, soda bottles and cans, debris, or other items becoming entangled in the accelerator or brake pedal. Floor mats and shoes can sometimes roll under the brake pedal and prevent it from being depressed properly. Similarly, a folded-over floor mat or trash item might retard the return of the accelerator when released.

Some people slip off their shoes while driving and simply allow them to lay in the floor in the foot well. This is especially true with respect to women who wear high-heeled dress shoes. The shoes can easily be slipped off. Stops and sharp turns can cause them to roll unnoticed under the driver's foot pedals.

## 14.13   NHTSA Study

Because of the public interest in SAIs, the National Highway Traffic Safety Administration conducted a study of ten vehicle types with above-average SAI complaints. The results of the study, entitled *An Examination of Sudden*

*Acceleration* (U.S. Dept. of Transportation, DOT-HS-807-367), were made public in 1989. The following is a summary of the study's findings:

1. A substantial number of SAIs involved accelerator linkages that "hung up" due to a mechanical problem. However, the problem caused the throttle to remain at its last position; it did not cause further acceleration.
2. A few cruise control failures were noted. However, in all cases, application of the brake would cause the cruise control to disengage.
3. Extensive testing of cruise controls in strong electromagnetic fields found no SAI causation.
4. In thousands of SAIs reported to the NHTSA, brake failure occurred in only a handful of cases. Actual brake system failure plays no significant role in SAIs.
5. For most SAIs, the most plausible cause of an open throttle condition while attempting to brake was pedal misapplication, which was perceived by the driver as brake failure.
6. It was hypothesized that the vehicle models with higher frequency SAIs had pedal placement or feel so that the driver could mistake one pedal for another causing pedal misapplication.
7. There were reported incidents where drivers thought they were starting their cars in "park" when they were actually in "neutral." When the driver acted to shift into "reverse," one notch back, the car would be in "drive." Often the selector was broken or unreadable.
8. Recommendations to change pedal placement and design were made to enhance driver recognition of the pedal being applied.

## 14.14   Maximum Climb

Since a typical mid-sized passenger car has a take-off acceleration of about 0.33g, this places a limit on the maximum grade that the car can climb.

When a box is laid on a frictionless inclined plane, the acceleration down the grade is as follows:

$$a = g \sin(\alpha) \qquad \text{(xviii)}$$

where $\alpha$ = the angle that the inclined plane makes with the horizontal.

In order for a vehicle to be able to go up an incline, the vehicle must be able to accelerate more than the downward acceleration along the incline, and must be able to grip the roadway.

The perpendicular force of the vehicle pressing down upon the inclined plane is given by the following:

$$F = W \cos(\alpha) \qquad\qquad (xix)$$

where W = weight of the vehicle.

In order for the vehicle to not simply slide down the incline, the friction between the tires and the surface must be greater than the sliding force parallel to the face of the plane.

$$Ff = Wf \cos(\alpha) > W \sin(\alpha) \qquad\qquad (xx)$$

where F = force applied normal to the inclined plane.

$$f > \tan(\alpha)$$

If the coefficient of friction was 0.7, then the angle of the incline would have to be less than 35 degrees for the truck to not slide off with locked brakes on all four wheels.

In order for the vehicle to be able to climb the incline, the sliding acceleration must be less than the acceleration the vehicle can generate.

$$a > g \sin(\alpha) \qquad\qquad (xxi)$$

where a = acceleration of vehicle and $\alpha$ = angle of incline.

If "a" is typically about 0.33 g for maximum initial acceleration, then "$\alpha$" must be less than about 19 degrees for the vehicle to make any headway up the hill.

It is noteworthy that in this case, the angle at which the vehicle stalls is much less than the angle at which it would lose traction with all four brakes locked.

There is still another item to consider in this situation. If the vehicle has four-wheel drive, then the above estimate of angle holds. If the vehicle is either front-wheel drive or rear-wheel drive, then an additional modification is required to account for the uneven distribution of drive through the wheels.

In a front- or rear-drive vehicle, traction is provided by only one set of wheels. The frictional resistance to slide when the brakes are released and the engine is engaged is as follows:

$$Wf(\cos\alpha) \, x > W(\sin\alpha) \qquad\qquad (xxii)$$

where x = fraction of vehicle weight carried by the drive axle.

$$f > (1/x)(\tan\alpha)$$

For front-wheel drive carrying 65% of the vehicle load with a coefficient of 0.7, the maximum angle is 24 degrees. For rear-wheel drive carrying 35% of the vehicle load with the same coefficient of friction, the maximum angle is 13.8 degrees. This simple calculation demonstrates one of the advantages of front-wheel drive.

It was shown that a vehicle that could produce 0.33g of forward acceleration could not make any headway up a slope greater than 19 degrees due to the limitations of power. The above has shown that if the vehicle is a rear drive with a typical loading of 35% of the vehicle's weight on the rear axle, it will not have sufficient traction to make headway unless the angle is less than 13.8 degrees. While it would have sufficient power, its back wheels would simply spin and the vehicle would slide down. The front-wheel drive, however, would.

This rear-wheel drive effect also applies to the parking brake, which in a passenger car or light truck, only locks the rear wheels. It is typical in passenger cars that the rear axle only carries 35% of the vehicle's weight. Thus, the resistance to slide would be only 35% of that which afforded when all four wheel brakes are engaged by the foot pedal. This can fool a driver into thinking that because the foot brakes can hold the car on an incline, so can the parking brakes.

In drum brakes, this problem is further exacerbated because the brakes may be directional; they may be designed to provided more braking effect in the forward direction than in the reverse direction. Thus, a parking brake that may hold a car when the car is pointed down the incline, may allow the car to roll when it is parked pointing up the incline.

## 14.15   Estimating Transmission Efficiency

A previous section briefly discussed the efficiency factors associated with getting power from the engine to the wheels, and some representative values for various gears in an automatic transmission were given. As was shown, the mechanical losses at slow speeds could be very high, while at high speeds the losses were much reduced.

Obtaining exact test information concerning mechanical efficiency of the drive train is often difficult, and doing the tests can be expensive. However, the values can be obtained indirectly from other performance parameters that are readily available.

Each vehicle manufactured or marketed in the U.S. must show EPA mileage ratings: one for highway driving and one for city driving. The highway rating is for flat highway driving at highway speeds. The city rating is for moderate speeds associated with city stop-and-go driving. Also, various consumer testing agencies publish similar fuel consumption figures. The ratio of these two values can be a useful indicator of the ratio of transmission efficiencies at the speeds tested.

For instance, in the previous example of the 1988 Ford Taurus, the EPA mileage ratings are 21 mpg city, and 29 mpg highway. The *Consumer Reports* figures are 15 and 29, respectively. This indicates that *Consumer Reports* tests for city driving at a lower average city speed than the EPA. The ratio of the EPA's figures are 0.72, and the *Consumer Reports* figures are 0.52. Since at highway speeds a typical transmission will have a mechanical efficiency of about 0.80, the EPA's figures indicate that at their city driving speeds, the transmission efficiency is 0.58.

Similarly, the *Consumer Reports* figures indicate an efficiency of 0.42 at their city driving speeds. The EPA's figures are closer to the values associated with the low end of third gear, while the figures from *Consumer Reports* are closer to the values for second gear. The combination of these figures provide three points from which a transmission efficiency curve can be constructed.

To demonstrate the consistency of these figures, consider the figures for a 1988 Buick Century with automatic transmission. The EPA figures are 20/27 and the *Consumer Reports'* are 16/29. The EPA ratio is 0.74 and the *Consumer Reports* ratio is 0.55. Using the same assumption that highway speed has a mechanical efficiency of about 0.80, then the EPA city driving efficiency is 0.59 and the *Consumer Reports* city driving efficiency is 0.44.

Manual transmissions are generally more efficient than automatic transmissions at the lower end, and about the same at the upper end, i.e., about 80–85% at highway speeds. For example, a 1989 Eagle Summit LX has an EPA of 29/35 and a *Consumer Reports* of 25/37. These are ratios of 0.83 and 0.68, respectively. Assuming a highway efficiency of 80%, the respective city efficiencies are 0.66 and 0.54.

Similarly, a 1989 Ford Escort LX with a manual transmission has an EPA 27/36 and a *Consumer Reports* of 23/38. The respective ratios are 0.75 and 0.60, and the respective transmission efficiencies are 0.60 and 0.48.

The Escort and the Eagle both have 5-speed manual transmissions, both weigh about 2340 pounds, both have similar weight distribution between the axles, and both have the same acceleration values for 0 to 30 mph and 0 to 60 mph (4.0 and 13.1 seconds, respectively). The Escort, however, has a four-cylinder, 1.9 liter displacement engine while the Eagle has a four-cylinder 1.5 displacement engine. Thus, because the Eagle has a smaller engine but has the same acceleration performance, it would be expected that it would have a better low-range mechanical efficiency. This is borne out in the efficiency estimates.

While it is not as good as test data, this method does provide reasonable estimates of transmission efficiencies.

## 14.16   Estimating Engine Thermal Efficiency

As a point of interest, an estimate of the thermal efficiency of the vehicle's engine can be made by comparing the energy input, vis-à-vis the engine fuel consumption rate, to the power output, as shown in Equation (iii).

A pound of gasoline has about 20,750 BTU (higher heating value) and a pound of diesel fuel has about 19,300 BTU. On a per gallon basis, this is 128,000 BTU and 144,000 BTU, respectively.

One horsepower is equivalent to 2545 BTU/hr. Since highway fuel consumption is usually figured at 55 mph, 1 hour of operation would use 55 mph/(x mpg), where "x" is the unknown mileage rating.

For example, a 1989 Nissan Sentra has been found to get 44 mpg at highway speeds on flat, expressway type roads. Its engine power rating is 90 hp. Converting the units, 44 mpg is equivalent to a consumption of 160,000 BTU/hr. If the Sentra has a frontal area of 20.5 ft$^2$, a "K" value of 0.01, a "C" value of 0.008 sec$^2$/ft, and a weight of 2275 pounds, then the frictional resistance at 55 mph would be about 103 lbf. Over the course of one mile, that would be 544,000 lbf-ft or 699 BTU. For one hour, that would be 38,400 BTU. The engine efficiency would then be output/input or 38,400/(0.08)(160,000) = 0.30 or 30%. (The 0.80 value is to account for transmission efficiency.)

The above calculated value is typical for an efficient Otto or air standard cycle spark ignition engine.

## 14.17   Peel-Out

When a vehicle initially accelerates in excess of the tractive ability of its tires, it will leave an obvious peel-out mark. The presence of this mark may often provide valuable information concerning what a vehicle was doing at a specific location. This can be especially useful in SAI cases (see Chapter 8), and certain intersection cases.

In the case of peel-outs, the torque applied to the tire by the engine exceeds the torque applied by the frictional resistance on the tire. This is expressed below.

$$T > fr(1/x)W \qquad\qquad \text{(xxiii)}$$

where T = torque engine applies to tire, $1/x$ = fraction of vehicle weight carried by tire, r = radius from center of wheel to where wheel contacts ground and is depressed, W = weight of vehicle, and f = coefficient of friction.

Equation (xxiii) explains why it is easier to peel out with a rear end drive car than a front end drive car. Since most cars have more weight distributed toward the front, a higher torque is required for a front-wheel drive vehicle to peel-out, all other factors being the same.

## 14.18   Lateral Tire Friction

When a car is braked or decelerated, it can skid or have slippage in the longitudinal direction, i.e., the direction of travel. However, it can also slide or slip laterally. Due to the orientation of the car when braking begins, due to the crown of the roadway, or due to side impacts, a vehicle can also slide sideways while skidding forward.

In general, when a vehicle has no longitudinal slip, the friction between the tire and the road across the lateral axis follows the same type of calculation as before. It is still the same tire, the same roadway, and the same amount of contact area. Some variations in the coefficient of friction may occur due to tread orientation or edge effects of the side of the tire, but essentially a tire responds laterally about the same as it does longitudinally.

On dry pavement, the longitudinal coefficient of friction is maximum when the longitudinal slip is 10–20%. It then diminishes more or less linearly until 100% slip occurs. When the pavement is wet, the maximum longitudinal friction occurs when longitudinal slip is 30–50%.

On the other hand, lateral friction is maximum when longitudinal slip is 5–10%, and then diminishes rapidly when longitudinal slip is greater than 15%. At a longitudinal slip of about 30–40%, the lateral friction coefficient may be half of what it was when the longitudinal slip was 0%. Between 80–100% longitudinal slip, the lateral friction coefficient may be reduced to 0.10–0.20, with 0.15 being a typical value for both radial and bias tires when the pavement is wet.

## 14.19   Bootlegger's Turn

A bootlegger's turn is when a car is turned around 180 degrees, generally on a two-lane road, while still being driven forward. Generally, to get a car turned around, it is necessary to stop the car, back up, cut the wheels sharply, and then proceed forward. A bootlegger's turn simply spins the car around to face the opposite direction while it is still moving.

As the name implies, a bootlegger's turn is so named because it was a maneuver used by moonshine runners to escape pursing "revenuers." Such quick turns would often catch the pursuers unawares, and they would over-run their quarry.

A bootlegger's turn is done as follows. While moving forward at a moderate or low speed, the emergency brake is suddenly applied so as to cause the rear wheels to slide. Just as the rear wheels begin to slide, the front wheels are steered sharply to the left or right. This will cause the rear end of the car to slide around. When the rear end of the car has spun sufficiently to where the car can take off in the opposite direction, the emergency brake is released and the accelerator is mashed down to make a quick getaway.

What makes this maneuver work is the fact that when the rear wheels begin to skid longitudinally, the lateral traction significantly drops to make side skidding very easy. However, because the front wheels are still rolling (and not significantly slipping), they retain their side traction. The car rotates around the front wheels, which are still able to hold to the road, while the rear wheels are more or less free to slide sideways.

If the rear wheels do not fully skid before the sharp turn is executed, there is a risk of overturning the car. This is because the lateral coefficient of friction will not have dropped off enough. Thus, instead of sliding around, the car might roll over. If the front wheels are allowed to skid also, which would happen if the driver used the foot brakes, the car would simply skid forward and not spin around sufficiently. For these reasons, it is not recommended that amateur drivers attempt this maneuver.

A similar maneuver called "doughnuts" uses the same principle of lower side traction due to longitudinal slip. In this maneuver, the driver proceeds forward at a moderate or slow speed usually in snow or on a loose gravel area. The driver mashes the accelerator to the floor so that the wheels spin or "peel-out." By keeping the front wheels sharply turned, the car will spin around in a circle pivoting about the front wheels making a "doughnut" figure. In this case, the longitudinal slip occurs due to the peel-out rather than a skid. By letting off the accelerator and allowing the rear wheels to "catch up," and straightening out the front wheels, the car can be driven out of the maneuver at will. This also is obviously not a recommended maneuver for an inexperienced driver.

Both maneuvers have been discussed because variations of these actions sometimes occur in an accident, often unintentionally by a panicked driver. During panic stops when longitudinal slip exceeds 50%, the concomitant reduction in lateral traction can result in loss of control of the vehicle, especially when rotational forces are present.

The loss of lateral traction resulting from longitudinal skidding is often a causative factor in the following.

- Truck accidents where jackknifing occurs.
- Car accidents where skidding precedes loss of steering control of the vehicle, especially at curves and corners.

- Motorcycle accidents where the driver loses control and lays the motorcycle down sometime after skidding initiates.

The primary advantage of an antilock brake system is that when the longitudinal slippage is limited to 30% or less, the reduction in lateral traction is held within sufficient limits to generally maintain control of the vehicle.

## Further Information and References

*Accident Reconstruction: Automobiles, Tractor-Semitrailers, Motorcycles and Pedestrians*, Society of Automotive Engineers, Warrendale, PA, February 1987. For more detailed information please see Further Information and References in the back of the book.

"The Amateur Scientist" by Jearl Walker, *Scientific American*, February 1989, pp. 104–107. For more detailed information please see Further Information and References in the back of the book.

*Consumer Reports*. For more detailed information please see Further Information and References in the back of the book.

*Detroit News*, Five part series of articles on SAIs, December 13–17, 1987.

*An Examination of Sudden Acceleration*, U.S.D.O.T., National Highway Traffic Safety Administration, Report Number DOT-HS-807-367, Washington DC, January 1989. For more detailed information please see Further Information and References in the back of the book.

The Insurance Institute for Highway Safety and the Highway Loss Data Institute are essentially the same, and are supported by the insurance industry. For more detailed information please see Further Information and References in the back of the book.

*Manual on Uniform Traffic Control Devices for Streets and Highways*, National Joint Committee on Uniform Traffic Control Devices, U.S. Department of Commerce, Bureau of Public Roads, Washington DC, June 1961. For more detailed information please see Further Information and References in the back of the book.

The Occurrence of Accelerator and Brake Pedal Actuation Errors During Simulated Driving, by S. B. Rogers and W. Wierwille, in *Human Factors*, 1988, 30(1), pp. 71–81.

"The Physics of Traffic Accidents" by Peter Knight, *Physics Education*, 10(1), January 1975, pp. 30–35.

*Research Dynamics of Vehicle Tires*, Vol. 4, by Andrew White, Research Center of Motor Vehicle Research of New Hampshire, Lee, New Hampshire, 1965. For more detailed information please see Further Information and References in the back of the book.

*SAE Handbook, Volume 4: On-Highway Vehicles and Off-Highway Machinery,* Society of Automotive Engineers, Warrendale, PA. For more detailed information please see Further Information and References in the back of the book.

"Unintended Acceleration" by Tom Lankard, *Autoweek,* January 19, 1987.

*The Way Things Work,* by David Macaulay, 1988, Houghton Mifflin, New York. For more detailed information please see Further Information and References in the back of the book.

*The Way Things Work: An Illustrated Encyclopedia of Technology,* 1967, Simon and Schuster, New York. For more detailed information please see Further Information and References in the back of the book.

*Work Zone Traffic Control, Standards and Guidelines,* U.S.D.O.T., Federal Highway Administration, Washington DC, 1985, also A.N.S.I. D6.1-1978. For more detailed information please see Further Information and References in the back of the book.

# Momentum Methods

# 15

The Eagle has landed.

— **Neil Armstrong,** *first landing on the moon, July 20, 1969.*

A new scientific truth does not triumph by convincing its opponents and making them see the light, but rather because its opponents eventually die, and a new generation grows up that is familiar with it.

— **Max Planck,** *Nobel Prize Winner in Physics, developer of quantum theory, 1858–1947.*

## 15.1   General

A milestone in the development of modern science was the publication of Newton's three laws of motion. In brief, the three laws are as follows:

I.  A body at rest tends to stay at rest and a body in motion tends to stay in motion unless acted upon by an external force.

II.  When a force is applied to a free body, the rate at which the momentum changes is proportional to the amount of force applied. The direction in the change of momentum caused by the force is that of the line of action of the force.

III.  For every action by a force, there is an equal but opposite reaction.

The above three principles form the basis for all nonrelativistic mechanics. Until vehicle velocities begin to reach about 3000 km/s, or about 1% the speed of light, the above three basic axioms of motion will do for every instance of vehicular accident analysis. Since the present relative velocities of impacting vehicles rarely exceed 200 km/hr or 0.0000185% the speed of light, there is little need of having to apply Lorentz transformations or other relativistic considerations to vehicular accident analysis.

Of course, this brings to mind the old joke about the pedantic motorist who told the judge that he did not run a red light; he was traveling fast enough that the red light appeared green to him. The judge, no stranger to

Doppler effect calculations himself, simply agreed with the motorist and charged him $1 for every mile per hour over the speed limit.

In reviewing Newton's three laws, it is apparent that the analysis of momentum changes and forces is central to their application. In vehicular accidents, impact with a second vehicle usually causes changes in the direction and speed of both the involved vehicles. The analysis of the momentum and forces in such cases can provide valuable information about the accident, especially pre-impact speeds.

Sometimes when the available accident data are insufficient for other types of analyses to work, there is sufficient information to apply momentum methods to analyze the accident. Also, the application of momentum methods is an independent way of corroborating results from other types of analyses.

## 15.2  Basic Momentum Equations

The basic equation of engineering mechanics is the ever faithful:

$$F_A = m_A \, a_A \qquad \text{(i)}$$

where $F_A$ = the applied force on item A, $m_A$ = the mass of item A, and $a_A$ = the resulting acceleration of item A.

Since the acceleration "a" is equal to the time derivative of the velocity, Equation (i) can be rewritten as follows to more closely parallel Newton's second law.

$$F_A = (d/dt)(mv)_A \qquad \text{(ii)}$$

where $v$ = velocity.

The term "$(mv)$" is referred to as the momentum.

By rearranging the terms in Equation (ii), the following is obtained.

$$F_A(dt) = m_A(dv_A) \qquad \text{(iii)}$$

In the above equation, it is assumed that mass is constant during the time when the velocity changes. Thus, the mass derivative term, "$v(dm)$," is discarded.

The indefinite integration of both sides of Equation (iii), as done in Equation (iv), produces a term called the impulse. If a constant force is applied to a body for a certain time, then the impulse on that body is simply the force multiplied by the time.

The amount of impulse applied to a free body is equal to the change of momentum that results from its application.

$$I_A = f[F_A(dt)] = m_A[v_{A2} - v_{A1}] \tag{iv}$$

Thus, a force applied on the body, "$m_A$," for a time duration "t" produces an impulse, "$I_A$." This impulse causes a change in the momentum of the body, which in this case equals the change in velocity vectors of the body.

Applying Newton's third law now allows a second body to be introduced into this analysis. If a second body is what is applying the force "$F_A$" to the first body, then the force on the second body will be equal but opposite in direction to that applied on the first body.

$$F_B = -F_A \tag{v}$$

where $F_B$ = is the equal but opposite force applied on the second body.

Since the force applied on body A is equal and opposite to the force applied on body B, and since the forces on both bodies are applied for the same amount of time, then the impulse on body A is equal but opposite in direction to the impulse on body B.

$$-I_A = I_B \tag{vi}$$

$$-f[F_A(dt)] = f[F_B(dt)]$$

$$m_A[v_{A1} - v_{A2}] = m_B[v_{B2} - v_{B1}]$$

By rearranging the above terms, the equation for the conservation of linear momentum for a two-body impact is obtained.

$$m_A v_{A1} + m_B v_{B1} = m_A v_{A2} + m_B v_{B2} \tag{vii}$$

The left-hand side of Equation (vii) is the total linear momentum prior to impact, and the right-hand side is the total linear momentum after impact. Equation (vii) is the fundamental equation used in momentum analysis of vehicular accidents.

It should be noted that all the terms involving velocity, acceleration, etc. in the above equations are vectors and should be treated accordingly.

## 15.3  Properties of an Elastic Collision

Properly defined, a fully elastic collision between two bodies is one in which the deformation of each body obeys Hooke's law. Hooke's law states that within certain limits, the strain of a material is directly proportional to the applied stress causing the strain.

However, a more practical definition of an elastic collision between two bodies is: a collision in which it is observed that the two bodies return to the same shape they had prior to the collision; they do not significantly deform because of the collision.

The second definition is more practical than the first because it can be generally determined by visual inspection. If a vehicle has sustained little or no permanent deformation due to the collision, then the collision can be assumed to be elastic, barring any special impact energy absorption equipment that restores itself. Importantly, if the second definition is applied and the collision is deemed elastic, then the first definition can be applied to analyze the impact.

In an elastic collision between two bodies, there are four stages to the collision.

1. Contact is made between the two bodies.
2. Deformation occurs on both bodies according to Hooke's law reaching some maximum.
3. Recoil or restitution occurs as the stored "spring" energy in each body is released.
4. Separation between the two bodies occurs.

If each body in the collision reacts like a linearly elastic spring, then the following Hookian rule is applied:

$$F = -ks \qquad \text{(viii)}$$

where $k$ = spring constant, $s$ = amount of deformation of the body (a vector), and $F$ = the applied force.

According to Newton's third law, the force of each body acting on the other is equal in magnitude, but opposite in direction. Thus,

$$-F_A = F_B \qquad \text{(ix)}$$

$$k_A s_A = -k_B s_B$$

The ratio of deformation between the two bodies is then as follows:

$$s_A/s_B = -k_B/k_A \qquad \text{(x)}$$

Since the time duration in which the two bodies are in contact with one another is the same, then the following also holds.

$$-I_A/I_B = \{m_Av_{A1} - m_Av_{A2}\}/\{m_Bv_{B2} - m_Bv_{B1}\} = k_As_A/-k_Bs_B \qquad (xi)$$

$$[m_Av_{A1} - m_Av_{A2}]/k_A = s_A \qquad (xii)$$

$$[m_Bv_{B2} - m_Bv_{B1}]/k_B = s_B$$

A review of Equations (x), (xi), and (xii) reveals two points. First, if the elastic constant "k" for both bodies is the same, then the resulting elastic deformation will be the same for both bodies. Secondly, the elastic deformation of each body is a function of its own change of momentum and elastic constant.

## 15.4   Coefficient of Restitution

During a two-body collision, there will be a point where the deformation of each body is at a maximum. When this point is reached, the two bodies will also have the same velocity. This common velocity is calculated by the following using the pre-impact momentum sum.

$$[m_A + m_B]u = m_Av_{A1} + m_Bv_{B1} \qquad (xiii)$$

$$u = [m_Av_{A1} + m_Bv_{B1}]/[m_A + m_B]$$

where u = the velocity at which both bodies are momentarily moving together.

This velocity can also be calculated from the momentum sum after the collision.

$$[m_A + m_B]u = m_Av_{A2} + m_Bv_{B2} \qquad (xiv)$$

$$u = [m_Av_{A2} + m_Bv_{B2}]/[m_A + m_B]$$

When the two bodies momentarily have the same velocity "u," the portion of the collision that occurs before it is considered the deformation portion of the collision, and the portion after it is considered the restitution portion. Therefore, the impulse for the deformation portion of the collision can be calculated using the definition of impulse given in Equation (iv).

$$I_{def} = m_A[v_{A1} - u] = m_B[u - v_{B1}] \qquad (xv)$$

where $I_{def}$ = impulse during the deformation portion of the collision.

Likewise, the impulse for the restitution portion of the collision can be calculated from the definition of impulse given in Equation (iv).

$$I_{res} = m_A[u - v_{A2}] = m_B[v_{B2} - u] \qquad \text{(xvi)}$$

The ratio of the restitution impulse to the deformation impulse is called the coefficient of restitution, "$\varepsilon$."

$$\varepsilon = I_{res}/I_{def} = [v_{B2} - v_{A2}]/[v_{A1} - v_{B1}] \qquad \text{(xvii)}$$

It should be noted that the collisions being considered are assumed to be central collisions. If a collision is oblique, then the component vectors have to be separated. Those component vectors that are involved in the central collision obey the relations for the coefficient of restitution as noted above. Those components that are not involved in the central collision are handled separately. The latter vector components are often involved in causing rotation or doing frictional work such as sideswiping.

## 15.5   Properties of a Plastic Collision

In a nonelastic or plastic collision between two bodies, there are also four basic stages of the collision:

1. Contact is made between the two bodies.
2. Deformation occurs on both bodies reaching some maximum amount.
3. Some recoil or restitution may occur as any stored "spring" energy in either body is released.
4. Separation may occur between the two bodies if recoil is sufficient.

The main differences between the four stages of a plastic collision vs. an elastic one are as follows:

1. The deformation does not follow Hooke's law, that is, it does not elastically deform. It is likely that some elastic deformation will initially occur, but it may be insignificant as compared to the amount of plastic deformation that occurs.
2. There may not be any significant recoil or restitution.
3. The two bodies may not have sufficient recoil to separate; they may stay "stuck" to one another after impact and move as a unit.

A perfectly elastic collision has a coefficient of restitution of 1.0, and a perfectly plastic collision has a coefficient of restitution of zero. Between those two extremes are combinations of the two, where both some elastic and some plastic deformation occurs. Obviously, the lower the magnitude of "ε," the less elastic deformation there is with respect to the plastic deformation.

Usually, any elastic deformation occurring in a collision takes place in the early part of contact between the two bodies. The plastic deformation then takes place after the elastic deformation range has been exhausted, and stops when the two bodies reach the "u" velocity [see Equation (xiii)]. Consequently, as the severity of the collision increases, the "ε" value approaches zero. Likewise, as the severity of the collision decreases, the "ε" value increases. Generally, the severity of the collision is proportional to the relative closing speed of the two vehicles.

The following coefficients of restitution, or "ε" values, have been found to occur in passenger vehicles either colliding with a fixed barrier or its equivalent. An example of a fixed barrier is a well-built brick wall that does not move or become damaged when it is impacted. Letting the subscripts "A" denote a vehicle and "B" denote a fixed barrier, then the coefficient of restitution of a fixed barrier is equal to the following:

$$\varepsilon = [v_{B2} - v_{A2}]/[v_{A1} - v_{B1}] = -v_{A2}/v_{A1},$$

because $v_{B2} = v_{B1} = 0$.

When forward or rearward velocities are 2.5 mph or less, "ε" may be considered to be 0.9 to 1.0. This is because U.S. federal law, as embodied in the National Traffic and Motor Vehicle Safety Act of 1966 with amendments, requires that bumpers on cars made after 1977 be able to withstand a 2.5 mph front or rear impact with a fixed barrier without significant damage occurring to the vehicle (49 CFR 581, 1987 edition).

For a time prior to 1978, vehicle bumpers were required by law to be able to withstand a 5 mph impact with a fixed barrier. However, this was rescinded in 1977, and the current, lower standard was then established. Prior to 1966, there were no legally mandated bumper standards. Thus, no similar "ε" value can be categorically assigned to vehicle bumpers made prior to 1966.

From experience, it has been found that at impact speeds of about 25 mph, an "ε" value of 0.2 is typical for most passenger cars. At impact speeds of about 35 mph, an "ε" value of 0.1 is typical, and at speeds exceeding 50 mph, an "ε" value of about 0.004 is typical.

Interestingly, when the above experience-derived values of "ε" are plotted against the vehicle impact speed, they tend to closely follow the following equation:

$$\varepsilon = e^{-0.065s} \tag{xviii}$$

where ε = the coefficient of restitution, s = the vehicle's speed in mph, and e = logarithmic "e," i.e., e = 2.718.

**Plate 15.1** Side impact. Note bumper imprint in crush damage along door.

## 15.6 Analysis of Forces during a Fixed Barrier Impact

In experimental tests, it has been found that when a 1980 Chevy Citation weighing 3130 lb was driven into a fixed barrier at 48 mph, the front end crushed 40.4 in deep across the front of the car.

Using Equation (xviii), it is found that this impact velocity would have a coefficient of restitution of about 0.044. Using this "e" value, the calculated rebound velocity would have been about 2 mph. For all practical purposes, this impact would be deemed a plastic collision.

The pre-impact momentum of the car would have been:

$$[(3130 \text{ lbf})/(32.17 \text{ ft/sec}^2)][(70.4 \text{ ft/sec})] = 6850 \text{ lbf-sec}.$$

The postimpact momentum of the car would have been:

$$[3130 \text{ lbf})/(32.17 \text{ ft/sec}^2)][3 \text{ ft/sec}] = 292 \text{ lbf-sec}.$$

The net change in momentum due to the impact was then 6558 lbf-sec.

After contact with the wall, the car came to a stop in 40.4 inches. Assuming that the velocity of the vehicle decreases linearity (i.e., constant deceleration), then the equation for the speed of the car during impact would be as follows:

$$70.4 \text{ ft/sec} - (1.743 \text{ ft/in-sec})x = V$$

where x = the amount of crush that has occurred, and V = the velocity at that point during the impact.

If the average velocity during the impact is simply "(1/2)(70.4 ft/sec)," then the time required to stop in 40.4 inches would be 0.0956 seconds. Thus, the actual impact is literally over in less than the blink of an eye.

Assuming that the force applied by the wall to stop the car was constant, then the impulse during the impact would be:

$$F(0.0956 \text{ sec}) = 6558 \text{ lbf-sec.}$$

$$F = 68{,}598 \text{ lbf.}$$

If the force would have been applied more or less evenly across the front of the car, and the front is assumed to be about 72 inches wide and about 30 inches in height, then an equivalent average pressure on the wall to resist the car impact would have been 32 psi.

The deceleration rate during the impact is estimated as follows:

$$d = (1/2)at^2$$

$$d = 40.4 \text{ in } t = 0.0956 \text{ sec}$$

$$a = 8841 \text{ in/sec}^2 = 737 \text{ ft/sec}^2 = 22.3 \text{ g's}$$

From this simple analysis, a few items are noted. First, if the crush depth is increased, then the deceleration can be decreased and the time of impact increased. For example, if the crush depth had been 50 in instead of the 40.4 in, then the time of impact would have been 0.118 sec, and the deceleration would have been 18.6 g's instead of 22.3 g's. The 9.6 in increase in crush depth, a 24% increase, would have reduced the deceleration by 17%. This is significant. The degree of injury sustained in a vehicle impact by the occupants is a direct function of the impact deceleration.

## 15.7   Energy Losses and "ε"

When two bodies collide, some of the total kinetic energy is dissipated in plastic deformation. The amount of energy loss is given by the following.

$$(1/2)(m_A v_{A1}^2) + (1/2)(m_B v_{B1}^2) = E_{Loss} + (1/2)(m_A v_{A2}^2)$$
$$+ (1/2)(m_B v_{B2}^2) \tag{xix}$$

$$E_{Loss} = (1/2)m_A(v_{A1}^2 - v_{A2}^2) + (1/2)m_B(v_{B1}^2 - v_{B2}^2)$$

In the above equation, the "$v^2$" terms are considered either vector dot products, or simply the absolute values of the velocity vectors squared.

The coefficient of restitution, "$\varepsilon$," is given by the following:

$$\varepsilon = I_{res}/I_{def} = [v_{B2} - v_{A2}]/[v_{A1} - v_{B1}].$$

Rearranging the above relation for "$\varepsilon$" to favor postimpact velocity terms gives the following:

$$v_{B2} = \varepsilon[v_{A1} - v_{B1}] + v_{A2} \quad \text{and} \tag{xx}$$

$$v_{A2} = v_{B2} - \varepsilon[v_{A1} - v_{B1}].$$

Using the above value for "$v_{A2}$," the following is generated.

$$(v_{A1}^2 - v_{A2}^2) = v_{A1}^2 - [v_{B2}^2 - 2(\varepsilon)(v_{A1} - v_{B1})(v_{B2}) + \varepsilon^2(v_{A1} - v_{B1})^2] \tag{xxi}$$
$$(v_{A1}^2 - v_{A2}^2) = v_{A1}^2(1 - \varepsilon^2) - v_{B2}^2 + v_{A1}v_{B2}(2\varepsilon) -$$
$$v_{B1}v_{B2}(2\varepsilon) + v_{A1}v_{B1}(2\varepsilon^2) - v_{B1}^2(\varepsilon^2)$$

Similarly, using the value in Equation (xx) for "$v_{B2}$," the following is generated.

$$(v_{B1}^2 - v_{B2}^2) = v_{B1}^2(1 - \varepsilon^2) - v_{A2}^2 + v_{B1}v_{A2}(2\varepsilon) - \tag{xxii}$$
$$v_{A1}v_{A2}(2\varepsilon) + v_{A1}v_{B1}(2\varepsilon^2) - v_{A1}^2(\varepsilon^2)$$

Substituting the above velocity difference terms, Equations (xxi) and (xxii), into Equation (xix) gives the following for the energy loss

$$E_{Loss} = (1/2)(m_A)[v_{A1}^2(1 - \varepsilon^2) - v_{B2}^2 + v_{A1}v_{B2}(2\varepsilon) - v_{B1}v_{B2}(2\varepsilon) \tag{xxiii}$$
$$+ v_{A1}v_{B1}(2\varepsilon^2) - v_{B1}^2(\varepsilon^2)] +$$
$$(1/2)(m_B)[v_{B1}^2(1 - \varepsilon^2) - v_{A2}^2 + v_{B1}v_{A2}(2\varepsilon) - v_{A1}v_{A2}(2\varepsilon)$$
$$+ v_{A1}v_{B1}(2\varepsilon^2) - v_{A1}^2(\varepsilon^2)]$$

Rearranging Equation (xxiii) to sort out the "$\varepsilon$" terms, and substituting to favor postimpact velocities, gives the following expression:

$$E_{Loss} = [(1/\varepsilon^2) - 1][(m_Am_B)/2(m_A + m_B)][v_{B2} - v_{A2}]^2 \tag{xxiv}$$

Similarly, if pre-impact terms are favored, the following expression can be derived:

$$E_{Loss} = [(1 - \varepsilon^2)/2][(m_A m_B)/(m_A + m_B)][v_{A1} - v_{B1}]^2 \qquad (xxv)$$

When $\varepsilon = 1$, both Equations (xxiv) and (xxv) become zero, indicating again that a fully elastic collision is one in which there is no energy loss. When $\varepsilon = 0$, Equation (xxx) is indeterminate due to division by zero. However, Equation (xxv) is solvable.

To be a half-plastic and half-elastic collision, "$\varepsilon$" must equal 0.707. In a passenger car colliding with a fixed barrier, such a half-plastic and half-elastic collision would occur at a speed of about 5.33 mph. A collision with only 10% elastic energy would have an "$\varepsilon$" of 0.316 and would occur at a speed of about 17.7 mph. It is readily apparent, then, that in most vehicular collisions, the predominant deformation mode is plastic. It is also apparent that in most vehicular collisions, recoil is a small effect.

## 15.8   Center of Gravity

In all of the foregoing analyses, it has been tacitly assumed that the masses of the bodies were point masses. They were considered to be like minute billiard balls, bouncing off one another. When such equations are applied to vehicles involved in collisions, they then apply to the centers of mass of the vehicles.

Thus, all of the previous equations were based on the premise of straight central impact, where the line of action between the bodies was through the center of gravity of each body.

The distribution of load on a vehicle is not necessarily such that the center of mass is at the geometric center of the vehicle. If "Q" is the load carried by the rear axle and "R" the load carried by front axle, then the center of gravity (C.G.) in a two-axle vehicle will be found as follows.

$$a = Qb/(Q + R) = Qb/W = b/(1 + R/Q) \qquad (xxvi)$$

where a = distance from the front axle to the center of gravity of the vehicle, b = distance between axles, and W = total weight of the vehicle.

For example, in a car with a 106-in wheel base and 60% of the curb weight carried by the front axle, the C.G. is located 42.4 in away from the front axle in a direction toward the rear of the car. Alternately, the C.G. is located 13.6 in away from the midpoint of the wheelbase toward the front of the car.

The above, of course, assumes that the C.G. is found on a line that bisects the length of the vehicle. For most purposes, this will be sufficient. If, however, there is some type of asymmetrical load on the vehicle, perhaps a load shift during the collision, it might be necessary to determine the C.G. using the point mass method. In this method, the weight carried under each tire is determined, and the coordinates of the center of gravity are determined by the following:

$$x_c = [\Sigma m_i x_i]/[\Sigma m_i] \quad y_c = [\Sigma m_i y_i]/[\Sigma m_i] \qquad \text{(xxvii)}$$

where $x_c$ and $y_c$ are the coordinates of the C.G.

Typically, the point of origin from which the "x" and "y" distances of the mass points are measured is one of the tire contact points. This causes the "x" and "y" values for that tire to be zero, thus simplifying some of the calculations.

Sometimes it is important also to know the position of the C.G. with respect to the "z" axis, that is, the vertical axis. However, while the center of gravity with respect to the horizontal plane and "x and y" coordinates is relatively easy to locate, the "z" coordinate requires more effort.

Perhaps the simplest, although not the easiest method of determining the C.G. with respect to all three coordinates, utilizes a crane. The vehicle is lifted with a three-point or four-point harness and is suspended by a single cable attached to the harness. Preferably, the vehicle should be somewhat tilted or otherwise not parallel with the ground. While suspended, a photograph of the vehicle is taken from a set position. A second lift is then similarly done except the vehicle is tilted differently from the first lift. With the vehicle facing the same side to the camera, a second photograph is taken.

The single cable that holds up the vehicle acts as a plumb line in the photographs. The line of action of the cable runs directly through the C.G. of the vehicle. Thus, the line of this cable should be graphically extended so that it runs down through the image of the vehicle. When the two photographs are superimposed, the point where the two plumb lines intersect on the vehicle is the three-coordinate C.G. If the photographs were taken with the same side facing the camera, it is an easy matter to locate this point on the photographs and then determine where it is on the car itself.

Unfortunately, an apparatus for lifting cars and taking their pictures is not always handy, or cheap. Thus, an alternate method may need to be employed that has a lessor equipment requirement.

Knowing that the "z" component of the C.G. will be directly above the "x and y" components, then the "z" component can be found by jacking up the vehicle with a scale under the jack to read the load. The calculations are easiest when the jack is placed in line with either the longitudinal or transverse axes of the C.G. By jacking up the vehicle until the initial load on the jack significantly reduces, and noting the angle of the vehicle, the "z" component can be found by simple statics.

If a person continued to jack the car up until it is balanced on the two wheels making contact, the C.G. would be directly over the axis of the two wheels, and the load on the jack would be zero.

By taking a photograph with the view parallel to the axis of the contacting wheels, and drawing a plumb line on the photograph that originates at the axis of the contacting wheels, the line would pass through the C.G. Assuming that the "z" component lines directly above the "x and y" C.G. coordinates, the intersection of the drawn plumb line with the line that bisects the vehicle is the C.G. The "z" component then can be scaled from the photograph.

As a rule of thumb, most late model passenger cars have a "z" coordinate located about 2 feet from the ground. Regular sized pickup trucks, e.g., Ford 100, have a "z" coordinate located about 2.5 feet above the ground.

## 15.9 Moment of Inertia

The moment of inertia of a vehicle can be determined experimentally by suspending the vehicle from a single cable and variously spinning it and measuring the rate of rotation. This experimental method takes advantage of the relation:

$$F \times r = I\alpha \qquad \text{(xxviii)}$$

where $F$ = the applied force, $r$ = distance from the C.G., $I$ = moment of inertia, and $\alpha$ = angular acceleration.

In this case, the cross product between two vectors is used, which is denoted by the "×."

In general a constant force is applied at one of the far ends of the vehicle, perpendicular to the lever arm between the point of application and C.G. of the vehicle. The final angular velocity, "$\omega_{f}$," is measured and its value is divided by the time the force was applied to obtain "$\alpha$," the angular acceleration. Thus, the moment of inertia is found by the following:

$$I = [F \times r]/[\omega_{f}t]. \qquad \text{(xxix)}$$

In Section 15.8 the method for finding the center of gravity by suspending the vehicle from a cable was discussed. Since the additional effort necessary to find "I" is rather small in comparison to the trouble of the setup, it is recommended that they both be done at the same time. However, if a person is unable to undertake hoisting a car at the end of a cable and spinning it, there are several methods that can be used to provide reasonable estimates of "I."

The first method assumes that the distribution of mass in the vehicle is more or less even over the whole projected x-y area and that the center of gravity is near the geometric center.

$$m_{areal} = m/A \qquad\qquad (xxx)$$

where A = area of the vehicle projected onto the horizontal plane, m = mass of the vehicle, and $m_{areal}$ = mass per unit area.

For a rectangular object the moment of inertia is:

$$I_{xx} = m_{areal}bh^3/12 = m_{areal}Ah^2/12 = mh^2/12 \qquad (xxxi)$$

where b = width, h = length, and $I_{xx}$ = moment of inertia about x-x axis.

Using the above method, for a 1988 Buick Century, which is 189 in long and 69 in wide, with a curb weight of 2950 lbs, its moment of inertia would be as follows:

$$m = 2950 \text{ lbf}/(386 \text{ in/sec}^2) = 7.642 \text{ lb-sec}^2/\text{in}$$

$$I_{xx} = (7.642 \text{ lb-sec}^2/\text{in})(189 \text{ in})^2/12 = 22,750 \text{ lb-sec}^2\text{-in.}$$

A second method considers the loads carried by the axles as point masses located half the wheelbase from geometric center.

$$I_{xx} = (x)m(l/2)^2 + (1-x)m(l/2)^2 = ml^2/4 \qquad (xxxii)$$

where l = length of wheel base, x = fraction of mass carried by front axle, and (1-x) = fraction of mass carried by rear axle.

Using the same 1988 Buick Century, whose wheelbase is 105 in, the "I" value is calculated as follows.

$$I_{xx} = (0.65)(7.642 \text{ lb-sec}^2/\text{in})(105 \text{ in}/2)^2 + (0.35)(\text{lb-sec}^2\text{-in})(105 \text{ in}/2)^2$$

$$I_{xx} = 21,063 \text{ lb-sec}^2\text{-in.}$$

A third method assumes that the engine is primarily responsible for the uneven distribution of load between axles. Thus, the mass of the engine is deducted from the total vehicle mass and an "I" value is computed for the "engine-less" vehicle. The engine mass is then added in as a point mass.

$$I_{xx} = 2(1-x)mh^2/12 + (2x-1)m(l/2)^2 \qquad \text{(xxxiii)}$$

where x = fraction of load on front axle, h = overall length of vehicle, l = wheel base length, and m = mass of vehicle.

Again, using the 1988 Buick Century, the following "I" value is computed.

$$I_{xx} = 2(0.35)(7.642 \text{ lb-sec}^2\text{-in})(189 \text{ in})^2/12 +$$
$$(0.30)(7.642 \text{ lb-sec}^2\text{-in})(105 \text{ in}/2)^2$$

$$I_{xx} = 15{,}923 \text{ lb-sec}^2\text{-in} + 6{,}319 \text{ lb-sec}^2\text{-in} = 22{,}242 \text{ lb-sec}^2\text{-in.}$$

When a vehicle is carrying an unusually heavy load in the trunk or elsewhere, the additional mass can be figured in the computation for "I" in the manner used for the engine mass.

Of the three methods, method three seems to work best for passenger cars, where the mass distribution is fairly even except for the engine. The first method seems to work best for truck trailers and vehicles where the general mass distribution is relatively even and there are no large concentrations of mass. Method two works well when there are heavy loads over the axles, like in a loaded pickup truck (engine over front axle, load over rear). However, as demonstrated by the 1988 Buick Century calculations, all three methods give similar results.

One additional item to note is that the "I" values discussed above assumed that the x-x axis was in the geometric center of the vehicle. As such, it was assumed that rotation would be about the center of the vehicle. If, however, rotation occurs where the center of rotation is about the rear axle or the front axle, then the parallel axis theorem will have to be applied to properly adjust the "I" to the point of rotation. The parallel axis theorem is given below.

$$I_{ii} = I_{xx} + mr^2 \qquad \text{(xxxiv)}$$

where $I_{ii}$ = moment of inertia about axis i-i, r = distance between axis i-i and axis x-x, and m = mass of the vehicle.

## 15.10   Torque

Again, starting with the basic equation of engineering:

$$F = ma.$$

Let us now apply the force not to an idealized ball or point of mass, but to a stiff object with some length. Further, let the force be applied at a point so that the line of action of the force does not pass through the center of mass. In that case, then the following occurs.

$$T = F \times r = (ma) \times r = m(a \times r) \qquad \text{(xxxv)}$$

where $T$ = torque, $r$ = vector distance from center of mass, $F$ = force applied at "$r$," $m$ = total mass of item, and $a$ = acceleration in the direction of the force.

As before, the "$\times$" denotes a cross-product vector multiplication.

The rotation of a body around its center of mass can be constant or variable. When the rotation is constant, the rotation rate is usually given in radians per second. When the rotation rate varies, the rate of change is quantified by the angular acceleration, which is given in radians per second. The relationship between angular velocity, "$w$," and angular acceleration, "$\alpha$," is given as follows.

$$\alpha = \Delta\omega/\Delta t \qquad \text{(xxxvi)}$$

where $\Delta$ = an incremental change and $t$ = time.

The tangential velocity at distance "$r$" from the center of mass is given by the following:

$$v_T = \omega \times r. \qquad \text{(xxxvii)}$$

Similarly, the tangential acceleration at distance "$r$" from the center of mass is given by the following:

$$\alpha_T = (d/dt)(\omega \times r) = \alpha \times r. \qquad \text{(xxxviii)}$$

Substituting that back into Equation (xxxv) gives the following:

$$T = F \times r = m(\alpha \times r) \times r \qquad \text{(xxxix)}$$

Collecting terms and simplifying gives:

$$T = F \times r = (mr^2)\alpha = I_{xx}\alpha \qquad \text{(xl)}$$

where $I_{xx}$ = the moment of inertia of a point mass about the axis of rotation.

If the mass is distributed over an area instead of at a point, the moment of inertia about the center of mass is used, provided the rotation is about the center of mass. If the rotation is about some other axis, then the parallel axis theorem is applied [Equation (xl)] to determine the moment of inertia about that particular axis.

## 15.11  Angular Momentum Equations

Like its linear acceleration counterpart, rotational acceleration is the time derivative of rotational velocity. Like its linear counterpart, force, torque is also equal to a time derivative of velocity.

$$T = I_{xx}\,(d/dt)(\omega) = (mr^2)\,(d/dt)(\omega) = m\,[(d/dt)(\omega)] \times r \times r \qquad \text{(xli)}$$

In the above, it is assumed that the "$I_{xx}$" is a point mass at a distance "r" from the axis of rotation.

Since $(d/dt)(\omega) \times r = (d/dt)(v_T)$, then

$$T\,(dt) = I_{xx}\,(d\omega) = m(dv_T) \times r \qquad \text{(xlii)}$$

where $v_T$ = tangential velocity at point "r" on the body.

The indefinite integration of both sides of Equation (xlii), as shown below, produces a term called the rotational impulse.

$$R = f[I_{xx}(d\omega)] = f[m(dv_T) \times r] \qquad \text{(xliii)}$$

$$R = I_{xx}[\omega_2 - \omega_1] = [m(v_{T2} - v_{T1}) \times r]$$

where R = the rotational impulse.

The term "$I\omega$" is referred to as the angular momentum.

Equation (xliii) is enormously useful in dealing with vehicular collisions that produce rotation. The left side of the equation indicates that the rotational impulse is simply the change in angular momentum. The right side of the equation indicates that the rotational pulse is equal to the change in linear momentum at point "r" about the rotational axis.

To demonstrate the usefulness of Equation (xliii) consider the following collision. Car "A" is traveling due south. Car "B" is traveling due east, but is stopped at an intersection waiting for the light to change. Car "B" creeps forward slowly into the intersection while it is waiting for the light to change. Car "A" drives through the intersection and impacts Car "B" at its front left corner, causing it to spin or yaw about the rear wheels. Car "A" does not spin, but is stopped by the collision at the point of impact.

The forward linear momentum of Car "A" has been transferred to Car "B" causing it to spin. The momentum of Car "A" was "$m_A v_A$." This momentum was then transferred to Car "B" at a point "r" from the axis of the rear wheels. The relationship of the spin, or angular momentum, of Car "B" to the loss of the forward linear momentum of Car "A" is given by:

$$(I_{Bxx} + m_B r^2)\omega = m_A v_A \times r$$

where r = distance from the rear axle to the front left corner of Car "B," $I_{Bxx}$ = moment of inertia about the center of mass of Car "B," $m_A v_A$ = forward moment of inertia of Car "A," and $\omega$ = spin or yaw of Car "B" just after impact is completed.

It should be noted that the speed "$v_A$" of Car "A" is tangential to the y-y or long axis of Car "B." If the impact had been at an angle other than 90°, it would be necessary to take the normal component of the velocity vector with respect to Car B's y-y axis. Alternately, this is expressed as:

$$v_A \times [r/(|r|)] = |v_A| \sin \phi = v_{AT}$$

where $\phi$ = the acute angle between the axis y-y of Car "B" and the velocity vector of Car "A," and $[r/(|r|)]$ = the unit directional vector "r".

## 15.12 Solution of Velocities Using the Coefficient of Restitution

As already noted, Equation (xvii) defines the coefficient of restitution as:

$$\varepsilon = I_{res}/I_{def} = [v_{B2} - v_{A2}]/[v_{A1} - v_{B1}].$$

If the above expression for "$\varepsilon$" is alternately solved for "$v_{A1}$" and "$v_{B1}$," and the results are then substituted back into the conservation of momentum equation, Equation (vi), then the following equations are obtained:

$$v_{A1} = [(\varepsilon m_A - m_B)/\varepsilon(m_A + m_B)]v_{A2} + \qquad \text{(xliv)}$$
$$[(\varepsilon m_B + m_B)/\varepsilon(m_A + m_B)]v_{B2}$$

$$v_{B1} = [(\varepsilon m_A + m_A)/\varepsilon(m_A + m_B)]v_{A2} + \qquad \text{(xlv)}$$
$$[(\varepsilon m_B - m_A)/\varepsilon(m_A + m_B)]v_{B2.}$$

Equations (xliv) and (xlv) indicate that if the velocities just after collision are known, and the "$\varepsilon$" value of the collision is known, then the individual velocities of each body prior to impact can be calculated. Inspection of these equations finds that both expressions are linear vector equations; both "$v_{A1}$" and "$v_{B1}$" are resultants of the addition of two vector quantities. This solution form lends itself well to graphic solution.

An interesting point is worth noting about Equations (xliv) and (xlv). If the two equations are written in the following form:

$$v_{A1} = [a]v_{A2} + [b]v_{B2} \qquad\qquad (xlvi)$$

$$v_{B1} = [c]v_{A2} + [d]v_{B2}$$

where a, b, c, and d are scalars, then the following relation holds between the scalars,

$$a + b = c + d = 1. \qquad\qquad (xlvii)$$

This relationship can be very useful in checking the computations of the scalars, and can be used as a calculation shortcut. Using this relationship, only one scalar in each equation needs to be computed. The other scalar can be quickly determined by inspection. The relationship can also save programming lines if the equations are programmed for solution on a computer.

If $\varepsilon = 1$, a fully elastic collision, it is seen that Equations (xliv) and (xlv) reduce to the following:

$$v_{A1} = [(m_A - m_B)/(m_A + m_B)]v_{A2} + [(2m_B)/(m_A + m_B)]v_{B2} \quad (xlviii)$$

$$v_{B1} = [(2m_A)/(m_A + m_B)]v_{A2} + [(m_B - m_A)/(m_A + m_B)]v_{B2} \qquad (xlix)$$

In this special case where the collision is wholly elastic, the scalar coefficients in Equation (xlvi) not only satisfy $a + b = c + d = 1$, but also the following:

$$a = -d \qquad\qquad (1)$$

$$b + c = 2$$

$$c - a = 1$$

$$b - d = 1$$

If $\varepsilon = 0$, a fully plastic collision, it is seen that Equations (xliv) and (xlv) are undefined due to the division by zero. However, when $\varepsilon = 1$, it is automatically known that $v_{A2} = v_{B2} = u$.

In a similar fashion, if Equation (xvii), the relation for the coefficient of friction, is alternately solved for "$v_{A2}$" and "$v_{B2}$," and the results are substituted back into the equation for the conservation of momentum, Equation (vii), then the following is obtained:

$$v_{A2} = [(m_A - \varepsilon m_B)/(m_A + m_B)]v_{A1} + [(m_B + \varepsilon m_B)/(m_A + m_B)]v_{B1} \quad \text{(li)}$$

$$v_{B2} = [(m_A + \varepsilon m_A)/(m_A + m_B)]v_{A1} + [(m_B - \varepsilon m_A)/(m_A + m_B)]v_{B1} \quad \text{(lii)}$$

Equations (li) and (lii) indicate that if the velocities just prior to collision are known, and if the "$\varepsilon$" value for the collision is known, then the individual velocities after impact can be calculated. This is also a very useful analytical tool.

The velocities of the two vehicles prior to impact, as reported by the drivers or witnesses, can be substituted into "$v_{A1}$" and "$V_{B1}$." Then, the after-impact velocities can be solved. The calculated after-impact velocities can then be compared to the actual accident scene information. If they match reasonably well, the drivers' accounts are validated. If they do not match, then one or both of the claimed pre-impact velocities are incorrect.

As before, these equations can also be arranged in the form:

$$v_{A2} = [w]v_{A1} + [x]v_{B1} \quad \text{(liii)}$$

$$v_{B2} = [y]v_{A1} + [z]v_{B1}.$$

As before, it is found that the following relationship holds for the scalars:

$$w + x = y + z = 1. \quad \text{(liv)}$$

In a fully elastic collision, Equations (li) and (lii) simplify to the following:

$$v_{A2} = [(m_A - m_B)/(m_A + m_B)]v_{A1} + [(2m_B)/(m_A + m_B)]v_{B1} \quad \text{(lv)}$$

$$v_{B2} = [(2m_A)/(m_A + m_B)]v_{A1} + [(m_B - m_A)/(m_A + m_B)]v_{B1} \quad \text{(lvi)}$$

In a fully elastic collision, it is seen that in addition to Equation (liv), the additional relations hold between the scalars:

$$w = -z \qquad \text{(lvii)}$$

$$x + y = 2$$

$$y - w = 1$$

$$x - z = 1.$$

The reader will undoubtedly note the similarity of the relations in Equation (lvii) with those of Equation (l).

Similarly, when the collision is fully plastic and $\varepsilon = 0$, the following holds:

$$v_{A2} = v_{B2} = [(m_A)/(m_A + m_B)]v_{A1} + [(m_B)/(m_A + m_B)]v_{B1} \qquad \text{(lviii)}$$

## 15.13  Estimation of a Collision Coefficient of Restitution from Fixed Barrier Data

When a vehicle impacts a fixed barrier, it is presumed that the fixed barrier does not crush or otherwise absorb any significant amount of kinetic energy from the vehicle. The fixed barrier pushes back against the vehicle during impact with a force equal to that being applied by the vehicle. In essence, since no work can be done on the fixed barrier, the vehicle does work on itself.

Earlier, Equation (xxv) was derived to show the relationship between energy dissipated by the collision and the coefficient of restitution. For convenience, the equation is shown again below.

$$E_{Loss} = [(1 - \varepsilon^2)/2][(m_A m_B)/(m_A + m_B)][v_{A1} - v_{B1}]^2 \qquad \text{(xxv)}$$

If it is assumed that a fixed barrier essentially has an "infinite" or immovable mass, and the wall has no initial speed prior to the collision, then the equation reduces to the following:

$$E_{Loss} = [(1 - \varepsilon^2)/2][m_A][v_{A1}^2]. \qquad \text{(lix)}$$

When $\varepsilon = 0$, a fully plastic collision, all of the initial kinetic energy is dissipated as work in crushing the vehicle. When $\varepsilon = 1.0$, none of the initial kinetic energy is dissipated as work; all of it goes into the rebound of the vehicle.

While not exactly contributing to the point at hand, the following aside may be interesting to the reader. By substituting Equation (xviii) into (lix), a function for collision energy loss against a fixed-barrier collision can be derived that is solely dependent upon initial speed:

$$E_{Loss} = [(1 - e^{-0.130s})/2][m_A][s_{A1}{}^2] \qquad \text{(lx)}$$

where s = velocity of Vehicle "A" given in mph and e = 2.718 etc.

An alternate version of the above equation where velocity is given in ft/sec is as follows:

$$E_{Loss} = [(1 - e^{-0.089v})/2][m_A][v_{A1}{}^2]. \qquad \text{(lxi)}$$

However, getting back to the point of this section, if it is assumed that crush caused by collision with the fixed barrier is proportional to the change in kinetic energy of the vehicle, then:

$$s_A \cdot F = (1/2)m_A(\Delta v_A{}^2) = U = E_{Loss} \qquad \text{(lxii)}$$

where U = work done in crushing the vehicle, $s_A$ = amount of crush in the vehicle, F = the average force applied to the vehicle during collision with the fixed barrier, and $\Delta v_A{}^2 = v_{A2}{}^2 - v_{A1}{}^2$.

When two vehicles collide head on, the total work done in crushing is as follows:

$$U = (1/2)m_A(\Delta v_A{}^2) + (1/2)m_B(\Delta v_B{}^2) = s_A \cdot F + s_B \cdot F. \qquad \text{(lxiii)}$$

The force "F" applied between the two vehicles during collision is the same in magnitude but opposite in sign.

Now, if in the above equation, Vehicle "B" is sitting still with respect to Vehicle "A" (i.e., $\Delta v_B{}^2 = 0$), then the crush damage on both vehicles would be due to the action of Vehicle "A" alone. As can be seen in comparing Equations (lxii) and (lxiii), this means that the crush on Vehicle "A" in colliding with Vehicle "B" would be about half that of Vehicle "A" colliding with a fixed barrier at the same speed.

$$1/2 = v_{Aeq}{}^2/v_{A1}{}^2 \quad \text{or} \quad v_{Aeq} = (0.707)v_{A1} \qquad \text{(lxiv)}$$

where $v_{Aeq}$ = impact velocity of "A" with a fixed barrier and $v_{A1}$ = impact velocity of "A" with Vehicle "B," which is sitting still.

Thus, the coefficient of restitution for the above collision situation can be estimated by converting "$(0.707)v_{A1}$" into mph units, and substituting into Equation (xviii).

In a similar fashion, equivalent fixed-barrier impact velocities can be derived for other situations. For example, if the two vehicles have about the same mass and collide with the same velocity head on, then the equivalent fixed-barrier impact velocity is equal to the actual impact velocity.

# 15.14   Discussion of Coefficient of Restitution Methods

In general, the pre-impact velocities in a two-vehicle collision can be solved by proceeding as follows:

1. A reasonable "guess" as to the velocities of the vehicles is made. It is important to note that the "guess" is for the central impact velocities. If the collision was oblique, then it may be necessary to separate the central velocity vector components and handle the others separately.
2. A tentative value for "ε" is calculated from the first guess.
3. Using the tentative value for "ε," Equations (xliv) and (xlv) are solved.
4. A new value of "ε" is calculated based on the outcome of Equations (xliv) and (xlv).
5. If the new value of "ε" matches the tentative value of "e" first assumed, and the calculated velocities reasonably match the "guessed" values, then a reasonable solution has been obtained.
6. If the variance is too great, then a new "guess" of the velocities is made and the process is repeated until a reasonable convergence occurs.

However, in solving pre-impact velocities using the above method, Equations (xliv) and (xlv) become extremely sensitive to small variations in the value of "ε" when "ε" is very small. Of course, when "e" is small, the collision is mostly plastic and has occurred at a relatively high speed.

Thus, when "ε" is very small and subject to imprecision, an alternate approach may be advisable.

1. Using Equations (li) and (lii), the pre-impact velocities are assumed, or taken to be the values as stated by the drivers or witnesses.
2. An "ε" value is determined from equivalent barrier impact velocity computations, or simply assumed.
3. The postimpact velocities are calculated.
4. From the actual police report data, the postimpact velocities are determined. These velocity vectors are then compared to the calculated velocity vectors. If they match reasonably, a satisfactory solution has been found.
5. If the postimpact velocities do not match reasonably, new preimpact velocity vectors are assumed and the process is repeated until convergence is achieved.

## Further Information and References

*Automobile Side Impact Collisions – Series II* by Severy, Mathewson and Siegel, University of California at Los Angeles, SAE SP-232. For more detailed information please see Further Information and References in the back of the book.

*Mechanical Design and Systems Handbook*, Harold Rothbart, Ed., Chapter 16, McGraw-Hill, New York, 1964. For more detailed information please see Further Information and References in the back of the book.

*Motor Vehicle Accident Reconstruction and Cause Analysis*, by Rudolf Limpert, 2nd ed., The Michie Company, Charlottesville, Virginia, 1984. For more detailed information please see Further Information and References in the back of the book.

*Symposium on Vehicle Crashworthiness Including Impact Biomechanics*, Tong, Ni, and Lantz, Eds., American Society of Mechanical Engineers, AMD-Vol. 79, BED-Vol. 1, 1986. For more detailed information please see Further Information and References in the back of the book.

*The Traffic Accident Investigation Manual*, by Baker and Fricke, 9th ed., Northwestern University Traffic Institute, 1986. For more detailed information please see Further Information and References in the back of the book.

# Energy Methods

# 16

We the undersigned, visitors to Bermuda, venture respectfully to express the opinion that the admission of automobiles to the island would alter the whole character of the place, in a way which would seem to us very serious indeed.
— **Woodrow Wilson,** *February 1, 1908*

## 16.1   General

In previous chapters some aspects of energy methods have already been discussed. In the chapter about simple skids, it was shown how the initial kinetic energy of a vehicle can be dissipated by irreversible work done during braking. In the chapters about simple skids and simple vehicular falls, the interplay between kinetic energy and potential energy was examined. In the previous chapter about momentum methods, it was shown how the irreversible work done in crushing a vehicle during impact relates to the degree of elasticity or plasticity of the collision. In all of these applications, the fundamental assumption has been the conservation of energy.

This chapter discusses in a more general way how the first law of thermodynamics, the conservation of energy, applies to vehicular accident reconstruction. It examines how the irreversible work done in crushing a vehicle can be quantified and related back to the pre-impact velocities of the interacting vehicles. Quantifying the various ways in which energy is dissipated in an accident to determine the pre-impact kinetic energies of the vehicles, and thus their pre-impact velocities, is known as the energy method.

In general, the application of the energy method is very simple. The total energy of a system prior to an accident is equated to the total energy of the system after the accident plus the irreversible work done during the accident interval.

$$E_1 = E_2 + U \tag{i}$$

where $E_1$ = total energy of the system, kinetic and potential, prior to the accident, $E_2$ = total energy of the system, kinetic and potential, after the accident, and U = total amount of irreversible work done during the accident, between states "1" and "2."

Prior to an accident, the interacting vehicles have a certain total amount of kinetic and potential energy, "$E_1$." The kinetic energy will be associated with the vehicles' velocity and mass. The potential energy will be associated with the vehicles' elevation. In the course of the accident, some of the total energy will be dissipated as irreversible work, "$U$." Also, some of the kinetic energy might be converted into potential energy or vice versa.

The initial kinetic energies of the interacting vehicles, and therefore their initial velocities, can be determined if:

a. The amount of irreversible work is quantifiable.
b. The initial and end states for potential energy are known.
c. The kinetic energies possessed by the interacting vehicles after impact is known.

Two items are particularly noteworthy in the application of energy methods. The first is that, in principle, kinetic energy can be reversibly converted into potential energy and vice versa. In practice, however, at least some of the energy in the conversion is lost due to frictional, that is, irreversible, work being done.

For example, if a car rolls downhill, its potential energy is converted into kinetic energy. However, due to rolling resistance, air resistance, tire hysteresis, and bearing friction, some of the kinetic energy is dissipated. Thus, the conversion from potential to kinetic (or vice versa) is not fully reversible. In many situations, though, the simplifying assumption of reversibility provides a reasonable model.

The second item is that in accident reconstruction, irreversible work is done by the conversion of kinetic energy. Potential energy is not directly converted into irreversible work. Potential energy must first be converted into kinetic energy before it can then be dissipated as irreversible work.

For example, if the elevation of a car changes such that its potential energy is decreased, the front end of the car does not, by itself, become crushed. However, if the same car lowers its potential energy by rolling downhill, the car gains speed as the potential energy is converted into kinetic energy. If the car then collides with a brick wall, the front end will be crushed as the kinetic energy is converted into irreversible work.

The energy method is an extemely powerful tool for the solution of pre-impact velocities. In conjunction with the application of the momentum equations, the two methodologies form the basis for the solution of most vehicular accidents. Sometimes information that one methodology is missing to complete a solution, can be supplied by the other. Sometimes both methodologies can be independently used to solve the same problem. When both methodologies independently converge on the same solution, it is taken as a strong indicator that the solution is correct.

## 16.2   Some Theoretical Underpinnings

First, let us consider what is called a conservative system. A conservative system is one in which there is complete reversibility between kinetic and potential energy; there is no irreversible work. The sum of the kinetic and potential energies of the system is the total energy.

$$E = KE + PE = constant \qquad (ii)$$

where E = the total energy of system, KE = the kinetic energy, and PE = the potential energy.

Taking the differential of the above gives:

$$dE = d(KE + PE) = d(KE) + d(PE) = 0 \qquad (iii)$$

Now for the moment, consider just the kinetic energy term, "KE." In a fully conservative system, the kinetic energy of an object depends on its velocity, which in turn depends on its position. In most vehicular accident cases, it is practical to apply the standard Cartesian "x-y-z" orthogonal coordinate system. In functional notation then, the function for kinetic energy is given by the following:

$$KE = KE(x,y,z,u,v,w) \qquad (iv)$$

where x = displacement in the x direction, y = displacement in the y direction, z = displacement in the z direction (elevation), u = dx/dt, v = dy/dt, and w = dz/dt.

The total differential of "KE" is then as follows:

$$d(KE) = [\delta(KE)/\delta x]dx + [\delta(KE)/\delta y]dy + [\delta(KE)/\delta z]dz + \qquad (v)$$
$$[\delta(KE)/\delta u]du + [\delta(KE)/\delta v]dv + [\delta(KE)/\delta w]dw.$$

From previous considerations, it is known that:

$$KE = (1/2)m[u^2 + v^2 + w^2] \qquad (vi)$$

If the gradient of the above scalar function is taken with respect to velocity, then the following vector relationship is derived:

$$[\delta(KE)/\delta u]i + [\delta(KE)/\delta v]j + [\delta(KE)/\delta w]k = (1/2)m[2u + 2v + 2w] \quad \text{(vii)}$$

where i, j, and k are unit vectors in the x, y, and z directions, respectively.

If the above vector relationship is dot multiplied by the velocity vector, $V = u + v + w$, then the following is obtained:

$$[\delta(KE)/\delta u]u + [\delta(KE)/\delta v]v + [\delta(KE)/\delta w]w = \quad \text{(viii)}$$
$$(1/2)m[2u^2 + 2v^2 + 2w^2].$$

Inspection of Equation (viii) finds that the right side of the equation is simply twice the kinetic energy, or 2(KE). Thus, the following is noted:

$$2(KE) = [\delta(KE)/\delta u]u + [\delta(KE)/\delta v]v + [\delta(KE)/\delta w]w \quad \text{(ix)}$$

If Equation (ix) is differentiated, then the following additional relation is obtained:

$$2[d(KE)] = d[\delta(KE)/\delta u]u + d[\delta(KE)/\delta v]v + d[\delta(KE)/\delta w]w \quad \text{(x)}$$
$$+ [\delta(KE)/\delta u]du + [\delta(KE)/\delta v]dv + [\delta(KE)/\delta w]dw.$$

Subtracting Equation (v) from Equation (x) gives the following:

$$d(KE) = d[\delta(KE)/\delta u]u + d[\delta(KE)/\delta v]v + d[\delta(KE)/\delta w]w \quad \text{(xi)}$$
$$+ [\delta(KE)/\delta u]du + [\delta(KE)/\delta v]dv + [\delta(KE)/\delta w]dw.$$
$$- [\delta(KE)/\delta x]dx - [\delta(KE)/\delta y]dy - [\delta(KE)/\delta z]dz$$
$$- [\delta(KE)/\delta u]du - [\delta(KE)/\delta v]dv - [\delta(KE)/\delta w]dw$$

$$d(KE) = d[\delta(KE)/\delta u]u + d[\delta(KE)/\delta v]v + d[\delta(KE)/\delta w]w$$
$$-[\delta(KE)/\delta x]dx - [\delta(KE)/\delta y]dy - [\delta(KE)/\delta z]dz$$

Substituting "dx/dt," "dy/dt," and "dz/dt" for "u," "v," and "w," respectively, and shifting the "dt" term to the differential operator, the following final form is obtained:

$$d(KE) = [d/dt][\delta(KE)/\delta u](dx) + [d/dt][\delta(KE)/\delta v](dy) + \quad \text{(xii)}$$
$$[d/dt][\delta(KE)/\delta w](dz) - [\delta(KE)/\delta x]dx - [\delta(KE)/\delta y]dy - [\delta(KE)/\delta z]dz$$

$$d(KE) = \{[d/dt][\delta(KE)/\delta u] - [\delta(KE)/\delta x]\}(dx) +$$
$$\{[d/dt][[\delta(KE)/\delta v] - [\delta(KE)/\delta y]\}(dy) +$$
$$\{[d/dt][\delta(KE)/\delta w] - [\delta(KE)/\delta z]\}(dz)$$

By doing the above manipulations, the terms "du," "dv," and "dw," which were present in Equation (v), have been eliminated.

Consider now the term for potential energy, "PE." The potential energy of a conservative system is simply a function of position. In fact, if the potential energy under consideration is simply that of gravity, then "PE" is simply a function of one coordinate, "z."

$$PE = PE(z) \tag{xiii}$$

The differential of "PE" is then:

$$d(PE) = [\delta(PE)/\delta z](dz) \tag{xiv}$$

Combining Equation (iii) with Equations (xii) and (xiv) gives the following:

$$\begin{aligned} dE = d(KE) + d(PE) &= \{[d/dt][\delta(KE)/\delta u] - [\delta(KE)/\delta x]\}(dx) + \quad \text{(xv)} \\ &\{[d/dt][\delta(KE)/\delta v] - [\delta(KE)/\delta y]\}(dy) + \\ &\{[d/dt][\delta(KE)/\delta w] - [\delta(KE)/\delta z] + [\delta(PE)/\delta z]\}(dz) = 0 \end{aligned}$$

Since the "x-y-z" coordinates are independent of one another, then the above can be broken apart into the following independent coordinate equations:

$$[d/dt][\delta(KE)/\delta u] - [\delta(KE)/\delta x] = 0 \tag{xvi}$$

$$[d/dt][\delta(KE)/\delta v] - [\delta(KE)/\delta y] = 0$$

$$[d/dt][\delta(KE)/\delta w] - [\delta(KE)/\delta z] + [\delta(PE)/\delta z] = 0.$$

Equations (xvi) are generally called Lagrange's equations for a conservative system. In many texts, the following substitution is usually made:

$$L = KE - PE \text{ or } KE = L + PE \tag{xvii}$$

where L is called the Lagrangian term.

The substitution of $KE = L + PE$ changes Equations (xvi) to the following:

$$[d/dt][\delta(L)/\delta u] - [\delta(L)/\delta x] = 0 \tag{xviii}$$

$$[d/dt][\delta(L)/\delta v] - [\delta(L)/\delta y] = 0 \quad \text{Note: } \delta(PE)/\delta u = 0$$

$$\delta(PE)/\delta v = 0$$

$$[d/dt][\delta(L)/\delta w] - [\delta(L)/\delta z] = 0. \quad \delta(PE)/\delta w = 0$$

$$\delta(PE)/\delta x = 0$$

$$\delta(PE)/\delta y = 0.$$

It is worth noting that at this point the term "$[\delta(L)/\delta u]$" is simply the momentum of the system.

$$[\delta(L)/\delta u] = \delta[KE - PE]\delta u = mu \text{ remembering that } \delta(PE)/\delta u = 0.$$

In fact, the gradient of the kinetic energy term, as was taken in Equation (vii), is simply the vector representation of the kinetic energy term:

$$grad(KE) = [\delta(KE)/\delta u]i + [\delta(KE)/\delta v]j + [\delta(KE)/\delta w]k = m[u + v + w].$$

It is worth noting that the momentum equations, already discussed in Chapter 14, are imbedded in the above energy equations of a conservative system.

The Lagrangian equations in Equation (xviii) can be further generalized by the introduction of generalized coordinates. In this way, Equations (xviii) can be applied to any orthogonal coordinate system. While for nearly all vehicular accidents, the Cartesian coordinate system is convenient, occasionally the cylindrical coordinate system may be useful. In such a system,

$$x = r \cos q, y = r \sin q, r = (x^2 + y^2)^{1/2}, \text{ and } z = z.$$

With appropriate substitutions, the Cartesian-oriented Lagrangian equations in Equation (xviii) can be converted into a cylindrical coordinate system.

At this point, Equations (xviii) are merely an elegant way of describing the "sloshing" of energy from kinetic to potential and vice versa in a fully conservative system. In celestial mechanics, the Lagrangian equations are a cornerstone with respect to central forces and orbital motion. Unfortunately, in vehicular accident reconstruction, about the only situation to which Equations (xviii) apply are vehicles coasting up or down hills or perhaps a car in orbit. To make the Lagrangian equations useful for vehicular accident reconstruction, it is necessary to introduce irreversible work into the equation. This can be done as follows:

$$d(KE + PE) = dU \tag{xix}$$

where $U$ = irreversible work.

Work is defined as force applied through a distance, or "$dU = F \cdot dx$." The differential "$dU$" is the dot product of a force, "$F$," and an infinitesimal distance "$dx$" through which the force is applied.

In the above, the differential "$dU$" is the result of one specific force being applied in a correspondingly specific direction. In a vehicular accident, however, there can be many ways in which irreversible work can be done (e.g., skidding, crush, rolling friction). Because of this, the above term for irreversible work is expanded to include many such forces and their corresponding coordinate directions of application.

$$dU = \Sigma \ [F_i \cdot dx + G_i \cdot dy + H_i \cdot dz] \qquad \text{(xx)}$$

where F, G, and H = forces in the respective x, y, and z directions.

If Equation (xx) is substituted into the right side of Equation (xix), and Equation (xv) is substituted into the left side of Equation (xix), and the results are separated into respective coordinate equations, then the following is obtained:

$$\{[d/dt][\delta(KE)/\delta u] - [\delta(KE)/\delta x]\}(dx) = \Sigma \ [F_i \cdot dx] \qquad \text{(xxi)}$$

$$\{[d/dt][\delta(KE)/\delta v] - [\delta(KE)/\delta y]\}(dy) = \Sigma \ [G_i \cdot dy]$$

$$\{[d/dt][\delta(KE)/\delta w] - [\delta(KE)/\delta z] + [\delta(PE)/\delta z]\}(dz) = \Sigma \ [H_i \cdot dz]$$

The reader will note that the Lagrangian term "L" was dropped in favor of using "KEs" and "PEs." This is because the potential energy term is only important in the "z" direction coordinate.

Finally, the integration of Equations (xxi) gives the following:

$$\int_{x_1}^{x_2} [\{[d/dt][\delta(KE)/\delta u] - [\delta(KE)/\delta x]\}(dx)] = \int_{x_1}^{x_2} [\Sigma F_i \cdot dx] \qquad \text{(xxii)}$$

$$\int_{y_1}^{y_2} [\{[d/dt][\delta(KE)/\delta v] - [\delta(KE)/\delta y]\}(dy)] = \int_{y_1}^{y_2} [\Sigma G_i \cdot dy]$$

$$\int_{z_1}^{z_2} [[d/dt][\delta(KE)/\delta w] - [\delta(KE)/\delta z] + [\delta(PE)/\delta z]\}(dz)] = \int_{z_1}^{z_2} [\Sigma H_i \cdot dz]$$

where the subscripts "2" and "1" refer to the preaccident and postaccident positions.

Inspection of Equations (xxii) finds the following.

- All the terms on the right are irreversible work.
- All the terms on the left are related to the kinetic or potential energy of the object.
- The first term on the left side is a work term, i.e., force x distance, where an external force is being applied on the item.
- The second term on the left side is the change in kinetic energy from position "1" to position "2."
- Irreversible work done by the object is done in the direction of travel of the object (in that coordinate frame).

If the usual terms for kinetic and potential energy are substituted into Equations (xxii), the following is obtained:

$$F_{x,ave}(x_2 - x_1) - (1/2)(m)(u_2^2 - u_1^2) = U_x \qquad \text{(xxiii)}$$

$$F_{y,ave}(y_2 - y_1) - (1/2)(m)(v_2^2 - v_1^2) = U_y$$

$$F_{z,ave}(z_2 - z_1) - (1/2)(m)(w_2^2 - w_1^2) + g(z_2 - z_1) = U_z.$$

In general, when no net external reversible forces are acting on a vehicle, the first terms on the left side of Equations (xxiii) involving force are zero. Since all of the terms in Equation (xxiii) are scalar, then the following is also true:

$$(1/2)(m)[(u_1^2 - u_2^2) + (v_1^2 - v_2^2) + (w_1^2 - w_2^2) + 2g(z_2 - z_1)] \qquad \text{(xxiv)}$$
$$= U_x + U_y + U_z.$$

Of course, the above is the equation for only one object, or one vehicle. For two vehicles colliding or nearly colliding with one another, the following holds:

$$(1/2)(m_A)[(u_{A1}^2 - u_{A2}^2) + (v_{A1}^2 - v_{A2}^2) + (w_{A1}^2 - w_{A2}^2) + 2g(z_{A2} - z_{A1})] \qquad \text{(xxv)}$$
$$+ (1/2)(m_B)[(u_{B1}^2 - u_{B2}^2) + (v_{B1}^2 - v_{B2}^2) + (w_{B1}^2 - w_{B2}^2) + 2g(z_{B2} - z_{B1})]$$
$$= U_{Ax} + U_{Ay} + U_{Az} + U_{Bx} + U_{By} + U_{Bz} = U_{Total}.$$

Equation (xxv) represents the general energy equations for a two-car accident in Cartesian coordinates. Equation (xxv) can also be represented in cylindrical coordinates with the following substitutions:

$$R_1^2 = [u_1^2 + v_1^2 + w_1^2] \qquad \text{(xxvi)}$$

$$R_2^2 = [u_2^2 + v_2^2 + w_2^2]$$

Rearranging terms and substituting gives the following:

$$(1/2)(m_A)[R_{A1}^2 - R_{A2}^2 + 2g(z_{A2} - z_{A1})] \qquad \text{(xxvii)}$$
$$+ (1/2)(m_B)[R_{B1}^2 - R_{B2}^2 + 2g(z_{B2} - z_{B1})]$$
$$= U_{AR} + U_{BR} = U_{Total}.$$

Using cylindrical coordinates removes the "x-y"-related velocity components in Equations (xxv) and simplifies the form of the energy equation. In essence, since Equation (xxvii) is scalar, it is only necessary to work with the absolute magnitudes of the velocities, and the absolute magnitudes of the irreversible work terms.

The above is very important. In the basic energy method, it is not necessary to know from which direction the vehicles came, in which direction they left, or in which direction the irreversible work was done. It is only important to know the absolute values of the various energy terms. In short, the energy method is more or less like balancing sums in an accountant's ledger.

Since the kinetic and potential energy terms on the left of Equations (xxv) or (xxvii) have already been discussed elsewhere in this text at some length, all that remains is to figure out how to quantify the irreversible work of the right-hand terms.

## 16.3   General Types of Irreversible Work

In vehicular accidents, the most common forms of irreversible work are as follows:

a.  Braking and skidding.
b.  Sheet metal crush of the vehicle(s).
c.  Rollovers.
d.  Crush or impact damage to items other than vehicles, such as telephone poles, bridge railings, sign posts, etc.

If a person wants to split hairs, items "b" and "c" above are actually types of sheet metal damage. However, the mechanics in a rollover are sufficiently different from typical impact crush to warrant its consideration separately. A fifth, miscellaneous category can be added that lumps together less common types of irreversible work. Some of these miscellaneous types of irreversible work include:

a. Driving into a deep pool of water.
b. Driving into saturated sand, mud, deep snow, or a freshly disked field.
c. Driving into a row of haystacks.
d. Allowing the air resistance, rolling friction, and engine drag to slow the vehicle down to a stop.

In general, the above miscellaneous irreversible work items involve slow-acting braking that may cause only limited damage to the vehicle itself. These cases can be difficult to quantify directly and may require special case-by-case consideration. Often, such cases require on-site experimentation to determine the pertinent parameters.

When a vehicle is being extracted from mud, sand, or a like substance, sometimes it is useful to place a load cell in the tow cable to measure the tow draw. By comparing this to the tow draw of the vehicle on level, solid pavement, the frictional resistance of the medium can be measured.

However, the above may not be sufficient for a pool of water. The resistance measured when a vehicle is slowly towed from a pool of water may not be the same as that experienced by the vehicle traveling through it at driving speeds. In general, the frictional resistance of water is a function of the forward velocity of the vehicle in much the same way as air resistance. A further complication may occur if the vehicle was traveling fast enough to hydroplane over the surface of the water before sinking. And of course, there is the obvious complication of dealing with water; water does not normally exhibit skid marks to mark a vehicle's path.

Fortunately (or unfortunately for the driver), when a vehicle is driven off a dock, pier, or similar, it usually has a sufficient angle of attack with the water to "bite" or "catch" the water, rather than skip or hydroplane over it. The point where the vehicle sinks will often be the terminus of the trajectory from the dock, pier, or the like. In such cases, the problem can be solved as a simple trajectory, as discussed in Chapter 3.

Air resistance, rolling resistance, and engine drag have already been discussed elsewhere. The braking and skidding category of irreversible work has also been discussed elsewhere in the text. Thus, the remaining items to consider are rollovers and crush damage.

## 16.4  Rollovers

A side rollover is one in which the vehicle initially rotates about the line formed by either the right-side or left-side wheels. Kinetic energy from the motion of the car is converted into work, which then lifts the center of gravity of the vehicle by rotation. When the vehicle is so lifted, its potential energy is increased. As the vehicle rotates or rolls over through 360 degrees, it will

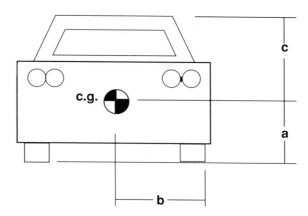

**Figure 16.1** Position of center of gravity.

pass through four balance or instability points. Each of these instability points are positions where the local potential energy of the vehicle is at a maximum.

If the conversion of kinetic energy to work is sufficient to lift the vehicle's center of gravity higher than the elevation of the first localized potential maximum, the vehicle will continue to roll over. If the conversion of kinetic energy is insufficient to lift the center of gravity above the first localized potential maximum, the vehicle will simply lift up on its side, and then fall back down onto its wheels.

Figure 16.1 shows the front view of a vehicle with the center of gravity located "a" distance from the pavement, and "b" distance from the outer edge of the tires. The first instability point, or localized potential maximum, of a vehicle experiencing a side rollover is given as follows:

$$h_1 = [a^2 + b^2]^{1/2} \qquad\qquad (\text{xxviii})$$

where $h_1$ = first localized maximum potential during side rollover.

It should be noted that "$h_1$" as given above is not exact. During tipping of the vehicle, the load carried by the tires and suspension on the side not tipped will double, causing some additional downward displacement or "squat." If the tipping is rapid, the downward displacement may be quite negligible since the suspension system may be damped such that it does not have time to respond. In any case, if needed, a "squat" version of Equation (xxviii) is given as follows:

$$h_1 = [a^2 + b^2]^{1/2} - (W/2k) \qquad\qquad (\text{xxix})$$

where $W/2$ = extra load being carried by the side wheels during tipping and $k$ = linear spring constant of the suspension system where $-(\Delta z)k$ = applied force.

For purposes of this section, Equation (xxviii) provides a sufficient approximation for "$h_1$."

The angle at which the vehicle must be tipped to be balanced at the first instability point is given by the following:

$$\alpha = \arctan(b/a). \qquad \text{(xxx)}$$

The angle "$\alpha$" is usually given in degrees.

In a typical car where a = 24 inches, b = 33 inches, and c = 30 inches, "$\alpha$" is about 54 degrees, and "$h_1$" is about 40.8 inches.

A second localized potential maximum occurs when the vehicle rolls from the 90-degree position (on its side) to its 180-degree position (on its top). If the vehicle is more or less box shaped, the second localized potential maximum is given by the following:

$$h_2 = [b^2 + c^2]^{1/2} \qquad \text{(xxxi)}$$

The position of the second localized potential maximum is given by the following:

$$\beta = 90 + \arctan(c/b) \qquad \text{(xxxii)}$$

If the same dimensions as noted above are substituted into Equations (xxxi) and (xxxii), it is found that "$h_2$" is 44.6 inches, and "b" is 132 degrees.

There are also third and fourth localized potential maximums that are symmetric to the first two. Using the same dimensions as before, the third point is located at an angle of 228 degrees with an "$h_3$" of 44.6 inches, and the fourth point is located at an angle of 306 degrees with an "$h_4$" of 40.8 inches.

Based on the same vehicle dimensions as given above, Table 16.1 shows the relative elevations of the center of gravity as the vehicle rolls over sideways through 360 degrees. It is assumed that no significant crush occurs during rollover to change the initial dimensions of the vehicle.

In essence, if the various values for the center of gravity elevations listed in Table 16.1 are plotted against the corresponding angles of rotation of the rollover, a plot of the potential energy vs. rollover angle is generated. Figure 16.2 is such a plot.

Inspection of the graph in Figure 16.2 finds several interesting features. First of all, the position of the car when it is upright is the position that minimizes the potential energy. This is the most stable position of the example vehicle. The second most stable position is when the car is on its roof. In order for the roof to be the most stable position, it would have to be crushed more than 6 inches.

**Table 16.1   Elevation of Center of Gravity — Rollover**

| Angular Position in Degrees | Elevation of Center of Gravity in Inches | Remarks |
|---|---|---|
| 0 | 24 | Upright |
| 54 | 41 | 1st max. elev. of c.g. |
| 90 | 33 | On side |
| 132 | 45 | 2nd max. elev. of c.g. |
| 180 | 30 | Upside down |
| 228 | 45 | 3rd max. elev. of c.g. |
| 270 | 33 | On other side |
| 306 | 41 | 4th max. elev. of c.g. |
| 360 | 24 | Upright |

During rollover, the center of gravity of the vehicle has four "peaks" of energy to climb and four "hills" of energy to slide down. The first peak is a change of 17 inches. After the first peak, there is an 8-inch slide to the side or 90 degree rest position. The second peak is a change of 12 inches from the 90-degree position, and a total change of 21 inches from the initial position of 0 degrees. After the second peak, there is a slide of 15 inches to the 180-degree position, i.e., upside down.

From experience in playing with old heaps in junk yards, the following "rules" with respect to the example vehicle appear to apply to rollovers:

1.  If there is just enough work performed to roll the car over the first peak, the car will then come to rest on its side. It takes 17 inches of

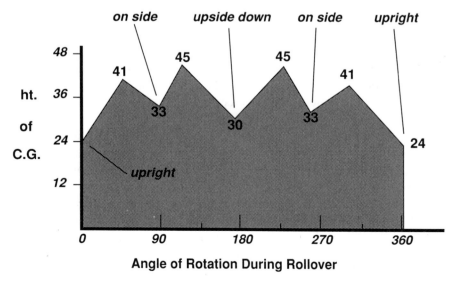

**Figure 16.2** Plot of center of gravity vs. rollover angle.

**Plate 16.1** Rollover damage.

energy to accomplish this. Eight inches of energy will be lost in crush,
damping, and friction. The other nine inches of energy goes to increasing the overall potential energy of the vehicle.

2.  If there is just enough work performed to roll the car onto its roof,
    another 12 inches of energy will be needed to go from the side position
    to the roof position. This is a total of 29 inches of energy expended
    so far.

3.  To continue rolling over from the upside-down position onto its other
    side, another 15 inches of energy is needed to clear the next peak. This
    is a total of 44 inches of energy expended so far.

4.  To finish the roll and land back upright, another 8 inches of energy
    is needed to clear the last peak. This is a total of about 52 inches of
    energy to complete a 360 degree rollover.

In the above scenario, it is assumed that the energy on the downhill side
of the peaks in Figure 16.2 is dissipated along the way by friction, crush,
damping, etc. For most instances, this is a reasonable approximation when
the vehicle rolls over more or less in contact with the ground.

Generalizing from the above, the following equations indicate what the
expenditure of energy will be for the various rest positions after a rollover.

**Case I** — Near rollover. The vehicle lifts up, but does not clear the first
instability point, and drops back down onto its wheels.

$$U < W(h_1 - a) \qquad\qquad\qquad (\text{xxxiii})$$

$$U < W([a^2 + b^2]^{1/2} - a)$$

where U = work expended in rollover action and W = weight of the vehicle.
**Case II** — Rollover that ends up on side position.

$$U = W(h_1 - a) \qquad\qquad (xxxiv)$$

**Case III** — Rollover that ends up on roof.

$$U = W([b^2 + c^2]^{1/2} - b + [a^2 + b^2]^{1/2} - a) \qquad (xxxv)$$

$$U = W(h_1 - a + h_2 - b)$$

**Case IV** — Rollover that ends up on the other side of the vehicle.

$$U = W([b^2 + c^2]^{1/2} - b + [a^2 + b^2]^{1/2} - a + [b^2 + c^2]^{1/2} - c) \quad (xxxvi)$$

$$U = W(h_1 - a + 2h_2 - b - c)$$

**Case V** — Rollover that ends upright again.

$$U = W([b^2 + c^2]^{1/2} - 2b + 2[a^2 + b^2]^{1/2} - a + [b^2 + c^2]^{1/2} - c) \quad (xxxvii)$$

$$U = W(2h_1 - a + 2h_2 - 2b - c)$$

Of course, it should be noted that the above equations have been developed for a car:

1. That is more or less a solid rectangle.
2. In which no significant crush has occurred to grossly change initial dimensions.
3. In which the "squat" due to load transfer to one side of the vehicle during the initial rollover is negligible.

The above equations would not exactly work for pickup trucks, since they might teeter front to back when they are upside down. Similarly, the tractors in tractor-trailer rigs are not as likely to turn upside down as a car would. They would likely roll over cocked, with ground contact simultaneously occurring at the front edge of the hood and the top edge of the roof. However, the principles demonstrated for developing equations for a solid rectangular car are the same for other shapes of vehicles.

If it is assumed that the same example vehicle is used as before, and that the sample vehicle has a weight of 2500 lbs, then the minimum velocity at which a rollover could occur is calculated as follows:

$$(1/2)mv^2 = W([a^2 + b^2]^{1/2} - a) \qquad\qquad \text{(xxxviii)}$$

$$v = \{2g([a^2 + b^2]^{1/2} - a)\}^{1/2}$$

$$v = 9.5 \text{ ft/sec.}$$

Note that the weight of the vehicle is partially canceled out by the mass term on the left, such that the equation ends up using "g." Equation (xxxviii) is independent of the vehicle's weight.

The above solution found in Equation (xxxviii) does not mean that a vehicle will turn over whenever it is traveling sideways at 9.5 ft/sec. However, if it was traveling sideways at 9.5 ft/sec and its side wheels caught a curb or something that momentarily held the wheels immobile, then rollover could occur. The velocity term in Equation (xxxviii) is the sideways or transverse speed of the center of gravity when the wheels are simultaneously prevented from having transverse motion.

If the same things about the example vehicle are assumed as before, but the rollover continues through 360 degrees until the vehicle is upright again, then the minimum velocity at which such a rollover could occur is calculated as follows:

$$(1/2)mv^2 = W(2h_1 - a + 2h_2 - 2b - c) \qquad\qquad \text{(xxxix)}$$

$$v = 16.5 \text{ ft/sec.}$$

The calculations in Equations (xxxviii) and (xxxix) demonstrate that it really does not take a lot of speed to have enough energy to roll a vehicle over. In other words, a sideways rollover does not expend a lot of energy. For example, the energy required for the 360-degree rollover of Equation (xxxix) is equivalent to just a 6-ft-long skid if the coefficient of friction is 0.7.

## 16.5   Flips

A flip is when a car or other type of vehicle rotates end over end. Usually, the rear end of the vehicle raises and rotates about the front wheels. Most of the time when a vehicle flips over, it is the rear end that raises up because most vehicles flip over when they are moving in a forward direction.

Like the rollover, a flip occurs when the center of gravity of the vehicle is traveling with a certain velocity while the front two (or possibly back two) wheels are momentarily motionless. Such a situation can occur if:

1. The vehicle hits a low retaining wall or embankment.
2. The vehicle drives off the road such that the front end of the car drops down and "catches" in the dirt.
3. The vehicle drives into a high curb.
4. The vehicle drives into a mound of gravel.

In each of the above cases, the front end of the car is momentarily stopped, while the rest of the car is still able to move by rotating about the front wheels. This occurs because the center of gravity of the car is located above the point of force that stops the car, as shown in Figure 16.3. If the center of gravity was in line with the point of force, the car would not raise up and rotate. If the center of gravity was below the point of force, the vehicle would "squat" during the impact as it tries to rotate into the ground. (In head-on collisions or similar, it is not unusual for one or both cars to "squat" enough to cause some tubeless tires to lose seal and deflate.)

Unlike a rollover, a flip is much simpler to calculate. While a rollover has four points of instability and four points of local stability (upright, right side, upside down, and left side), a flip has only two points of instability and two points of local stability. The two points of instability are the front and the back, that is, being stood on either end. The two points of stability are upright, and upside down. All these points, that is both the stable and unstable points, are basically spaced 90 degrees from one another.

While it may be arguable that in a vehicle with a squared-off front and rear there are theoretically four points of instability and four points of sta-

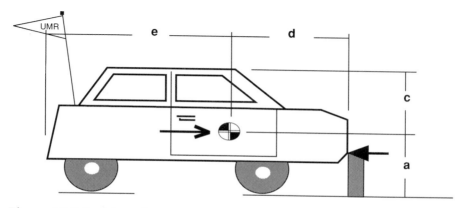

**Figure 16.3** Position of center of gravity.

bility to a flip, there is no practical significance to considering all four points. The first two instability points, or potential energy peaks, are so close to one another, and the intervening potential energy valley is so slight between them, that there is no practical significance to separating the three points. This is also true of the second two instability points and the intervening potential energy valley.

The above should be more or less intuitively obvious. Vehicles just don't regularly flip over and come to rest poised like balancing rocks standing on their ends. It is very difficult to stand a vehicle on either its front end or rear end and have it stay put. While this could possibly be done with a delicate touch and a very box-like vehicle, such a delicate touch does not occur in vehicular accidents. Vehicles usually just smash around like rogue elephants.

As with rollovers, the energy needed to flip a vehicle over is basically a function of the work it takes to lift its center of gravity. Assume that the same sample car as considered before is used and it has the same center of gravity, that is, a = 24 in., b = 33 in., and c = 30 in. Further, assume that as shown in Figure 16.3, the other dimensions locating the center of gravity from the side are d = 75 in. and e = 115 in. Then a table similar to Table 16.1 can be made showing the variation in center-of-gravity elevation as a vehicle flips forward.

As in the example vehicle, most cars have their center of gravity significantly closer to the front end of the car than to the rear. This means that it is significantly easier to flip a car over the front end than over the rear end. If the values in Table 16.2 were to be plotted like those in Table 16.1, the peaks would occur at 90 and 270 degrees, and the valleys would occur at 0 and 180 degrees.

While Table 16.2 shows the elevation of the center of gravity through a 360-degree flip, it is very unusual for a car to flip through one full rotation. In many cases, a car will first flip over onto its top, and if there is enough remaining kinetic energy, it may then rollover rather than continue with the flip. There is a tendency for rolling over even when there is sufficient energy available to continue the flip. The underlying reasons for this behavior are rooted in rotational dynamics and deserve a brief look even if they do not exactly relate to the topic at hand.

**Table 16.2   Elevation of Center of Gravity — Flip**

| Angular Position in Degrees | Elevation of Center of Gravity in Inches | Remarks |
|---|---|---|
| 0 | 24 | Upright |
| 90 | 75 | Standing on front end |
| 180 | 30 | Upside down |
| 270 | 115 | Standing on rear end |
| 360 | 24 | Upright |

A similar rotational effect occurs with a book that is significantly longer than it is wide. If a person tries to roll the book, that is rotate it about a line parallel to its binding, the long axis, the book will spin stably about the axis. If a person then attempts to rotate the book about a line parallel with its top edge across, the short axis, the book may begin to tumble or wobble as it rotates after about 180 degrees of spin. It is demonstrably harder to make the book rotate cleanly about its short axis than its long axis.

This effect is a consequence of Euler's equations of motion and the mass moments of inertia in the three principle axes. Euler's equations of motion, shown below, relate torque, angular velocity, angular acceleration, and the mass moments of inertia. They are applied to items like tops, gyroscopes, and other things that rotate in two or three dimensions. Without showing the derivations in a body that has symmetry about each of its three principle axes the following equations hold:

$$T_x = I_{xx}\alpha_x + (I_{zz} - I_{yy})\omega_z\omega_y \tag{xl}$$

$$T_y = I_{yy}\alpha_y + (I_{xx} - I_{zz})\omega_x\omega_z$$

$$T_z = I_{zz}\alpha_z + (I_{yy} - I_{xx})\omega_y\omega_x$$

where $T_i$ = moment or torque about the "i" axis, $I_{ii}$ = mass moment of inertia about the "i-i" axis, $\omega_i$ = angular velocity about the "i" axis, and $\alpha_i = d(\omega_i)/dt$, angular acceleration about the "i" axis.

Two items are apparent upon inspection of the above equations. First, if the mass moments of inertia are all equal, like it would be in a sphere, then all three equations simply become $I_{ii}a_i = T_i$. When all three mass moments of inertia are different, it is possible for linkage to occur among the rotational motions of the three axes. This linkage causes rotational motions to behave much differently than linear motions. This is the reason why a top precesses when it leans over while spinning, and is one of the reasons why it is easier to balance a moving bicycle than one that is stopped.

If the coordinates of the vehicle are oriented such that "x-x" is along a line from the right side to the left side of the vehicle, "y-y" is along a line from the front to the rear, and "z-z" is along a line from the bottom to the top, in general it is found that "$I_{xx}$" and "$I_{zz}$" are equal to one another. This is because the distribution of mass from the axis of rotation is about the same in both cases for most types of vehicles.

For most vehicles, the least mass moment of inertia is about the "y-y" axis, and it is usually significantly less than the other two mass moments of inertia along the principle axes. In the "$I_{yy}$" term, the line of rotation goes through more of the mass, which zeros out that portion of the mass in the moment of inertia calculation. Also, the longest distance from the axis is perhaps only 1/3 or 1/4 that of the other axes.

Thus, it is easier to rotate a car about its short axis than about its long axis. A greater rotational response occurs when a torque is applied to the "$I_{yy}$" term than would occur if applied to the other two moments of inertia.

By applying the above general findings to Equations (xl), the following simplified Euler's equation of motion are derived, which apply to many types of automobiles:

$$T_x = I_{xx}\alpha_x + (I_{zz})\omega_z\omega_y \qquad\qquad\text{(xli)}$$

$$T_y = I_{yy}\alpha_y$$

$$T_z = I_{zz}\alpha_z - (I_{xx})\omega_y\omega_x$$

$$\text{where } (I_{xx} - I_{zz}) \sim 0$$

$$I_{yy} << I_{xx} \text{ and } I_{zz}.$$

Inspection of the above equations finds that torque around the "y-y" axis is not interlinked to motion in the other axes. However, torque in either the "x-x" or "z-z" axis is interlinked to motions in the other axes.

In a rollover, rotation occurs about the "y-y" axis due to the application of a torque at the bottom of the tires on one side of the vehicle. The influence of small angular velocity rotations around the other two axes upon the "y" torque is nullified by the "$(I_{xx} - I_{zz})$" term, which is zero. However, because it is a rollover action, rotation in the "x-x" axis is zero anyway. This is because prior to the rollover occurring, the vehicle is in contact with the ground, which constrains any "x-x" axis rotation. Thus, in a rollover, rotation can typically only occur in the "y-y" and "z-z" axes. Angular velocities in both of these axes could cause a torque to be developed in the "x" direction if the angular velocity in the "z-z" axis were significant. But, this occurs infrequently. For these reasons, rollovers don't usually turn into flips unless the ground helps out.

In a flip, where rotation is about the "x-x" axis, the term "$(I_{zz} - I_{yy})$" is not zero. The term "$(I_{zz} - I_{yy})$" is the difference between the maximum and the minimum mass moments of inertia of the vehicle. Since "$I_{yy}$" is usually very small as compared to "$I_{zz}$," the "$I_{yy}$" term is neglected for simplicity. Also, there is no angular velocity about the "y-y" axis because the vehicle has been restrained from rotation in that direction by contact with the ground.

Thus, a vehicle can go into a forward flip with some previous rotation about the "z-z" axis. This could cause the vehicle to continue yawing during the flip, and land upside down in a partially or fully sideways position. Thus, if there is sufficient energy available, a vehicle could flip 180 degrees onto its top, and then go into a rollover.

In a spin, rotation is about the "z-z" axis (yaw). A vehicle that is yawing will normally not have rotations in the other two directions, since the vehicle is constrained by the ground. Thus, the "$\omega_x$" and "$\omega_y$" terms are zero. Because of this, except in unusual circumstances, only flips typically require the application of Euler's equations of motion.

Returning the main topic of this section, the minimum kinetic energy to cause a forward flip to occur, such that the car lands on its top, is given by the following:

$$(1/2)mv^2 = W(d - a) \qquad\qquad (xlii)$$

where m = mass of the vehicle, W= weight of the vehicle, v = velocity of the vehicle, and d and a are the coordinates of the center of gravity as shown in Figures 16.1 and 16.3.

Solving for the velocity term in Equation (xli) gives the following:

$$v = (2g[d - a])^{1/2} \qquad\qquad (xliii)$$

where v = minimum velocity to cause a forward flip.

For the example car, substituting in the dimensions finds that v = 16.54 ft/sec. Comparing this with the velocity required to initiate rollover finds that a forward flip requires about 74% more velocity to initiate than a rollover. In terms of energy, it requires about three times more energy to initiate a flip than it does to initiate a roll. This explains, in part, why rollovers are more common than flips.

In a forward flip, sometimes the vehicle will land on its top, the top will crush and crumple, and the motion of the vehicle will stop. However, because it takes so much more energy to flip a vehicle than roll it, there is more available energy on the downhill side from the first instability point. If the roof of the vehicle is rigid and elastic, it is possible that it can bounce sufficiently for the vehicle to then roll over onto its side. Whether it lands on its top or its side, the energy expended is still the same.

If the vehicle continues to roll after landing on its side, the rest of the roll is handled as discussed in the previous section on rollovers.

If, however, the vehicle does not roll over but continues to flip over through 360 degrees, then the following holds:

$$(1/2)mv^2 = W(d - a + e - c). \qquad\qquad (xliv)$$

Solving the above gives:

$$v = (2g[d - a + e - c])^{1/2}. \qquad\qquad (xlv)$$

Again, substituting in the dimensions from the example vehicle finds that the minimum speed to accomplish a 360-degree forward flip is 27.0 ft/sec. This is 64% more velocity than is required minimally to cause a 360-degree rollover. In terms of energy, it requires about 2.7 times more energy to cause a 360-degree flip than a 360-degree rollover.

However, whether it be a flip or a rollover, it is noteworthy that the minimum vehicular velocity required to cause either is low. Under the right circumstances, a vehicle traveling at even relatively slow speeds has sufficient velocity to either flip or roll over.

## 16.6  Modeling Vehicular Crush

Several types of conceptual models have been used to represent a vehicle that has sustained crush damage. Some computer programs that analyze vehicular collisions have modeled a vehicle like an elastic ball. In such a model, the collision is considered elastic, and the spring stiffness is the same from any direction of impact. This "billiard ball" model works reasonably well when the collision impulse is in the elastic response range of the vehicle, but quickly becomes inadequate when there is a high degree of plastic deformation

**Plate 16.2** Crush damage to side of car.

involved. As noted in Chapter 15, however, significant plastic deformation occurs at all but very modest vehicular speeds. The billiard ball model, however, does have some usefulness with respect to glancing impacts at moderate to high speeds where the central impact vector component is very small as compared to the parallel impact vector.

An alternative conceptual model used in many analyses is to assume that the deformation response of the vehicle is linear. This approach is then used to develop delta-V vs. crush depth relations. This is the conceptual model used in *EDCRASH*, a popular computer program used in analyzing collisions. In essence, this model assumes that the rate of crush deflection at the beginning of contact between the vehicles (or whatever) is equal to the closing velocity between the vehicles.

$$dx/dt = v_{A1} - v_{B1} = \Delta v \qquad \text{(xlvi)}$$

where $x$ = crush displacement, $v_{A1}$ = pre-impact speed of vehicle "A," and $v_{B1}$ = pre-impact speed of vehicle "B."

A crush constant "k" is then assigned to each vehicle such that the energies dissipated in crush by the respective vehicles are:

$$E_A = (1/2)k_A x_A^2 \quad \text{and} \quad E_B = (1/2)k_B x_B^2. \qquad \text{(xlvii)}$$

In this model, the crush response of the vehicle is basically treated like a linear spring. The energy dissipated in crush is given the same mathematical form as the stored energy in a compressed elastic spring that obeys Hooke's law.

Following the above model through, the function for energy consumed by crush in a vehicle is then:

$$E = (a^2/2b) + ac + b(c^2/2) \qquad \text{(xlviii)}$$

where $a$ and $b$ = arbitrary constants associated with each vehicle's relative stiffness and $c$ = average crush depth across the front of the vehicle, assuming that the collision was more or less across the front.

Because of the linear assumption of force vs. crush depth, several problems have been reported in the above model. At low speeds the delta-V parameter is often underestimated. At high delta-V speeds of more than about 50 mph, it overestimates delta-V (see *An Overview of the Way EDCRASH Computes Delta-V* by Day and Hargens, SAE paper 870045, p. 192).

The model used in this text presumes that a vehicle is like a reinforced thin-walled enclosed shell or a thin-walled column. This model appears to more closely resemble what actually occurs in a collision than the various "spring" models.

In a thin-walled column, when compressive load is applied at low levels, the column responds elastically in the same direction as the compression. As the load is increased, the column may undergo elastic buckling, where the shape of the column distorts in directions other than that of the applied force. However, in elastic buckling, the column is still able to snap back if the load is released. When the load is then further increased, inelastic buckling occurs. The column collapses by folding into itself, somewhat like an accordion, and the deformation is permanent.

In most structural engineering or "mechanics of materials" courses at the undergraduate level, failure of a structural component is considered to have occurred when elastic buckling initiates, and possibly even before that, when the part has deflected beyond its dimensional tolerances. The goal of good design is to avoid reaching a level of loading that causes either excessive dimensional distortion or rupture of the part, depending upon whether the part is composed of ductile or brittle material. Thus, while there is considerable study of what occurs in the regime below the buckling point, there is much less attention paid to what occurs after buckling has occurred.

## 16.7  Post-Buckling Behavior of Columns

The analysis of buckled elastic columns, which is sometimes known as the problem of the elastica, was first done by Euler more than 200 years ago. While the derivations are somewhat beyond the scope of this text, the results are important. In a simple column where each end is free to rotate and a load is applied at one end axially, the column will buckle elastically at the Euler buckling load, which is usually given as:

$$F_{cr} = K(\pi^2 EI)/L^2 \tag{xlix}$$

where $F_{cr}$ = force required for elastic buckling, i.e., critical load, E = modulus of elasticity, K = constant for the end conditions of the column, which in this case is 1, I = areal moment of inertia of cross-section of column, and L = length of column.

When both ends of the column are fixed, the value of "K" is 2. Thus, a column with fixed or clamped ends will have a Euler buckling load twice that of one that has ends that allow angular rotation.

Elastic analysis also shows that a column will sustain only small increases in load above the elastic buckling load point. After that, yielding occurs, if it has not already done so, as the column continues to buckle and collapse under the applied load. In essence, buckling theory indicates that there is no significant linear relationship between force and displacement after buckling occurs.

In fact, the Euler buckling load is simply a bifurcation point as discussed in chaos theory. Below the Euler buckling load, the relationship between load and displacement is linear and deterministic. At the Euler buckling load and slightly above it, the column can assume several deflection modalities: a half sine wave bowed outward, a half sine wave bowed inward, an S-shaped sine wave bowed outward, an S-shaped sine wave bowed inward, etc. The more complicated modalities are essentially higher frequency sinusoidal functions.

In short columns where the column is a thin-walled metal cylinder or similar structure, localized buckling is what causes the accordion-like folding or crushing of the sheet metal. The buckling occurs where the Euler buckling load is exceed on a localized basis. As long as the same load that caused the buckling to initiate is applied, more columns will feed into the buckling zone and become crushed. Once started, the process proceeds rapidly and only stops when either the load is removed, or when the column has wholly collapsed into itself.

The above can be easily verified experimentally using thin-walled aluminum soda cans. Most soda cans are nominally about 4 13/16 inches in height, and 2 9/16 inches in diameter. If a load is evenly applied around the top of an undented can and the can is sitting on a firm, flat surface, the can will compress and rebound elastically as long as the load is less than about 118 pounds. When the load is at or slightly exceeds 118 pounds, the can will collapse to a height of about 3 inches.

Because the can has reinforced ends, at about a height of 3 inches, the end conditions at the top and bottom of the can begin to exert influence. At that height it is necessary to increase the applied load to cause further collapse. As noted before, a column with fully fixed ends has a Euler buckling load twice that of one with fully pinned ends. A generalized graph of what occurs is shown in Figure 16.4.

However, even with the change in end conditions as noted, an aluminum soda can will collapse nearly 40% of its height in response to an applied constant force. Thus, the amount of work necessary to cause the can to crush no more than 40% of its original height is simply 118 pounds times the amount of crush. Since energy units are more often expressed in lbf-ft than lbf-in, this relationship can be restated as follows:

$$U = kc \qquad (1)$$

where U = amount of irreversible work done in crushing the can, c = the amount of crush in inches, and k = the force needed for buckling, which in this case is 118 lbf or 9.833 lbf-ft/in.

Usually, the "k" term is simply called the stiffness coefficient.

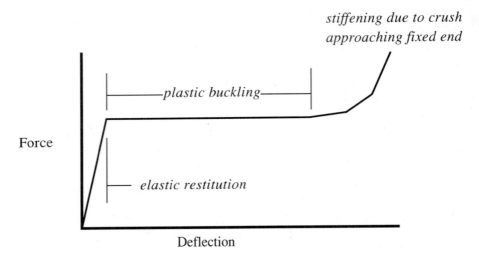

**Figure 16.4** Generalized graph of buckling in a short, thin-walled column.

## 16.8   Going from Soda Cans to the Old 'Can You Drive?'

Many people jokingly refer to their cars as "old tin cans." In principle, however, an empty soda can has similarities to a vehicle. Both are thin-walled
metal shells. Both exhibit buckling failure in response to loading before they
experience traditional compressive stress material rupture (i.e., they crush
before they break). Both have crush load thresholds that depend on the point
of application. The force need to crush a can by pushing down on its top is
different from the force needed to crush it by pushing on its side.

If the energy required for front end crush in a vehicle follows the general
form as shown in Equation (1), then the amount of crush expected in a fixed
barrier impact vs. impact speed is derived as follows:

$$U = kc \tag{li}$$

$$(1/2)mv_{eq}^2 = kc$$

$$v_{eq} = [2kc/m]^{1/2} = [2k/m]^{1/2}\,[c]^{1/2}$$

where $v_{eq}$ = impact speed with a fixed barrier.

If it is assumed that crush begins only above a certain impact speed,
where the vehicle's kinetic energy exceeds the elastic energy portion of its
force-deflection function, then the above can be modified to the following:

$$v_{eq} = v_o + [2k/m]^{1/2}\,[c]^{1/2} \tag{lii}$$

where $v_{eq}$ = impact velocity with a fixed barrier, $v_o$ = impact velocity above which damage occurs, c = average crush depth across front of car, and 2k/m = a constant associated with the properties of the vehicle.

In Equation (lii), if "c = 0," then "$v_{eq} \leq v_o$."

Equations (li) and (lii) above contain a square root of "c," the crush depth. Because of this, the graph of "$v_{eq}$" vs. "c" follows the general plot of a square root function like $y = [ax]^{1/2}$. Thus, the graph of "$v_{eq}$" vs. "c" would look linear at low values of "c," but would then flatten out somewhat at high impact velocities.

In more mathematical terms, the slope of the function $v_{eq} = f(c)$ is greater at low values of "c" than high values. If the derivative of Equation (lii) is taken, the reason for this becomes apparent.

$$dv_{eq}/dc = [k/2m]^{1/2} [1/c]^{1/2} \qquad \text{(liii)}$$

As "c" decreases below the value $[k/2m]^{1/2}$, the slope increases. As "c" increases more than the value $[k/2m]^{1/2}$, the slope decreases. In a vehicle with a "k" value of about 6000 lbf-ft/in or 500 lbf, and a mass of 106 lbf-sec²/ft, the term "[k/2m]" computes to a value of 1.54 ft or 18.4 inches. Thus, in this example, the slope of "$dv_{eq}/dc$" is 45 degrees when crush is 18.4 inches deep.

Equations (li) and (lii) are also interesting concerning the accuracy of calculating the fixed barrier impact velocity from the average crush depth. If the crush were 6 inches deep, the corresponding impact velocity computes to 26.06 ft/sec. A 1% increase in crush causes the computed speed to be 26.19 ft/sec, a 0.5% increase. Similarly if the crush were 30 inches deep, the corresponding velocity computes to 58.28 ft/sec. A 1% increase in crush causes the computed speed to be 58.57 ft/sec, a 0.5% increase. In short, the square root of "c" tends to halve the error of calculating "$v_{eq}$" from "c."

This is because of the following relation:

$$[1 + \Delta x]^{1/2} \sim 1 + (\Delta x/2) \quad \text{where} \quad \Delta x \ll 1.0.$$

In Equation (lii), it should be noted that the "$v_o$" term is essentially the velocity associated with overcoming the elastic energy portion of the vehicle impact. This should only be used if the elastic energy of rebound is not accounted for elsewhere. When the elastic rebound is accounted for, such as in calculating the energy expended in the rebound skid, the "$v_o$" term should be left out to avoid "double bookkeeping."

In general, because the elastic energy component is very small with respect to moderate- or high-speed collisions, there is no significant accuracy lost if the elastic energy term is simply ignored. For example, at moderate or higher impact speeds, a vehicle impacting a fixed barrier will expend about

99% of its kinetic energy in crush. If 1% of the total energy is ignored, the accuracy of the impact speed calculation will be affected as follows:

$$U = (1/2)mv^2 \tag{liv}$$

$$v = [2U/m]^{1/2}$$

$$(v_{approx})/(v_{actual}) = [0.99/1.00]^{1/2} = 0.995.$$

As can be seen by inspection and considering some of the other variables in these calculations, this is a very reasonable simplification for moderate- or high-speed impacts.

## 16.9   Evaluation of Actual Crash Data

In some crash tests, a 1980 Chevrolet Citation was run into a fixed barrier at 35, 40, and 48 mph. The residual crush across the front was measured to be 21.4, 27.9, and 40.4 inches, respectively. The weight of the Citation was 3130 pounds in each case. Table 16.3 was created using this information.

The data in Table 16.3 agrees very well with the model over a range of moderate speeds.

In similar fixed-barrier crash tests, a 1977 Honda Civic was frontally impacted at 19.1 and 26.8 mph. The weight of the Civic was 1641 pounds. The residual crush across the front was measured to be 9.5 and 14.3 inches, respectively. Table 16.4 depicts the resulting data.

The data in Table 16.4 agrees generally with the theory. Due to the form of the equation for velocity, the 13% variation from the average "k" translates to a variation of about 6–7% in the calculation of the impact speed.

**Table 16.3   1980 Citation Crash Data**

| Impact Speed | Kinetic Energy | Crush "c" | Calc. Stiffness "k" |
|---|---|---|---|
| 35 mph | 128,200 lb-ft | 21.4 in | 5990 lb-ft/in |
| 40 mph | 167,400 lb-ft | 27.9 in | 6001 lb-ft/in |
| 48 mph | 241,100 lb-ft | 40.4 in | 5968 lb-ft/in |
| | average "k" = 5986 lb-ft/in | | |
| | max. deviation from average = 0.3% | | |

Note:   Data from *Field Accidents: Data Collection, Analysis, Methodologies and Crash Injury Reconstruction*, 1985, paper 850437, "Barrier Equivalent Velocity, Delta-V and CRASH3 Stiffness in Automobile Collisions" by Hight, Hight, and Lent-Koop, Figure 16.4.

**Table 16.4   1977 Civic Crash Data**

| Impact Speed | Kinetic Energy | Crush "c" | Calc. Stiffness "k" |
|---|---|---|---|
| 19.1 mph | 20,011 lb-ft | 9.5 in | 2106 lb-ft/in |
| 26.8 mph | 39,397 lb-ft | 14.3 in | 2755 lb-ft/in |
| | average "k" = 2430 lb-ft/in | | |
| | max. deviation from average = 13% | | |

Note:  Data from "Elastic Properties of Selected Vehicles," by Navin, Mac-Nabb, and Miyasaki, SAE paper 880223.


**Table 16.5   1975 Civic Crash Data**

| Impact Speed | Kinetic Energy | Crush "c" | Calc. Stiffness "k" |
|---|---|---|---|
| 19.7 mph | 19,916 lb-ft | 10.0 in | 1992 lb-ft/in |
| 23.0 mph | 27,148 lb-ft | 12.5 in | 2172 lb-ft/in |
| | average "k" = 2082 lb-ft/in | | |
| | max. deviation from average = 4.3% | | |

Note:  Data from "Elastic Properties of Selected Vehicles," by Navin, Mac-Nabb, and Miyasaki, SAE paper 880223.

In two tests of a 1975 Honda Civic, it was found that the first one crushed 10.0 inches in a 19.7 mph frontal impact with a fixed barrier. In the second, a crush of 12.5 inches was reported in for a 23.0 mph impact. The vehicle weighed 1535 pounds. Table 16.5 depicts the data.

Table 16.5 also has excellent agreement with the theory, although the range of velocities involved are not very far apart.

## 16.10   Low Velocity Impacts — Accounting for the Elastic Component

A noted in the previous section, when the impact velocity is well above the point where plastic deformation occurs, that is crush, the elastic energy component can be ignored if it is accounted for in the rebound skid or similar. The idea is that any residual elastic energy will eventually be dissipated via other means. However, if it cannot be accounted for in that way, and it is necessary to obtain the impact speed using only the impact crush, then when the elastic energy constitutes perhaps 15% of the total impact kinetic energy, some allowance for it should be made.

For example, consider a car that impacts against a fixed barrier and only has 1 inch of deformation across the front. If the stiffness coefficient "k" is 5000 lb-ft/in, and the vehicle weighs 2600 pounds, then it might be estimated that the impact speed was as follows:

$$v_{eq} = [2(5000 \text{ lb-ft/in})(1 \text{ in})/(80.8 \text{ lb-sec}^2/\text{ft})]^{1/2} = 11 \text{ ft/sec}$$

However, since the impact speed is close to the elastic-plastic impact boundary, a more accurate estimate of the impact speed should include the elastic energy expended in rebound effects that are unaccounted for otherwise. If it is known that this model car will not show crush at fixed barrier impacts of less than 5 mph, then the elastic energy component is:

$$E_{elas} = (1/2)mv_o^2 = 2172 \text{ lb-ft.}$$

Combining this with the previous expression for "$v_{eq}$" gives the following:

$$v_{eq} = [2(5000 \text{ lb-ft} + 2172 \text{ lb-ft})/(80.8 \text{ lb-sec}^2/\text{ft})]^{1/2} = 13.3 \text{ ft/sec.}$$

Thus, allowing for the elastic component in this case increases the calculated impact velocity by about 21%.

## 16.11   Representative Stiffness Coefficients

When automobiles are arranged in categories of weight, it is found that the stiffness coefficients of the various vehicles are similar.

The following tables show some of these categories and their representative stiffness coefficients. The data was prepared from crash data, where impacts to a fixed barrier occurred at about 35 mph.

#### Table 16.6   Small Cars Stiffness Coefficients

| Vehicle | Weight (lb) | Stiffness Coefficient (lb-ft/in) |
|---|---|---|
| 1979 Honda Civic | 2180 | 4720 |
| 1979 Ford Fiesta | 2190 | 4040 |
| 1979 Plymouth Champ | 2310 | 4260 |
| 1979 Datson 210 | 2430 | 3960 |
| 1979 VW Rabbit | 2600 | 4860 |
| 1979 Toyota Corolla | 2650 | 5340 |
| 1979 Chevette | 2730 | 4619 |
| Average | 2441 | 4620 |
| Range | +12/−11% | +16/−14% |

Note:  Data from *Field Accidents: Data Collection, Analysis, Methodologies and Crash Injury Reconstruction*, 1985, paper 850437, "Barrier Equivalent Velocity, Delta-V and CRASH3 Stiffness in Automobile Collisions" by Hight, Hight, and Lent-Koop, Figure 16.4.

#### Table 16.7 Medium Cars Stiffness Coefficients

| Vehicle | Weight (lb) | Stiffness Coefficient (lb-ft/in) |
|---|---|---|
| 1979 Mustang | 3070 | 7610 |
| 1979 Mercury Capri | 3070 | 7178 |
| 1979 Chevrolet Monza | 3240 | 5970 |
| 1979 Volvo 242 | 3290 | 4600 |
| 1979 Ford Fairmont | 3300 | 6000 |
| 1982 Volvo DL | 3350 | 5040 |
| 1979 Volvo 244DL | 3370 | 4960 |
| Average | 3241 | 5908 |
| Range | +4/−5% | +28/−22% |

Note: Data from *Field Accidents: Data Collection, Analysis, Methodologies and Crash Injury Reconstruction*, 1985, paper 850437, "Barrier Equivalent Velocity, Delta-V and CRASH3 Stiffness in Automobile Collisions" by Hight, Hight, and Lent-Koop, Figure 16.4.

#### Table 16.8 Full Sized Cars Stiffness Coefficients

| Vehicle | Weight (lb) | Stiffness Coefficient (lb-ft/in) |
|---|---|---|
| 1980 AMC Concord | 3700 | 7460 |
| 1979 Plymouth Volare | 3820 | 7170 |
| 1979 Old Cutlass | 3820 | 5600 |
| 1979 BMW 528 | 3840 | 6400 |
| 1979 Ford Granada | 3950 | 6145 |
| 1979 Mercury Marquis | 4220 | 6300 |
| 1979 Ford LTD | 4370 | 6850 |
| 1979 Dodge St. Regis | 4460 | 6470 |
| 1979 Olds Regency | 4710 | 7355 |
| 1979 Ford LTD II | 4810 | 6000 |
| 1979 Lincoln Continental | 5360 | 7384 |
| Average | 4278 | 6649 |
| Range | +25/−14% | +12/−16% |

Note: Data from *Field Accidents: Data Collection, Analysis, Methodologies and Crash Injury Reconstruction*, 1985, paper 850437, "Barrier Equivalent Velocity, Delta-V and CRASH3 Stiffness in Automobile Collisions" by Hight, Hight, and Lent-Koop, Figure 16.4.

As the various tables show, the stiffness coefficient generally increases with the curb weight of the vehicle. This follows since heavier vehicles usually have more sheet metal in them making them structurally stronger. In fact, a rough estimate of the stiffness coefficient can be made by using the following empirical equation:

$$k = W(1.75 \text{ ft/in}) \tag{lv}$$

where $k$ = frontal stiffness coefficient and $W$ = curb weight of the vehicle in pounds.

The side stiffness coefficient for a car ranges from 9000 lb-ft/in for small cars to 11,500 lb-ft/in for large cars. The rear stiffness coefficient for a car ranges from 3500 lb-ft/in for older cars, to 6000 lb-ft/in for late model cars. When a motorcycle impacts the side of a car, the stiffness coefficient for the impact depth is reportedly about 1200 lb-ft/in for a 200-pound motorcycle. For a 350-pound motorcycle the stiffness coefficient is reportedly about 1750 lb-ft/in.

## 16.12   Some Additional Comments

As noted before, the stiffness coefficient is basically the force at which buckling occurs. For example, a car with a frontal stiffness coefficient of 6000 lb-ft/in, has a Euler buckling load of 500 lb. However, it should be recognized that this is the force that must be applied evenly across the front of the car. Thus, when the impact across the front is not even, an average depth is determined by making depth measurements at regular intervals and averaging them, as shown in Figure 16.5.

With regard to side crush, this is usually taken to mean the average crush over the area in which impact occurred, especially the area of the vehicle

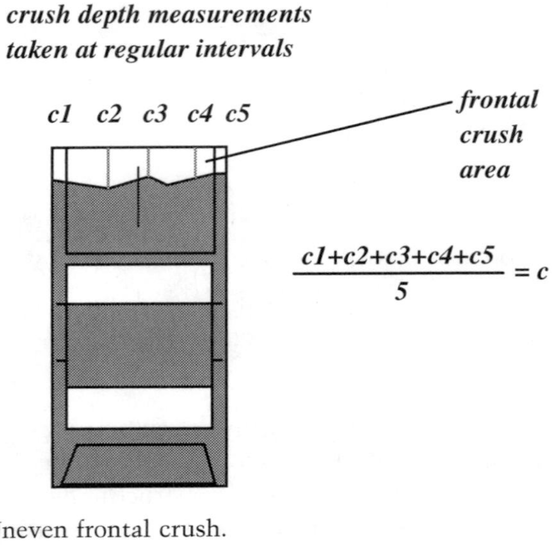

*crush depth measurements*
*taken at regular intervals*

*c1  c2  c3  c4  c5*                                    *frontal*
                                                        *crush*
                                                        *area*

$$\frac{c1+c2+c3+c4+c5}{5} = c$$

**Figure 16.5** Uneven frontal crush.

located between the front and rear wheel wells. This is often simply the width across the front of the impacting vehicle.

Because the elastic energy, whether it be a front, side, or rear impact, is usually only a very small component of the total energy consumed in an impact, a reasonable approximation in many instances is to assume that the energy expended in the crush of one vehicle is the same as that expended in the other.

## Further Information and References

*Atlas of Stress Strain Curves*, Howard Boxer, Ed., ASM International, Metals Park, Ohio, 1987. For more detailed information please see Further Information and References in the back of the book.

*Automobile Side Impact Collisions – Series II*, by Severy, Mathewson and Siegel, University of California at Los Angeles, SAE SP-232. For more detailed information please see Further Information and References in the back of the book.

*Basic Principles and Laws of Mechanics*, by Alfred Zajac, D.C. Heath and Co., Chicago, 1966. For more detailed information please see Further Information and References in the back of the book.

*A Crash Test Facility to Determine Automobile Crush Coefficients*, by Miyasaki, Navin, and MacNabb, University of British Columbia, SAE Paper 880224. For more detailed information please see Further Information and References in the back of the book.

"Differences between EDCRASH and CRASH3," by Day and Hargens, Engineering Dynamics Corp., SAE 850253. For more detailed information please see Further Information and References in the back of the book.

*Dynamics*, by Pestel and Thomson, McGraw-Hill, New York, 1968, pp. 349–353. For more detailed information please see Further Information and References in the back of the book.

"Energy Basis for Collision Severity," by K. L. Campbell, GM Safety Research and Development Laboratory, July 1974, SAE 740565. For more detailed information please see Further Information and References in the back of the book.

*Energy Methods in Applied Mechanics*, by Henry Langhaar, John Wiley & Sons, Inc., New York, 1962. For more detailed information please see Further Information and References in the back of the book.

*Field Accidents: Data Collection, Analysis, Methodologies, and Crash Injury Reconstructions*, Society of Automotive Engineers, February 1985. For more detailed information please see Further Information and References in the back of the book.

*Formulas for Stress and Strain*, by Raymond Roarck, McGraw-Hill, New York, 1943. For more detailed information please see Further Information and References in the back of the book.

The Insurance Institute for Highway Safety and the Highway Loss Data Institute are organizations supported by the insurance industry. For more detailed information please see Further Information and References in the back of the book.

*Measuring Protocol for Quantifying Vehicle Damage from and Energy Basis Point of View,* by Tumbas and Smith, SAE Paper 880072. For more detailed information please see Further Information and References in the back of the book.

*Motor Vehicle Accident Reconstruction and Cause Analysis,* by Rudolf Limpert, 2nd ed., 1984, The Michie Company, Charlottesville, Virginia. For more detailed information please see Further Information and References in the back of the book.

NHSTA. For more detailed information please see Further Information and References in the back of the book.

"An Overview of the Way EDSMAC Computes Delta-V," by Day and Hargens; Engineering Dynamics Corp., SAE 880069. For more detailed information please see Further Information and References in the back of the book.

*An Introduction to the Use of Generalized Coordinates in Mechanics and Physics,* by William Byerly, Dover Co., New York, 1965. For more detailed information please see Further Information and References in the back of the book.

*Symposium on Vehicle Crashworthiness Including Impact Biomechanics,* Tong, Ni, and Lantz, Eds., American Society of Mechanical Engineers, AMD-Vol. 79, BED-Vol. 1, 1986. For more detailed information please see Further Information and References in the back of the book.

# Curves and Turns

17

Personally, I never cared for fiction or story-books. What I like to read about are facts and statistics of any kind. If they are only facts about the raising of radishes, they interest me. Just now, for instance, before you came in I was reading an article about mathematics. Perfectly pure mathematics. My own knowledge of mathematics stops at twelve times twelve, but I enjoyed that article immensely. I didn't understand a word of it; but facts, or what a man believes to be facts, are always delightful.

— **Mark Twain,** *during an interview by Rudyard Kipling, 1889*

When in doubt, have two guys come through the door with guns.

— **Raymond Chandler,** *noire detective story writer, 1888–1959*

A problem well stated is a problem half solved.

— **Charles Kettering,** *inventor of the electric self-starter, 1876–1958*

## 17.1  Transverse Sliding on a Curve

When a vehicle executes a turn, the vehicle experiences a centripetal force in accordance to the following:

$$F_c = mv^2/R = m\omega^2 R \qquad (i)$$

where $F_c$ = centripetal force due to rotational motion, R = turning radius, v = tangential velocity of vehicle, i.e., forward speed of vehicle, m = mass of the vehicle, and $\omega$ = angular velocity.

When the centripetal force on the vehicle is significant, it can cause the vehicle to either slip off the roadway or turn over. If the centripetal force is less than is required to cause the vehicle to turn over, but more than the sideways frictional force of the tires, it will slide. Ignoring turnover for the moment, a slide will occur when:

$$F_c \geq F_f \tag{ii}$$

where $F_c$ = centrifugal force and $F_f$ = sideways frictional force.

Appropriate substitution into Equation (ii) above gives the following:

$$mv^2/R \geq Wf \tag{iii}$$

$$v^2/(gR) \geq f \quad \text{or} \quad v \geq [gRf]^{1/2}$$

where f = coefficient of friction, g = gravitational constant, and W = weight of vehicle.

The meaning of Equation (iii) is simple: if the value of "$v^2/(gR)$" is greater than the value of "f," the vehicle will lose traction and slide. If this occurs, the direction of slide will be the tangent to the curve at the point where the centripetal force exceeded the restraining sideways frictional force.

Equation (iii) explains why it is best to steer out of a curve rather than steer into it. During a turn, steering into the turn causes the value for "R," the radius of curvature, to decrease. A decrease in "R" in Equation (iii) causes the value of "$mv^2/R$" to increase. If the vehicle was already near the point of losing traction, a small decrease in "R" may be sufficient to increase the centripetal force to the point where it exceeds the frictional forces holding it. However, steering out of the curve causes the value of "R" to increase, which decreases the centripetal force.

Also, if a vehicle is taking a turn near the point at which sliding may occur, any lessening of the side coefficient of friction can cause the vehicle to slide out of the turn. Such reductions in the side coefficient of friction can not only be caused by variations in the pavement, but also by braking. When there is significant slip in the forward or longitudinal direction, the coefficient of friction in the transverse direction decreases. Thus, braking during a tight turn can cause a vehicle to lose side traction and slide out of the turn.

This is the reason why it is sometimes recommended that a driver accelerate slightly during a turn and not attempt any braking. By accelerating slightly through a turn, the amount of slip between the tires and the pavement is minimized, and the lateral coefficient of friction remains as high as possible.

Professional drivers, especially on dirt tracks, often use these principles to turn corners faster than otherwise. For example, when approaching a turn, the driver will cause the rear wheels to lose traction and the rear end of the vehicle will swing around. When the rear end has swung around to the point where the car is pointed toward the inner radius of the turn, the driver will steer out of the slide and then accelerate. In a sense, instead of steering through the turn, the driver literally rotates his vehicle about the front wheels. This allows a turn to be taken without slowing down. (Like all other such vehicular maneuvers described in this book, it is not recommended that this be done by an amateur.)

So far it has been assumed that the roadway is flat. In many cases, a curved roadway will be banked, that is, it is set such that the roadway has a certain pitch or angle in the transverse direction. If the roadway has been well designed, it will be banked such that if the curve is driven at the recommended speed, the centripetal force effects will be balanced out by the angle of the bank, or the superelevation, as it is sometimes called.

If the angle of the bank just cancels out the centripetal force, the following holds:

$$(mv^2/R)(\cos\theta) = W(\sin\theta) \qquad\qquad (iv)$$

where $\theta$ = angle of bank of the roadway.

Solving Equation (iv) for "$\theta$" yields:

$$\tan\theta = (v^2)/(gR) \qquad\qquad (v)$$

$$\theta = \text{Arctan}[(v^2)/(gR)].$$

If the angle is small, say less than about 8 degrees, then the small angle approximation can be applied where

$$\tan\theta \sim \theta \text{ when the angle is measured in radians.}$$

Of course, since the definition of measurement in radians is the amount of circumference of a circle divided by its radius, then

$$\theta = h/l$$

where h = the height of the roadway above level and l = the length across the roadway.

Thus, when the angle of bank is 8 degrees or less, the following holds:

$$h/l = \theta = (v^2)/(gRf). \qquad\qquad (vi)$$

If it is presumed that the roadway is banked positively (the angle of bank helps keep the vehicle from side sliding during the curve), then the general expression for slippage is as follows:

$$(mv^2/R)(\cos\theta) \geq Wf(\cos\theta) + W(\sin\theta) + (mv^2/R)(\sin\theta)f \qquad (vii)$$

$$v^2 \geq [gRf + gR(\tan\theta)]/[1 - f(\tan\theta)]$$

$$v^2 \geq [gRf + gR(h/l)]/[1 - f(h/l)]$$

Since the term "f(h/l)" is small, and becomes even smaller under the square root radical, the above expression for turns on banks of 8 degrees or less can be simplified to the following:

$$v \geq [gR(f + h/l)]^{1/2}. \tag{viii}$$

When the banking angle is 8 degrees, or "h/l = 0.14," the error in the above equation is about −5%, that is, Equation (viii) will compute a speed that is about 5% too low when the angle of bank is 8 degrees. Since most roadways are banked at angles much less than 8 degrees, the approximation is reasonably accurate for most purposes.

Sometimes an older roadway or a very poorly designed one will have negative banking. Negative banking is where the angle of the bank reduces the speed at which a turn can be made. When there is negative banking, Equation (vii) has a change in sign in the second and third terms on the right side of the inequality as follows:

$$(mv^2/R)(\cos \theta) \geq Wf(\cos \theta) - W(\sin \theta) - (mv^2/R)(\sin \theta)f \tag{ix}$$

Equation (ix) can be similarly reduced to the following:

$$v^2 \geq [gRf - gR(h/l)]/[1 + f(h/l)] \quad \text{or} \tag{x}$$

$$v \geq [gR(f - h/l)]^{1/2}$$

when f(h/l) is small.

Equation (x) shows that if the coefficient of friction "f" becomes less than "h/l," the vehicle can slide down the bank without having any forward motion at all.

The above relations can be useful in setting a lower bound limit to the speed of a vehicle if it slid out of the curve. If there is a question concerning whether the vehicle slid out of the curve due to excessive speed for the curve, or just by poor operation of the brakes at a lower speed, the skid marks can be examined. Skid marks which are initiated by a braking action will generally have longitudinal striations left by the tire treads. Skid marks left by lateral slide caused by excessive speed in rounding the curve, will usually have barber pole-appearing striations, showing that the tire was still rotating forward as the transverse sliding occurred. Such marks are usually call yaw marks or yaw skid marks.

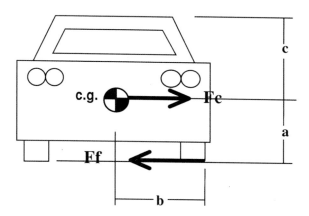

**Figure 17.1** Forces during a turn.

## 17.2   Turnovers

A turnover occurs during a turn or curve when the centripetal force becomes sufficient to overcome the forces that hold the vehicle on the ground. However, in order for a turnover to occur, the vehicle must maintain lateral traction so that lateral slip does not occur first. Referring to Figure 17.1, the moment and force equations that determine when turnover occurs are as follows:

$$(mv^2/R)a > W(b) \quad \text{and} \tag{xi}$$

$$Wf \geq (mv^2/R) \tag{}$$

where $W$ = weight of the vehicle, $m$ = mass of the vehicle, $g$ = acceleration due to gravity, $R$ = radius of curve, and $a$ and $b$ are as shown in Figure 17.1.

Solving Equations (xi) gives the following:

$$[gRf]^{1/2} > v > [gR(b/a)]^{1/2}. \tag{xii}$$

If the roadway is positively banked, the conditions for turnover are then modified, as discussed in the previous section, to the following:

$$[gR(f + h/l)]^{1/2} > v > [gR(b/a)]^{1/2}. \tag{xiii}$$

Similarly, if the roadway is negatively banked, the conditions for turnover are as follows:

$$[gR(f - h/l)]^{1/2} > v > [gR(b/a)]^{1/2}. \tag{xiv}$$

Equation (xiii) contains the basic information necessary to determine whether a vehicle will turn over first or slide first. Rearranging Equation (xiii) gives the following relation, which defines the conditions under which turnovers occur:

$$(f + h/l) > v^2 gR > (b/a) \qquad \text{(xv)}$$

$$\text{or} \quad (f + h/l) > (b/a).$$

A typical compact car will have an "a" value of about 2 feet, and a "b" value of about 2.58 ft. Thus, in order for such a car to turn over in making a curve, the coefficient of friction must be greater than 1.29. Since a typical coefficient of friction for a dry, very soft tire is no more than about .80, under most circumstances, a typical compact car will slide out of a turn rather than turnover. Of course, a coefficient of friction of 1.29 or greater can be accomplished by side impact with a curb or other type of low barrier. In such cases, lateral sliding is prevented by the barrier.

The maximum coefficient of friction for a soft truck tire on dry concrete can be as high as 0.94. This means that a truck or trailer with a "b/a" ratio of 0.94 or less could turn over in a turn. Since a common width of a truck or trailer is about 80 inches, it is possible for a truck or trailer whose center of gravity is 43 inches or more off the ground to turn over rather than slide. Since the floor height of commercial trailers is often 4 feet or more and cargo is stacked on top of the floor, commercial trailers can turn over in a turn rather than slide. This is especially true if the trailer is stacked with cargo to the ceiling and is carrying the maximum load. This greatly shifts the center of gravity upward.

Further, if the load in a trailer is not centered over the wheels, then the center of gravity will not be located halfway between the wheels as is assumed. If a heavy load, for example, were stacked along one side of the trailer, the "b" value might be much shorter than half the width of the trailer. This would further increase the tendency of the trailer to tip over during a turn.

## 17.3   Load Shifting

Often, when a trailer turns over on a curve, the driver blames "load shift" rather than excessive speed. "Load shift" is when the load being carried shifts position during the turn, causing the center of gravity to shift to an unfavorable position. The shifting load may even impact the wall of the trailer, which would further increase the tendency of the trailer to tip. In such cases, it is worth noting that the U.S. Department of Transportation (USDOT) regulations, as contained in the *Federal Motor Carrier Safety Regulations,* require that the load being carried be positioned in the most favorable position to minimize the center of gravity (Section 393.100). The regulations also require that the load be secured to prevent load shifting.

Because of the common use of load shifting to explain trailer turnovers on curves, the following USDOT regulation (Section 392.9) is quoted in whole:

> (a) General. No person shall drive a motor vehicle and a motor carrier shall not require or permit a person to drive a motor vehicle unless —
> (1) The vehicle's cargo is properly distributed and adequately secured as specified in 393.100–393.106 of this subchapter.

## 17.4   Side vs. Longitudinal Friction

The total amount of frictional resistance of a tire has been previously given as:

$$F_f = Wf \qquad (xvi)$$

where $F_f$ = the friction force resisting motion, f = coefficient of friction, and W = the load applied on the tire.

The above is generally true when only one resistive force is at work, like a longitudinal or transverse resistive force. However, when both side (transverse) and longitudinal resistive forces are at work, the total frictional force is the vector sum of the two component frictional forces.

For example, let "$F_{long}$" be a frictional force being applied in the longitudinal direction, and "$F_{side}$" be a frictional force being applied along the side or transverse direction. In that case, the following would hold:

$$[F_{side}^2 + F_{long}^2]^{1/2} = F_f = Wf \quad \text{or alternately} \qquad (xvii)$$

$$[f_{long}^2 + f_{side}^2]^{1/2} = f.$$

Thus, if a vehicle is accelerating forward and is using 80% of its available maximum traction to accomplish this, then there would only be 60% of the maximum traction available for side frictional forces.

$$[(0.64)F_f^2 + F_{side}^2]^{1/2} = Wf = F_f$$

$$F_{side} = 0.60F_f$$

Equation (xvii) is often called the friction circle. When longitudinal braking and side friction or traction are being considered simultaneously, however, the friction circle relationship is not exact. Laboratory and field measurements have found that the side friction is usually a little less than would be predicted strictly by Equation (xvii). This is due to tire deformation mechanics. For this reason, the following empirical modification to Equation (xvii) is employed when braking is involved:

$$[(1.2)f_{side}^2 + f_{long}^2]^{1/2} = f. \tag{xviii}$$

Using the above relationship, if a tire-pavement coefficient of friction is 0.75, and the vehicle is making a turn such that the side friction just equals the centripetal force, then a braking action causing a 0.2g slow down ($f_{long}$ = 0.2) will cause the side coefficient of friction to drop from 0.685 to 0.660, a 3.6% decrease. Since prior to the braking action the side forces just matched the centripetal forces, then the slight decrease in side friction caused by the braking will cause the vehicle to slide out of the turn.

## 17.5  Cornering and Side Slip

When a vehicle executes a turn, the front wheels are rotated a certain angle, called the steering angle. Unless the vehicle is traveling at a low speed, the steering angle and the actual direction of travel of the vehicle will not be exactly the same because of tire slip in both the longitudinal and transverse directions. Usually, the steering angle will be a little greater than the actual angle of turn. The difference between the two angles is called the angle of slip.

One way to imagine what is going on, is to think about the car going straight forward and then turning. Unless the vehicle is traveling at a very slow speed, the forward momentum of the vehicle will tend to continue to push the vehicle straight ahead, while the front wheels try to turn the vehicle left or right. This results in some slip across the contact "patch" between the tire and the ground in both the longitudinal and transverse directions. Thus, the tire will exert both side and longitudinal forces.

The angle of slip effect can be easily observed by executing the same tight turn at several speeds. At very slow speeds, a car can make a tighter turn at the same steering angle than at higher speeds. In fact, with the steering wheel turned to its maximum left or right position, it can be seen that a car will spiral outward with an increasing radius of curvature as the forward speed is slowly increased.

Most ordinary turns develop slip angles of 5 degrees or less. In this range, the side forces developed by a tire follow the linear relation:

$$F_{side} = K\lambda \tag{xix}$$

where $\lambda$ = slip angle and K = turning stiffness constant for a specific load, tire type, camber, etc.

Very hard turns at relatively high speeds can have slip angles of 10 to 17 degrees. Unfortunately, above 5 degrees, the relationship between slip angle and side force is not linear. The side force curve tends to flatten out, like a square root curve.

This flattening out effect at high slip angles is sometimes observed at dirt track races, where a driver takes a corner at too sharp a turn. Instead of turning, the car will often just continue forward until it slows down enough for the tires to "bite" the surface again.

In general, radial tires have higher "K" values than bias tires. Increased camber angles of the tires will reduce "K." Tire inflation pressure and tire load will also increase the value of "K."

For passenger cars, "K" may range from 50 to 200 lbf/deg with an average value of about 100 lbf/deg. For trucks, "K" may range from 200 to 500 lbf/deg. with an average value of about 300 lbf/deg.

## 17.6 Turning Resistance

When a vehicle executes a turn, a resistance to forward motion is produced by the slip angle of the tires. If the slip angle is "$\lambda$" and the side coefficient of friction is "$f_{side}$," then the relationship between slip angle, side friction, and forward friction is as follows:

$$f = f_{side} (W_{front}/W)(\sin\lambda) \qquad (xx)$$

where $W_{front}$ = load carried by the front wheels and W = total load of the vehicle carried by both axles.

When the front wheel slip angle is small, the rear wheels have a negligible slip angle, and Equation (xx) is appropriate. However, if the turn is large enough where slip angle in the rear wheels must be taken into account, Equation (xx) must be expanded to include the rear wheels as follows:

$$f = f_{side} [(W_{front}/W)(\sin\lambda_{front}) + (W_{rear}/W)(\sin\lambda_{rear})] \qquad (xxi)$$

For example, if a 3000 lbf vehicle has a cornering stiffness of 100 lbf/deg, the front wheels are turning at a slip angle of 3 degrees and the rear wheels are turning at a slip angle of 1 degree, and the front axle carries 60% of the total load, then the forward turning resistance is as follows:

$$F_{side} = K\lambda = (100 \text{ lbf/deg})(3 \text{ deg}) + (100 \text{ lbf/deg})(1 \text{ deg}) = 400 \text{ lbf}$$

$f_{side} = F_{side}/W = 400 \text{ lbf}/3000 \text{ lbf} = 0.133$

$f = (0.133)[(0.60)(\sin 3) + (0.40)(\sin 1)] = 0.005$

In this case, when the total weight of the vehicle is taken into consideration, the total resistive force in the direction of travel caused by turning is:

$$F_f = Wf = (3000 \text{ lbf})(0.005) = 15 \text{ lbs.}$$

This is not particularly significant. In general, at low to moderate slip angles, the effects of cornering with respect to resistance to forward motion are negligible.

## 17.7  Turning Radius

The following discussion is limited to two-axle vehicles.

At slow speeds, the turning radius, as measured to the longitudinal axis of the vehicle, is given by the following:

$$R = L/(j_f - j_r) \qquad \text{(xxii)}$$

where R = turning radius, L = distance between axles (wheel base), j = steering angle (in radians) as measured from longitudinal axis, subscript r denotes rear, and subscript f denotes front.

In most vehicles, "$j_r$" is zero, that is, the rear wheels don't steer. However, there have been a limited number of production passenger vehicles that have steerable rear wheels, and there are also a limited number of special-purpose industrial vehicles that have steerable rear wheels. In such cases, "$j_r$" has a value, usually, but not always, the same as "$j_f$" at normal driving speeds.

The angle turned by the steering wheel usually is not the same angle that is turned by the front wheels. The ratio between the steering wheel angle and the front wheels steering angle is called the steering ratio.

$$\theta = \alpha/\varphi \qquad \text{(xxiii)}$$

where $\alpha$ = steering wheel angle and $\varphi$ = steering angle of front wheels.

At speeds where turning slip occurs, the turning radius is given by:

$$R = L/(\varphi_f - \lambda_f - \varphi_r + \lambda_r) \qquad \text{(xxiv)}$$

where L = wheel base, $\varphi$ = steering angle of front wheels (in radians), $\lambda$ = slip angle (in radians), subscript r denotes rear, and subscript f denotes front.

An inspection of Equation (xxiv) finds that there are three general regimes:

1. $\lambda_f - \lambda_r = 0$. Neutral steering. The effects of slip angle at the front wheels is canceled by the slip angle of the rear wheels.
2. $\lambda_f > \lambda_r$. Understeering. The turning radius is larger than that indicated by the steering angle.
3. $\lambda_f < \lambda_r$. Oversteering. The turning radius is smaller than that indicated by the steering angle.

In a vehicle that oversteers, the rear tires will tend to slide before the front tires. In a vehicle that understeers, the front tires tend to slide before the rear tires. In a vehicle with neutral steering, the front and rear tires tend to slide at the same time. Many production vehicles are designed to have a slight oversteer bias.

A vehicle's understeering or oversteering tendencies may be important in making curves near the critical speed for either slip or turnover. For example, in a vehicle that oversteers, if the curve is being taken at an excessive speed, release of the accelerator (and therefore reduction of the slip angle) automatically causes the vehicle to turn at a slightly larger radius without adjustment of the steering angle.

## 17.8   Measuring Roadway Curvature

Referring to Figure 17.2, it can be seen that:

$$y^2 + (L/2)^2 = R^2. \qquad \text{(xxv)}$$

It is also noted that:

$$y = R - x \qquad \text{(xxvi)}$$

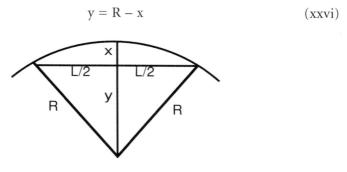

**Figure 17.2** Circle segment.

Combining Equations (xxv) and (xxvi) gives the following:

$$(R - x)_2 + (L/2)^2 = R^2 \qquad \text{(xxvii)}$$

$$R = L^2/(8x) + x/2.$$

From a practical standpoint, it is convenient to use a standard length to measure "L," the chord length, and then measure "x," the depth of the chord. If, for example, a 50-ft tape is used as the standard, Equation (xxvii) becomes:

$$R = (2500)/(8x) + x/2 = 312.5/x + x/2 \qquad \text{(xxviii)}$$

where all measurements are in feet.

## 17.9  Motorcycle Turns

The height of the center of gravity of a motorcycle and rider is often about 24 inches off the ground. When a motorcycle makes a turn, the lean of the rider and motorcycle must equalize the centripetal force in the following way:

$$(mv^2/R)a(\cos \theta) = Wa(\sin \theta) \qquad \text{(xxix)}$$

where $\theta$ = angle of lean (as measured from the vertical), W = weight of the motorcycle and rider, and a = height of the center of gravity when fully upright.

Equation (xxix) reduces to the following:

$$\tan \theta = v^2/gR \qquad \text{(xxx)}$$

$$\theta = \text{Arctan}(v^2/gR).$$

While professional riders may lean as much as 45 degrees from the vertical to take a turn, most riders usually don't lean more than about 20–25 degrees from the vertical. Inexperienced riders may lean less than 15 degrees from the vertical.

## Further Information and References

*Motor Vehicle Accident Reconstruction and Cause Analysis*, by Rudolf Limpert, 2nd ed., 1984, The Michie Co., Charlottesville, Virginia. For more detailed information please see Further Information and References in the back of the book.

*The Traffic Accident Investigation Manual,* by Baker and Fricke, 9th ed., 1986, North-western University Traffic Institute, LCCC No. 86-606-16. For more detailed information please see Further Information and References in the back of the book.

# Visual Perception and Motorcycle Accidents

18

Lift up your eyes, and behold them that come from the north: where is the flock that was given thee, thy beautiful flock?

— Jeremiah 13:20

He came out of nowhere; I checked before I made the turn and everything was clear.

— **Routine statement made after a motorcycle/car accident.**

## 18.1  General

The following chapter about motorcycle accidents is somewhat different than the other chapters. In part, it demonstrates that some accidents have a built-in tendency to occur once a certain threshold is crossed: even when everyone involved is apparently doing the right thing.

A large fraction of accidents involving motorcycles and four-wheel vehicles share a number of common characteristics including the following.

- The accident involved a motorcycle and a car, often a large car or truck.
- The accident occurred on a straight section of roadway, often a simple two-lane road. The car and motorcycle were headed toward each other.
- The driver of the car or truck was usually over 40 years of age.
- The driver of the car or truck was making a left turn, and the motorcyclist was proceeding straight ahead.
- The accident occurred when the motorcycle impacted the left turning car.
- The driver of the car often indicated that he or she did not see an approaching motorcycle. There was often no visual obstructions to impede either driver's view of the other.

To determine the underlying cause of this oft-repeated accident pattern, a small study was conducted. The following are the results of that study.

343

## 18.2  Background Information

According to the 1990 Statistical Abstract of the United States, in 1987, there were about 181 million registered vehicles in the U.S. This includes cars, trucks, and buses. Of this total, 139 million were automobiles. Additionally, there were 5.1 million registered motorcycles. In terms of the total number of registered vehicles, motorcycles constituted about 2.7%, cars constituted 75%, and trucks constituted 22.6%.

On the average in 1987, a registered motorcycle was driven only 17.6% as far in a year as the average registered car, truck, or bus. The total road miles of motorcycles were only 0.5% that of the total for cars, trucks, and buses.

Considering the statistics, it might be expected that the number of accidents involving motorcycles would be comparatively low. Based on the single factor of total miles driven, it might be expected that the share of accidents involving motorcycles would be perhaps 0.5%, give or take some statistical variance.

The actual accident figures, however, are as follows. The total number of car accidents in 1985 was 27.7 million. Truck accident totals were 6.1 million, and the total for motorcycle accidents was 440,000. Of the total of 34.2 million vehicle accidents, 81% were car accidents, 17.8% were truck accidents, and 1.3% were motorcycle accidents.

Further, the total number of deaths in 1986 resulting within 30 days of a vehicular accident was 46,100. Persons in vehicles accounted for 33,700, pedestrians accounted for 6,800, motorcyclists accounted for 4,600, and bicyclists accounted for 900. Of the deaths related to vehicular accidents, motorcyclists accounted for 10% of the total.

In short, the statistics indicate that on a per-driven-mile basis:

- motorcyclists have 2.6 times more accidents.
- motorcyclists have 20 times more fatalities than drivers of other types of vehicles.

The reason for the higher number of fatalities per road mile of motorcycles vs. other vehicles is readily apparent. In almost all severe collisions, the motorcyclist is ejected from the motorcycle and dissipates his or her kinetic energy by impacting the ground or slamming against another object, often the other vehicle. In this respect, the higher per-mile fatality rate of motorcycles vs. cars simply reflects the relative safety of the vehicle configuration once an accident has occurred. In short, it is safer to be in a car than on a motorcycle when an accident occurs.

However, the reason motorcycles have more accidents per driven road mile is not so readily apparent. Some argue that a motorcyclist should be able to more easily avoid an accident due to the greater maneuverability of a motorcycle. Some may argue that since fewer infirm persons drive motorcycles than cars, and that motorcyclists tend to be young males, the average motorcyclist would have better reflexes, and would therefore react faster to accident situations.

As noted previously, a large proportion of motorcycle accidents involve motorcycles that are driving straight ahead and collide with cars making left turns. This may be a left turn into a driveway, or into a cross street. In nearly all these cases, the driver of the car indicates that he or she never saw the approaching motorcycle, or did not see it until it was too late.

Thus, the perceptions of the motorcycles and motorcyclists by drivers of the other vehicles appear to be key factors in the higher accident rate of motorcycles, not the maneuverability of the motorcycles or the reflexive actions of the motorcyclists.

## 18.3  Headlight Perception

When a driver observes an approaching vehicle at night, the most readily apparent visual cue is the angular separation of the two headlights located on either side of the front of the vehicle. Since the distance from the left to the right headlight is more or less standardized, an experienced driver can estimate the distance between his car and the oncoming vehicle by the amount of angular separation between the headlights. In order to estimate the closing speed, the driver observes the oncoming vehicle for a few moments to note how quickly the angular separation increases.

For example, a 1975 Ford Mustang has headlights that are 52 inches apart. At a distance of 500 feet, the angular separation between the two headlights is 0.497 degrees. At 250 feet, the angular separation is 0.993 degrees. At 125, the angular separation is 1.985 degrees.

The angular separation is given by the following equation:

$$\alpha = \arctan (x /d) \qquad (i)$$

where $x$ = distance from center of left headlight to right headlight, $d$ = closing distance between vehicles, and $\alpha$ = angular separation in degrees.

Due to the small angles involved, the angular separation is, for all practical purposes, a linear function as follows:

$$\alpha = 57.3(x/d) \qquad (ii)$$

Even as close as 50 feet, the error of the linear approximation given in Equation (ii) is less than 1%.

Thus, when the headlights are twice as far apart as first observed, the oncoming car has halved the closing distance. This linear relationship is easy, and appeals to common intuitive notions of speed and distance. For most purposes, the small variations of headlight spacing among car models are negligible.

However, a motorcycle does not have two headlights spaced apart; it typically has only one front headlight. The angular separation of headlights as observed on an approaching car or truck is not there to cue a driver of the distance and speed of an approaching motorcycle. The only cue available is the light intensity of the single headlight.

Light diminishes in intensity as the square of the distance. In other words, a light with an intensity of 24 foot-candles at 50 feet will have an intensity of 6 foot-candles at 100 feet. The relationship is expressed as follows:

$$I = k/d^2 \tag{iii}$$

where I = light intensity, k = arbitrary constant, and d = the closing distance.

In order for a driver to estimate the closing distance of an approaching single-headlight-equipped motorcycle, he or she must be able to compare the light intensity of the headlight as it is observed with some memorized value. This is difficult.

Some brands of headlights appear brighter than others because of the type of glass in the cover lens, or the efficiency of the reflector behind the filament. Some headlights produce slightly different colors of light that affect their perceived intensity. The aim of the headlight will also affect the perceived brightness, and the aim of the headlight will be affected by the slope and curvature of the road. Furthermore, road dirt on the headlight cover lens will affect its brightness as well as atmospheric conditions such as fog, rain, snow, haze, etc.

In order to estimate the speed of an approaching motorcycle, the driver of the other vehicle must compare the intensity of the headlight at one instant with another instant and apply the inverse-square rule. As compared to estimating the speed of an oncoming car from the angular separation of its lights, the driver must now use a different and less reliable visual cue to determine distance, and must apply a more difficult and less familiar mathematical rule to estimate speed.

As was noted before, only 2.7% of registered vehicles in 1987 were motorcycles, and motorcycles constituted only 0.5% of all road miles driven. Most drivers are very experienced in observing two-headlight vehicles at night, but significantly less experienced in observing approaching motorcycles. In cold climates, motorcycles may not even be encountered at all during the winter months.

Thus, as compared to vehicles with two headlights, a driver is more likely to misjudge the closing distance and speed of a motorcycle at night because:

- the visual cue is completely different.
- the rule that applies to the cue to estimate speed is different.
- the experience factor in applying the above two items is low.

## 18.4  Daylight Perception

The width of many cars is about 72 inches. The average width of a motorcycle without extensive faring is about 30 inches. Thus, the width of a car is about 2.4 times that of a motorcycle.

A person with 20/20 vision will be able to readily distinguish and read 4-point type at a reading distance of 12 inches. A 4-point type letter is 4/72 inches tall. Therefore, a person with 20/20 vision can distinguish and read type that subtends an arc of about 1/4 degree, which is about half the width of a full moon.

A person with 20/50 vision can readily distinguish and read materials at 20 feet that a person with 20/20 vision can distinguish and read at 50 feet. Some states allow persons with vision correctable to 20/50 drive.

A person with 20/20 vision will fully recognize a vehicle during daylight hours when it subtends a horizontal arc of at least 1/4 degree. For a car, this occurs when the car is 1375 feet away, or about a quarter mile. For a motorcycle, this will occur when the motorcycle is 573 feet away.

A person with 20/50 vision would not be able to distinguish the same car until it subtended a horizontal arc of 5/8 degree. This would occur when the car is 550 feet away, or when a motorcycle is 229 feet away.

Most two-lane highways are posted for no more than 55 mph, or 80.7 ft/sec. If two vehicles approach one another at the maximum legal speed, the closing velocity is 110 mph or 161.4 ft/sec. A driver with 20/50 vision would discern an oncoming car 3.4 seconds before they met, but would only discern a motorcycle 1.4 seconds before they met.

Generally, younger drivers have faster reaction times than older drivers. A reaction time of 1/2 second for young drivers and 1 second or more for older drivers is typical. If both an oncoming motorcycle and a car are traveling at 55 mph, an older car driver with 20/50 vision may barely have enough time for recognition and reaction before the two vehicles meet.

As has been noted, a significant portion of car and motorcycle accidents involve an oncoming motorcycle and a left turning car. Since a modest car is about 16 feet long, a left turn from stop will take about 19 feet of arc for the front end of the car to reach the left shoulder and another 16 feet to get the car clear of the roadway. A moderate acceleration for a medium car from stop is 5 ft/sec². A simple calculation shows that the required time for a modest car to fully execute a left turn and be clear of the left lane is about 3.74 seconds.

If the driver of the car has 20/50 vision, he or she will discern a car at a distance of 550 feet away, and a motorcycle at 229 feet. This means that an oncoming car can be approaching at speed of 100 mph or less, and will be seen soon enough by the driver to avoid making a left turn in front of him. However, an oncoming motorcycle could collide with the left turning car if the motorcycle were exceeding 61 ft/sec or 42 mph. The driver may not see the motorcycle in time prior to making the left turn, and the motorcycle may close the distance before the car can fully execute the turn.

The problem is further exacerbated during hard braking, because motorcycles require more driving technique, especially while turning. In fact, even under normal circumstances, a motorcycle typically takes longer to stop than a car. While the motorcyclist may see the left turning car well in advance, he or she may not perceive that the driver is about to execute a left turn until he or she is too close to affect a safe stop.

A typical hard braking deceleration for a motorcycle is about 16 ft/sec². At 42 mph, a motorcycle will require about 2.76 seconds to come to a complete stop. This time does not include any reaction time. If the motorcyclist traveling at 42 mph does not perceive within 1 second that the car is initiating a left turn in front of him or her, he or she will not be able to fully stop the motorcycle before impacting the car. The motorcyclist must then make an evasive turn of some sort to avoid making contact with the car.

Thus, while the above kinematic analysis indicates that car drivers with 20/20 vision will likely be able to discern a motorcycle well in advance of any problem, a car driver with 20/50 vision may not discern an approaching motorcycle in time. It may be no coincidence that the ratio of average car width to average motorcycle width is about the same as the ratio of motorcycle accidents per mile to other vehicle accidents per mile.

## 18.5  Review of the Factors in Common

In Section 18.1 of this chapter, several items were listed as being common to this type of motorcycle and car accident. At this point all of them have been explicitly explained except two: the age of the driver of the car and the size of the car.

Larger size cars or trucks take longer to get fully across the roadway when making a left turn. The longer length requires a little more time to clear off the roadway, and takes up more broadside space across the roadway, making it a larger obstacle for the motorcyclist to avoid. Thus, the motorcyclist is more likely to collide with a long vehicle than a short one.

With respect to the age of the driver, it is known that some persons experience a significant drop in visual acuity in their mid-forties or so. Reflexes also slow down in middle age. Up until the mid-forties, a person's visual acuity may be very stable, changing very little for many years. There would be little need for a change in glasses or perhaps even a need for glasses. Often, persons who experience such a loss in vision are slow to recognize or admit the problem.

Some states require that driver's licenses be renewed every three years; others require renewal after six years. Usually, a license renewal requires a vision test when renewal is made. A driver who has had a significant drop in visual acuity will likely have the problem detected at that time. However, if the drop in visual acuity occurs just after a license renewal, a driver could go perhaps 3–6 years before having to correct the problem.

It was previously demonstrated that the failure to recognize an approaching motorcycle in time can occur when the car driver's vision is 20/50 and the speed of the motorcycle exceeds 42 mph. In some states, visual acuity correctable to 20/50 is legal. Thus, the driver of the car may have visual acuity sufficient to satisfy legal requirements, but insufficient to avoid this type of accident.

## 18.6  Difficulty Finding a Solution

It might be argued that something should be done to correct this problem and save lives. However, the obvious solutions are not easy to deal with. Should people with 20/50 vision not be allowed to drive, or make left turns? Should motorcycles be outlawed, or perhaps forced to have two headlights spaced apart like a car? Should speed limits for motorcycles be lowered below 42 mph in areas where left turns by cars occur frequently? None of these solutions are politically acceptable, or even practical.

It has been a difficult legal struggle to simply get motorcyclists to wear helmets, much less alter the basic design of their vehicles. And, how may registered voters are car drivers that are also middle age or older? How well would they appreciate a law that would not allow them to drive unless they have vision correctable to 20/20?

In short, the problem appears to be very difficult, if not intractable. Fortunately, the number of persons injured in this type of accident is relatively small as compared to the number of deaths and injuries from other types of accidents.

## Further Information and References

*Bicycle Accident Reconstruction: A Guide for the Attorney and Forensic Engineer*, by James Green, P.E., 2nd ed., Lawyers and Judges Publishing Co., Tucson, AZ, 1992. For more detailed information please see Further Information and References in the back of the book.

*Statistical Abstract of the United States*, U.S. Printing Office. For more detailed information please see Further Information and References in the back of the book.

# Interpreting Lamp Filament Damages 19

But, soft: what light through yonder window breaks?
— *Romeo and Juliet*, Act II, Scene II, *William Shakespeare, 1564–1616*

## 19.1   General

Many vehicular and marine accidents occur at night. In such instances, it is not uncommon for one party involved in the accident to claim that the headlights, taillights, running lights, warning lights, brake lights, or turn signals of the other vehicle were not on or did not work at the time of the accident. Of course, if the lights were not on at the time of the accident, there is a legitimate issue concerning the visibility of the unlit vehicle and the prudence of the person driving it in that condition.

Even when the accident has not occurred during nighttime hours, there can be reasons for wanting to know if certain lights were on at the time of the accident. For example, in some cars, when the car is in reverse gear, backup lights will come on. If it can be shown that the backup lights were on at the time of the accident, then the position of the gear selector can be verified at the time of the accident. Likewise, if a driver was braking at the time of impact, the brake lights would be expected to be operating. Thus, if it can be shown that the brake lights were on at the time of impact, then the driver's statement about braking can be verified by physical evidence.

## 19.2   Filaments

Commercial incandescent light filaments are nearly universally made of tungsten. Tungsten is a relatively heavy metal with an atomic number of 74, and an atomic weight of 183.85. At room temperature, tungsten is a hard, brittle gray solid with a specific gravity of 19.3. As a comparison, iron has a specific gravity of 7.87. Tungsten is a transition metal and falls into the same periodic element family as chromium and molybdenum.

One of tungsten's attributes with respect to light filaments is that it has the highest melting point of all the metals, 6179°F. It is electrically conductive

**Plate 19.1** Taillight bulb. One filament sheered away and one exhibits "hot shock."

**Plate 19.2** Halogen headlight bulb. Filament disheveled due to "hot shock" during front-end impact.

and at elevated temperatures can be drawn into wire. This combination of characteristics makes it an excellent material for incandescent light filaments.

In an incandescent lightbulb, an electrical current is passed through the filament. The filament is a resistive load, and heats up due to "$I^2R$" effects. When the filament has heated up sufficiently, it emits visible light. Unlike many materials, tungsten's melting temperature is higher than the temperatures at which incandescence occurs, which ranges from 4000° to 5500°F.

## 19.3   Oxidation of Tungsten

Tungsten will oxidize in air at 752°F to form tungsten trioxide.

$$2W + 3O_2 \rightarrow 2WO_3 \tag{i}$$

As the temperature increases, the rate of oxidation rapidly increases. Also, extremely fine tungsten is highly flammable and can ignite spontaneously in air.

Since the oxidation temperature of tungsten is well below the temperature at which incandescence occurs, to keep the tungsten from simply burning up, it must be encased in a bulb containing a nonreacting atmosphere, such as nitrogen or carbon dioxide. While tungsten's reactiveness with oxygen may seem like a complication with respect to the manufacture of lightbulbs, it can be useful in determining whether a light was on at the time of an accident.

If the glass bulb around a lighted filament is broken open so that air can reach the filament, the filament will literally burn with yellow flame and white smoke. The resulting product, tungsten trioxide, will not dissolve in water. This is fortunate, because simple rainwater will not chemically dissolve the tungsten trioxide away.

Sometimes the tungsten trioxide smoke residue can be observed on the headlight reflector as a whitish substance months after the accident, even if the vehicle has not been sheltered from the weather. However, tungsten trioxide will dissolve in caustic alkali solutions.

Tungsten trioxide is normally considered canary yellow at room temperature. When heated, it becomes dark orange. It regains its original yellow coloring on cooling. Due to its yellow color, tungsten trioxide is sometimes used as a coloring pigment in ceramics. Tungsten trioxide has a specific gravity of 7.16, and has a melting temperature of 2683°F.

If the tungsten wire has been exposed to air, but not sufficiently for an extensive oxidation reaction to occur, the filament will noticeably darken. This will be in contrast to the normally silvery, shiny appearance of the filament. The darkening is caused by light interference of a thin layer of tungsten trioxide on the surface.

Thus, if the bulb envelope has broken open during the accident, a person may reasonably conclude that the light was on at the time of the accident if:

- A yellow flame was observed at the light just after the accident.
- A yellowish or whitish powder or smoke stain is observed on the reflector, in the bulb remains, or elsewhere around the light.
- The filament wire is noticeably darkened.

Chemical analysis of the residual powder or smoke residue can further confirm that the material observed is tungsten trioxide. In this regard, the identification of tungsten trioxide by chemical or physical means is a relatively unique marker. There are few common uses for tungsten trioxide, except for coloring ceramics. Unless a person crashes into a ceramic shop carrying yellow pottery, the identification of tungsten trioxide near a broken light bulb is a very certain indication that the light was on when the bulb envelop was broken open.

In some cases, it might be claimed that the lights were on when the envelope was broken open, but that the accident severed the wiring just before the bulb was broken open. Thus, the filament would not be affected. This argument is specious.

The oxidation temperature of tungsten, 752°F, is well below the temperature range of incandescence. When a light bulb is turned off, it takes some time for the filament to cool down below the oxidation temperature. Since most accident events are measured in seconds or fractions of seconds, there is generally not enough time for a filament to cool down to avoid any oxidation effects.

Further, if the bulb is relatively cool with respect to the oxidation temperature, or little air reached the filament before it cooled, the filament coil will simply exhibit discoloration in proportion to its temperature when air reached it. Dark colors in the wire, like black or purple, indicate a higher temperature than lighter colors like green or brown.

A rough indication of filament discoloration vs. initial filament temperature is given in Table 19.1.

**Table 19.1 Filament Discoloration vs. Temperature**

| Observed Color | Initial Temperature of Filament |
| --- | --- |
| Green, brown | 1100°F |
| Purple, black | 1200–1300°F |
| Yellow powder forms | +1400°F |
| White powder forms | +4000°F |

Of course, if the bulb envelope is broken open, the total lack of any tungsten trioxide formation or coil darkening indicates that the filament temperature was below the oxidation point, and that the light had been off at least a short while prior to the accident.

## 19.4   Brittleness in Tungsten

At ambient temperatures, tungsten is very brittle. It exhibits no practical ductility. The yield point and the rupture point are practically the same. The material does not significantly stretch prior to breaking. This characteristic can be put to good use in determining if the lights were off, or at least relatively cold, at the time of the accident.

In accidents where the impact deceleration is high, the filaments may break apart. If they break apart with no discernible stretching of the filament coils, the filament was cold at the time of the impact, i.e., it was not on. Examination of the fractured filament ends under magnification would also show brittle type fracture with no significant indications of ductility, such as necking, cup and cone fracture, etc.

## 19.5   Ductility in Tungsten

At elevated temperatures, tungsten becomes ductile. This characteristic allows it to be drawn into wires and filaments in the first place. Table 19.2 shows the relationship of temperature to the true plastic strain in tungsten.

True stress and true strain are the stresses and strains that occur in the material instantaneously, and are calculated using the actual cross-sectional area rather than the nominal cross-section area. True strain is generally calculated as follows:

$$\varepsilon = 2 \ln (d_o/d) \tag{ii}$$

where $\varepsilon$ = true stain, d = instantaneous diameter, and $d_o$ = nominal diameter.

**Table 19.2   True Plastic Strain vs. Temperature — Tungsten**

| Temperature (°F) | Stress (psi) | True Stain (in/in) |
|---|---|---|
| 600 | 65,000 | 0.1 |
| 800 | 65,000 | 0.2 |
| 1000 | 65,000 | 0.3 |
| 1200 | 65,000 | 0.4 |

**Plate 19.3** Tailer bulb with filament distention was lit at time of impact.

The temperature at which tungsten changes from brittle to ductile is about 645°F. This is not too much less than the lower temperature bound for oxidation to occur.

If an impact occurs when the light is on and the bulb envelope is not broken open, it is possible to still confirm that the light was on due to ductile stretch of the filament coil, sometimes called "hot shock." Like a soft spring, the filament will simply stretch out in response to the impact. If the filament breaks, the fractured ends will exhibit necking down, reduced cross-sectional area, or cup and cone fracture patterns.

By testing several lightbulbs of the same make and model at various deceleration levels, it is possible to roughly calibrate the amount of coil stretch to the deceleration experienced by the bulb at impact. However, this is not to be taken directly as the deceleration experienced by the vehicle as a whole. The localized deceleration rate in the vicinity of the bulb can differ from the deceleration of the vehicle as a whole, and the former should not be taken as an automatic measure of the latter.

Similarly, the direction of distention of the filament coil is not an automatic indication of the direction of impact. The localized impact forces in the vicinity of the bulb can be different than that of the vehicle as a whole. Also, the filament coil can experience "rebound" of the coil, which would cause it to distend in the opposite direction. In cases where there are multiple impacts, the filament may have several superimposed distentions, making correlation to any one impact extremely tenuous.

However, if information is scant, some general indications of deceleration magnitude and impact direction can be learned from a careful study of the filaments as long as the above caveats are understood.

In cases where there are two filaments in a bulb, such as in a headlight with both high and low beams, the lighted filament will indirectly heat up the unlighted filament. Thus, the lighted filament will stretch the most, or perhaps simply break. Such markers can be very useful in establishing whether the high beams or low beams were on at the time of the accident.

As a useful informational item, the low beam filament in a sealed beam is usually the upper one, the one closest to the back dazzle shade. In the duplo-type headlight, where the bulb is inserted through the back of the headlight, the upper filament is also the low beam. It is also nearest to the back dazzle shade.

## 19.6   Turn Signals

Turn signals normally turn off and on about one or two times a second. While in a signal mode, the cooling down time of a turn signal is no more than about 0.5 seconds. Oxidation of a turn signal light filament still occurs up to a second or more after the filament is turned off. Darkening and stretching can still occur several seconds after the turn signal is turned off. All of the markers used in steady light type bulbs can be applied to turn signals and flashers to determine whether they were on at the time of the accident.

## 19.7   Other Applications

The filament stretching effect can be used to determine if any of the interior lights in a vehicle were on at the time of the accident, such as dome lights, map light, etc. It is also useful in nighttime marine collisions to determine which of the running lights would have been working at the time of impact.

Of course, different make and model lights will respond differently to impact declarations. While some will stretch readily, others may not discernibly distend under the same deceleration, and as inferred before, the localized deceleration can significantly differ from point to point in the vehicle.

## 19.8   Melted Glass

When a glass bulb envelope breaks open, tiny shards are produced. If some of these shards land on the hot filament, they may partially melt and stick to the filament. Glass typically melts at 1200°F or more.

Examination of the filament under magnification is often required to observe this effect. Finding such melted glass particles on the filament can confirm that the filament was at least hotter than 1200°F when the glass envelope was broken. Since incandescence occurs at temperatures well above 1200°F, the presence of melted glass on the filament cannot automatically be taken to mean that the light was on; it only means that the light had been on just prior to the accident and had not yet cooled to less than 1200°F.

## 19.9   Sources of Error

In cases where a vehicle has experienced a previous impact, the light filaments may exhibit stretch effects from the prior accident. In combination with new stretching, the combined direction or stretch may give an erroneous impact direction.

Repeated hard road shocks from potholes can also cause filaments to stretch in other directions. Very old filaments can appear stretched due to normal, high temperature creep effects, sometimes call "age sag." Thus, the age and use of the vehicle should be noted when interpreting impact direction from stretched filaments. Age sag and slight distentions from hard pothole impacts could be misinterpreted as "hot shock" effects.

However, some types of bulbs, usually taillight-type bulbs, will exhibit a darkened envelope as they age. This is due to the tungsten metal boiling out of the filament and depositing on the inside of the envelope. Pitting of the filament also indicates age. When a filament burns out due to age, the filament separates. At the point of separation, electrical arcing damage such as beading or necking down may be observed under magnification. Age sag or slight distentions from long wear can be discriminated from "hot shock" effects by the presence of the above secondary age characteristics.

Occasionally after an impact, an uninformed person will turn on the lights to see what still works. This can cause an otherwise unburned filament in a cracked bulb to burn up or darken. To check the circuitry after an accident, the use of a low voltage ohmmeter is recommended rather than turning on equipment.

In relatively low-velocity impacts where the envelope is not broken, there may not have been sufficient deceleration to cause filament deformation or breakage. In such cases, the lack of filament markers can mean that the impact deceleration was just too low to come into play. It does not automatically mean that the lights were off. However, if the impact was too low to affect the filaments, it will be corroborated with correspondingly low damage levels.

Sometimes when the general impact is low, and the lights in question do not exhibit any observable "hot shock" effects, it is possible to observe some "hot shock" effects in other lights that are nearest the point of impact on the vehicle. If the lights showing "hot shock" effects normally turn on when the lights in question turn on, the operation of one set of lights can be inferred from the other. As a rule, the lights closest to the point of impact exhibit the greatest degree of "hot shock" effects. As the distance from the point of impact increases, the degree of observable "hot shock" effects diminishes rapidly.

## Further Information and References

*Atlas of Stress Strain Curves*, Howard Boxer, Ed., ASM International, Metals Park, Ohio, 1987. For more detailed information please see Further Information and References in the back of the book.

*The Condensed Chemical Dictionary*, 9th ed., Gessner Hawley, Ed., Van Nostrand Reinhold Co., New York, 1977. For more detailed information please see Further Information and References in the back of the book.

*Handbook of Chemistry and Physics*, 51st ed., 1970–1971, CRC Press, Boca Raton, FL. For more detailed information please see Further Information and References in the back of the book.

*Handbook of Physics*, edited by Condon and Odishaw, McGraw-Hill, New York, 1967. For more detailed information please see Further Information and References in the back of the book.

*Materials Handbook*, by Brady and Clauser, 11th ed., McGraw-Hill, New York, 1977. For more detailed information please see Further Information and References in the back of the book.

*Response of Brake Light Filaments to Impact*, by Dydo, Bixel, Wiechel, and Stansifer of SEA Inc., and Guenther of Ohio State University, SAE Paper 880234. For more detailed information please see Further Information and References in the back of the book.

*The Traffic-Accident Investigation Manual*, by Baker and Fricke, 9th ed., Northwestern University, 1986. For more detailed information please see Further Information and References in the back of the book.

Van Nostrand's Scientific Encyclopedia, 5th ed., Van Nostrand Reinhold Co., New York, 1976. For more detailed information please see Further Information and References in the back of the book.

# Automotive Fires

# 20

Will you love me when my carburetor's busted?
Will you love me when my windshield's broke in-two?
Will you love me when my brakes can't be adjusted,
And my muffler, it goes zoop-poop-poopity-doo?
Will you love me when the radiator's leakin',
And my spark plugs have lost their self-respect?
When the nuts and bolts are fallin',
And the junkyard is a-callin',
Will you love me when my fliver is a wreck?

— **Thomas R. Noon,** *refrain from "The Fliver Song," circa 1941, 1925–1998*

## 20.1 General

Fires in automobiles, trucks, and vans fall into one of the following six categories of causes, which are listed roughly in order of importance or frequency.

1. **Fuel-related fires.** The fire is caused by fuel leakage onto hot components, usually in the engine compartment, or along the exhaust system.
2. **Electrical-related fires.** This includes short circuits, overheated wiring, and electrical malfunction of components such as fans, blowers, heaters, etc.
3. **Arson.**
4. **Garage fires, or similar.** This is where the garage in which the car is stored catches fire and burns the car or truck along with it. In some cases, the cause of the fire in the garage is wholly unrelated to the vehicle. In other cases, there may be a causal link, e.g., gasoline fumes being ignited by a drop light laying on the floor.
5. **Dropped cigarettes and smoking materials.**
6. **All other causes.** This includes fires in brake linings, fires due to hydraulic fluid leakage, battery explosions, various types of mechanical failures, contact of the catalytic converter with combustibles, and so on.

Statistically, the first three categories represent the majority of all fires involving vehicles. Surprisingly, fires in vehicles due to collisions and impacts are low in number. Car accidents that involve fire breaking out typically constitute less than 1% of such accidents. Despite the dramatic imagery created in movies and novels, few cars actually blow up and catch on fire when they wreck.

## 20.2   Vehicle Arson and Incendiary Fires

There are nearly as many fires, arsons, and incendiary fires in personal vehicles as there are involving homes. The most common motives for incendiary fires in personal vehicles include the following.

1.  **Cash flow problems.** The owner cannot afford the payments and is in danger of default.
2.  **The value of the car is "upside down."** This is when the remaining payments on the car total more than the present value of the car. For example, a car that has been poorly cared for and abused may be shot after just 2 years. However, if the owner financed the car for 4 or even 5 years, he may still have thousands of dollars left to pay on a piece of junk.
3.  **A lemon.** The car has simply been nothing but trouble for the owner since it was purchased. Burning the car removes the millstone from the owner's neck, and allows the owner to get a new one that does not spend all its time in the shop.
4.  **Theft.** It is common practice for some carjackers and joyriders to simply set fire to a vehicle in a remote area after they are finished with it. The car was likely not stolen by professionals, but by gang members or adolescents who simply used it for a good time, or perhaps even a gang initiation rite.

Of course, the above list is not complete. It just represents the most common motives. Verification of either of the first two potential motives can be checked relatively easy by routine background credit checks by an adjuster, private investigator, or law enforcement authority. In the case of the "lemon" motive, maintenance records at the dealership or local garage used by the suspect are generally readily available.

Be careful of the fourth category, theft. Sometimes people who wish to rid themselves of a car will fake a theft. While it is best to leave the investigative work in determining whether a theft has been committed to the proper law enforcement authorities, the investigator should be cognizant that faked thefts do occur.

Incendiary automobile fires whose primary motive is revenge or jealousy do occur, but are infrequent when compared to other causes.

One common element in incendiary vehicle fires is that there are often either no witnesses, or just one witness, the owner. When there is no witness, the fire will often occur in remote areas, or hidden areas like secluded garages, where passersby cannot observe the initiation of the fire. In a "no witness" incendiary car fire, the owner will often have no idea how the fire started, and will often say that he had not used the car for some time prior to the fire. This is to give the car enough time to burn completely.

When there is just one witness, the owner, often he will state that the fire started while he was driving the vehicle and that he had to abandon the vehicle and walk "x" amount of distance to get to a telephone while the vehicle was burning. This action also provides the time for the vehicle to be fully consumed by the fire. It would do an arsonist no good if the fire was seen, reported, and put out promptly, leaving the car repairable.

One of the more common "dead giveaways" of an arson is when the point of origin of the fire is located in the seats, floor, or upholstery of the car's interior. In late model cars, most of the interior materials are made of fire-resistant or fire-treated materials to prevent fire from quickly propagating through an occupied car. Most of the materials now used will not sustain flames on their own, and require an outside source of heat. However, this was not true in the past. In older cars, the cloth and coverings, seat stuffing, and carpets could support combustion, and sometimes would burn fiercely.

When a cigarette is dropped onto the seat of a late model car, the cigarette will generally burn a hole in the shape of an inverted cone in the seat, and then go out when the cigarette is exhausted. Cigarette fires often have points of origin under the seats, where errant cigarettes have rolled, in the crack of the seats, or in the ashtrays. Car ashtrays loaded with old butts can make a nice fire in the middle of the dashboard.

A common method of incendiary fire making the rounds, is to squirt lighter fluid into the vent openings of the dashboard of the car. It is an attempt to make the fire appear to be a short in the wiring under the dashboard. (Check the fuses. Sometimes, they aren't even blown because the circuits in the particular area were not operating at the time of the fire.) The arsonist then lights the accelerant and closes the door, or lights it with the door window left open an inch or two to provide air. If the car can be examined right after the fire, residual lighter fluid can usually be easily found having dripped onto the carpet below the fire area.

Another common technique is to squirt gas in the engine compartment and light it, closing the hood quickly. This method is also easy to detect

**Plate 20.1** Fire in pickup front seat due to cigarette lodged between seat cushion and seat back.

because the fire origin will be nonspecific, the severest burn damage will cover a broad pattern, and no particular fuel-related parts will be more severely burned than others.

## 20.3 Fuel-Related Fires

By far the largest category of automotive fires usually involves fuel leaking onto a hot portion of the engine. The engine block, and especially the exhaust manifold and exhaust, is a ready source of ignition energy after the engine has run for a time. Fuel can leak onto the engine from many locations. In modern engines, there is not only the main fuel line from the fuel tank to the engine, but there are also fuel return lines associated with the vapor control system, the emissions system, and the separate fuel lines from the injectors to the cylinders.

In newer vehicles, one of the more common places for leaks to occur is at line connections. If the various connections on the fuel lines are not properly fastened, fuel can leak from the connection. Many manufacturers' fuel system recalls are related to the failure of clips or fasteners to properly secure the fuel line and prevent it from leaking. In older cars, especially with rubber or elastomer type fuel lines, the fuel lines may crack at bends and corners. Fuel can leak from these cracks.

In some vehicles, metal fuel lines are used. Metal lines, especially steel lines, seem to be superior to other types because burning gasoline does not melt steel. Rubberized lines will be burned with the fuel, and will act a little like a fusee when fuel has caught fire. The fire will follow the leak back to the point of leakage, fire will engulf the rubber line, and the line will eventually break open the rubber line and allow the fire to burn the remaining fuel inside the line, following it down the hose. Aluminum lines easily melt during a fuel fire. Copper lines can melt when there is direct contact between the copper and the burning fuel, which is common.

To minimize the transfer of vibrations from the engine to the car body, some manufacturers will often use a short piece of rubber hose between the connection at the engine, and the metal line. These short pieces of rubber hose are often where fuel leaks occur. This is because they are often disconnected and reconnected during mechanical work, hence there is an opportunity for error in reconnecting the pieces. Also, the rubberized portions of the line are often subject to more fatigue after long periods of service, or when the engine has had a high amount of vibration.

Since a drip can often follow a line several inches before dropping onto the engine or manifold, the point of leakage is not always the same as the point of ignition of the fuel. It is usual for the fire to follow the leakage back to the source, leaving a fire trail from the point of ignition back to the point of leakage.

When fuel fires occur in the engine compartment, as most do, the fire damage pattern on the hood can be very helpful. Since many fuel fires occur in the upper portions of the engine compartment, the flames will often directly impinge on the underside of the hood. The hood will often be a sort of plan view of the fire and fire spread in the engine compartment. The hottest areas of fire will be demarcated by complete loss of paint, primer, and any galvanization coatings. Cooler areas will have correspondingly less severe damage to the finish.

In a carbureted engine, if a fire begins at a connection beside the carburetor due to leakage, the carburetor will be melted away on that side. Most carburetors are now made of aluminum, and thus melt easily when in contact with burning fuel. If the fire began inside the carburetor, due to a backfire or flooded condition, the carburetor will have collapsed away from its center, similar to the way the sides of a large wax candle collapse away from the hot wick. If the fire is simply hot, and has begun elsewhere, the carburetor will simply whiten or oxidize on the side from which the heat originates.

If the air filter canister to the carburetor is made of steel, it can also be used to discriminate between a fire originating at a connection to the carburetor, or from within the throat of the carburetor. Fire in the carburetor throat will shoot up into the air filter canister and perhaps directly burn the filter. Fire alongside the carburetor will burn away on the underside of the canister, leave the center of the canister more or less intact, and cook or char the filter.

In some cases, the part that actually leaked fuel, will be the missing part. Many parts within the fuel system in a modern engine are made of aluminum, plastic, or rubber elastomer. Direct contact with burning gasoline will usually destroy those parts, turning them into so much melted char or debris.

In a modern vehicle, the engine firewall does not keep fire from entering the passenger areas from the engine compartment as well as it used to. In recognition of this fact, some car companies no longer use the term firewall, and instead use the term bulkhead. This is because there are so many openings in the bulkhead for ventilation, hoses, wiring, and the like. Most of these openings are for plastic ventilation ducts, rubberized hoses, and plastic insulated wiring. When fire reaches these parts, they quickly collapse or slough away, leaving the hole open for the passage of fire into the passenger area. Thus, it is usual for a severe engine fire to pass through the bulkhead into the interior space in the vehicle.

In vehicles with electric fuel pumps which run off the battery, it is possible for the fuel pumps to continue to operate for a time while the fire is burning, even if the engine has stopped. This usually occurs when the fire begins while the car is being driven, and the driver gets out but does not shut off the engine. Often, the driver does this to investigate the source of some smoke observed in the vehicle interior coming through the floorboards.

When a car engine has been run for a while and then turned off, it is usual for the engine block and manifold to actually increase in temperature right after the engine has been turned off. This is because when the engine is turned off, the oil stops circulating, the coolant stops circulating, the radiator fan may be turned off, and the car is no longer moving. The last two items supply forced convection cooling to the radiator. Thus, the heat from the last engine firing is trapped in the engine, and is not carried away.

To alleviate this temperature rise, which shortens the life of an engine, many modern cars have a thermostat enabling the radiator fan to operate even though the engine has been turned off. This does not completely get rid of the temperature rise, but it can significantly reduce it.

This temperature rise effect is important to understand because many car fires start when the car is stopped, even though the leak may have been occurring while the car was being driven. While the car is being driven, the engine runs cooler, and the leaked fuel may be simply blown away from the engine by moving air. Most cars and trucks are designed so that moving air will pass through the radiator and wash over the engine to carry away heat.

When a car is stopped, the engine block and manifold temperatures rise, and the leak is no longer blown away by moving air. It can drip intact on the hot engine block, manifold, or exhaust and ignite. Most importantly, with no moving air, the fire can backtrack and follow the leakage path back to the source to get things really going. It is for this reason that vehicle fires often break out just after the vehicle's engine has been turned off, or the driver has just stopped the vehicle.

In some cases, if the leak is in the main fuel line, it may cause a small drop in the fuel line pressure. This drop in pressure may be sufficient to cause the engine to sputter or stall when climbing a hill, passing, or otherwise operating at a higher load. If the driver recalls when this occurred, it may provide a useful clue as to when the leak began.

**Plate 20.2** Pattern on hood indicates fire spread in engine compartment.

## 20.4   Other Fire Loads under the Hood

While fuel-related fires easily outnumber the rest, fires under the hood can also be caused by some of the other flammable liquids leaking onto hot surfaces. This includes the following.

- Brake fluid squirting from a bad seal or line connection.
- Hydraulic fluid leaking from a seal in the power assist systems.
- Engine oil from a leaky valve cover.
- Transmission oil boiling out of the dip stick in overheating automatic transmissions.
- Air conditioner compressor oil from a leak in the high pressure side of the refrigerant loop.
- Some types of engine coolants leaking from a radiator hose.

For example, ethylene glycol, one of the common coolants used in radiators, has a flash point of 241°F and ignites at 775°F. While it is not nearly as hazardous with respect to fire as gasoline, it still can burn and take out a car under the right circumstances. Some types of rubberized hoses may even catch fire if the hoses are displaced and contact the manifold or other hot portions of the engine block.

## 20.5   Electrical Fires

A majority of electrical fires in a vehicle are caused by the following.

- Wires that have become cut by abrasion and have grounded out.
- Wires that have become displaced and have contacted the manifold or hot portions of the engine block and grounded out.
- Wires that have simply overheated due to overload where the fuses have been bypassed (usually done in home-installed aftermarket equipment).
- Appliances that have failed and have inadequate fusing to catch the fault (also popular in home-installed aftermarket equipment like CBs, radios, and tape decks).

Often, the damages associated with an electrical fire are located on either side of the bulkhead or fire wall. This is, of course, because this is where most of the wiring harnesses are located. Because of the way equipment in the engine compartment is packed together, it is possible for an electrically caused fire to turn into a fuel system fed fire. If the fuel system fed fire is severe enough, it may hide or obscure the electrical origins of the fire.

An electrical fire under the dashboard will usually create inside-out damages, that is, damages that are most severe on the interior, and whose severity diminishes toward the exterior or perimeter of the area. Because the wiring is usually run just under the dashboard, the area will have a hollowed out appearance.

Most of the plastic materials used in modern automobile dashboards do not readily support a flame or fire. However, they will melt, char, and emit copious dense smoke, and may support flames and fire as long as the electrical fault can supply heat energy.

An examination of the fuses will often provide information as to which grouping of equipment was involved in the fault. Then, the point of origin has to be tracked down like any other electrical fault. However, in an automobile, a ground fault can occur any time the bare wiring can make contact with the metal frame or parts of a car.

The metal frame is the return wire for most circuits. The negative side of all the circuits in the vehicle are run through the metal frame as a common ground return. Thus, once heat reaches a group of wires in a harness that is positioned on a metal fender or the bulkhead, and the protective insulation melts away, it is possible for nearly all the wires in the harness to ground fault at more or less the same time at the same place.

The home installation of aftermarket equipment often causes electrical fires in vehicles. The handyman is often unfamiliar with the circuits, and will simply tap into the most convenient wire he can find. Usually, the fuse block is bypassed, so the new item does not have the benefit of overcurrent protection. Further, the wires to the new item will be loose, and often simply tucked loosely under the carpet (where they can be stepped on repeatedly), or laid over metal pieces that may eventually abrade and cut into the insulation.

Because a car is constantly vibrated when rolling along, loose wiring has a much higher probability of abrading itself on nearby edges or rough surfaces. This is one reason why the manufacturer groups the wiring in harnesses. Sometimes, when repair work is done on an electrical circuit, it is considered too much trouble for the mechanic to put the particular wire back into the bundle or harness as before. He may simply tuck it in at a convenient spot, or tape it to something nearby to "hold it down."

Of course, loose wiring in the engine compartment may also eventually come to rest on either the manifold or the engine block. Since much of the wiring is simply sheathed in relatively low temperature plastic insulation, proximity to the hot items may be enough to damage the insulation and allow the circuit to ground fault.

## 20.6  Mechanical and Other Causes

Fires in cars are also, but less frequently, caused by mechanical failure. Wheel bearings, brakes, and exhaust components are the most common culprits.

When the brakes are involved, usually the individual wheel brake cylinders have lost the seals, and brake fluid has leaked into the drum or disk where the high temperatures generated by braking friction ignite the fuel. The point of origin at the wheel is usually easy to spot.

Most axles and wheels use a heavy oil or grease for lubrication. The failure of wheel bearings can allow these lubricants to leak or flow into the brakes, or may itself generate sufficient heat for ignition to occur.

Some fires are blamed on the catalytic converter. Most are in one of two groups.

1. Where the fire is caused by the catalytic converter contacting weeds or dry grass when hot, and igniting materials exterior to the car. In the process, the vehicle is burned up by the grass or flammables underneath it.
2. The fuel mixture of the engine is too rich, and clogs up the catalytic converter causing it to operate very hot. A fire can then originate within the converter itself.

Of course, a catalytic converter actually is a place where fire is occurring. The converter actually "slow burns" unburned fuel in the exhaust stream so that fewer unburned carbon compounds exit the exhaust pipe. This slow burn is accomplished by the use of a surface catalyst, usually a platinum-type compound, which causes the fuel to oxidize at temperatures and conditions less than it would otherwise.

Because of this, catalytic converters are coolest when the vehicle is driving along with the engine efficiency high, which normally occurs when the vehicle is driving on the highway. The moving air under the vehicle also helps keep it cool.

Catalytic converters operate hottest when the car is sitting still. The engine is operating under very poor thermodynamic efficiency, and there is no moving air for forced convection to occur. If the vehicle is then parked and left running in a dense, dry, and tall growth of weeds or foliage, the vegetative material may not only insulate the bottom of the vehicle, prevent air flow on the underside of the vehicle and further worsen convection, but may also directly contact the hot metal skin of the converter. The combination of these factors may then lead to ignition of the combustible vegetation under the vehicle.

Another cause of fire that occasionally occurs is improper refueling. Careless fueling of the vehicle can often leave a small amount of gasoline trapped in the refueling port. This can then drop down onto the exhaust pipe and ignite. The flames then engulf the fuel wetted areas, which are usually around the refueling port or door. This is why gas stations always post the sign, "turn off engine when refueling." This type of fire is perhaps a little more prevalent now than in years past because of the proliferation of self-service gas stations. Many drivers are not particularly careful in the handling of gasoline around the pumps. Seeing drivers smoke while refueling is also hazardous.

## Further Information and References

*ATF Arson Investigation Guide*, published by the Department of the Treasury. For more detailed information please see Further Information and References in the back of the book.

*Fire Investigation Handbook*, U.S. Department of Commerce, National Bureau of Standards, NBS Handbook 134, Francis Brannigan, Ed., August 1980. For more detailed information please see Further Information and References in the back of the book.

*Investigation of Motor Vehicle Fires*, by Lee Cole, Lee Books, 1992. For more detailed information please see Further Information and References in the back of the book.

*Manual for the Investigation of Automobile Fires*, National Automobile Theft Bureau. For more detailed information please see Further Information and References in the back of the book.

*A Pocket Guide to Arson and Fire Investigation*, Factory Mutual Engineering Corp., 3rd ed., 1992. For more detailed information please see Further Information and References in the back of the book.

# Hail Damage

# 21

## 21.1   General

Hail is a type of hydrometeor, which simply means that it is a piece of water that falls from the sky. Rain, snow, and sleet are also hydrometeors. Hail occurs in association with strong convective cloud systems, that is, cumulonimbus clouds.

A cumulonimbus cloud is the classic towering thunderstorm cloud and has low, middle, and upper altitude formations. The lower portion of the cloud may form as low as 500 feet above the ground. The cloud appears billowy, dense, and tall, and may have cloud colors ranging from white to black. The top of the cloud may spread and flatten out over a large area giving rise to the description of thunderhead or anvil cloud.

Starting from the ground, the kinetic air temperature of the troposphere drops linearly with altitude at a rate of about 6.5°C per kilometer. This continues until an altitude of about 11 kilometers is reached, which is the approximate top of the troposphere. At that altitude, the kinetic air temperature stabilizes at about −56.5°C. From 11 kilometers upwards to about 25 kilometers, the temperature stays relatively constant at about −56.5°C. This region is called the tropopause. Above 25 kilometers, in the stratosphere, the kinetic air temperature rises again. Thunderstorms and hail do not form in the stratosphere. So with respect to hail formation, we lose interest in the vertical structure of the atmosphere higher than about 12 to 14 km.

**Plate 21.1** Burnish mark on steel A/C cabinet.

Most cumulonimbus clouds "top" off and flatten where the kinetic air temperature stabilizes at −56.5°C, that is, at the boundary between the troposphere and the tropopause. This is because the vertical rise in the cloud formation is motivated by the temperature vs. altitude gradient. The greater buoyancy of the warm, moist air with respect to the colder surrounding air causes it to float upward. This is why prior to flattening out on top, cumulonimbus clouds often appear in groups that look like puffy vertical columns, or columns of chimney smoke.

When the temperature gradient of the ambient air approaches or becomes zero, that is when $\delta T/\delta x = 0$, the impetus for upward cloud formation ceases. The conveyor belt-like upward movement of the convective portion of the cloud then "piles up" at an altitude of about 11 kilometers or so, and spreads out laterally because it can't rise any higher. The cloud appears to mash up against the lower boundary of the tropopause and forms a flat top.

When a properly prepared cross-section of a hailstone is viewed, concentric layers of ice are visible, which give it an onion skin appearance. There are usually milky-colored layers of ice separated by clear ones. Up to 22 layers of ice have been observed in hailstones. This layered appearance supports the theory of multiple incursions with respect to the formation of hail. In this theory, hailstones are assumed to alternately ascend and descend by convective updrafts and downdrafts through freezing and nonfreezing zones of the cloud formation.

For example, if the ground has a summery temperature of about 25°C, the air temperature becomes freezing at an altitude of about 4 kilometers. If water droplets that have formed at an altitude lower than 4 kilometers are caught in an updraft, they can readily freeze when they are lofted to an altitude of 4 kilometers or more. If the frozen droplets are then allowed to drop down to slightly below 4 kilometers, they can be rewetted, and the process can be repeated until a layered hailstone is built up. Of course, some hailstones increase in size by colliding with other hailstones and assimilating them into their structure.

Thus, the formation of hail requires a tall cloud formation that spans both freezing and nonfreezing zones of the atmosphere. Of course, ample moisture content must be present, especially in the lower portion of the cloud, so that the hailstone has the raw materials to accrete to an appreciable size.

If hail damage to a structure or vehicle is claimed, one of the first verification checks a person should do is determine if thunderstorm activity was present in the area at the same time the hail damage reportedly occurred. It is also prudent to check if the cloud height or ceiling well exceeded 4 kilometers or about 13,000 feet.

These are the reasons why hail fall is not associated with relatively flat or short cloud formations, especially those wholly below 4 kilometers in altitude. The fact that hail formation requires specific conditions means that while hail requires a thunderstorm or a thunderstorm system in order to form, not all thunderstorms produce hail. In fact, as will be shown, most don't.

Since a hailstone is kept aloft during its formation by a combination of updrafts and downdrafts, the maximum hailstone size is a function of the maximum updraft velocity. That is, the formation of large hailstones requires correspondingly high velocity updrafts.

Soft hail, which is sometimes called snow pellets or graupel, is also formed in association with convective cloud systems. Soft hail, however, is usually less than 1/4 inch in diameter and can be crushed easily between the fingers. Soft hail is significantly less dense than regular hail and contains a large amount of trapped air. Soft hail typically occurs in association with winter showers. In these showers much of the cloud formation is below freezing and the available moisture is limited. Soft hail is too small and fragile to cause damage to property.

## 21.2  Hail Size

The average hail fall duration in the Midwest is about 5 minutes and covers an area approximately 20 square miles, which is why it is always prudent to check for collateral damage to nearby similar buildings and structures when a hail damage claim requires verification. Hail falls, however, can last as long as 15 minutes.

The relative size range of hail within a particular hail fall area is usually no more than 1 to 3, that is, the largest hailstone will usually be no more than about three times the size of the smallest hailstone. Thus, the largest hailstones in a particular hail fall area can be readily estimated from the size of the smallest hailstones found there, and vice versa.

Hailstones range in size from less than that of a pea to several inches in diameter. A severe thunderstorm is defined as one that produces a tornado, horizontal surface winds of at least 58 mph, or one that produces hail at least 3/4 inches in diameter. A thunderstorm is said to approach the "severe" category when the associated horizontal surface winds are at least 40 mph, or there is hail with at least a diameter of 1/2 inch.

The largest hailstone recorded in the U.S. was 17.5 inches in circumference. It was found September 3, 1979, in Coffeyville, Kansas. The previous record holder had a maximum circumference of 17 inches, reportedly weighed about 1.5 pounds and fell in Potter, Nebraska, in November 1928.

Stories of hailstones as large as bowling balls have been anecdotally reported, but none have actually been found or otherwise verified. Such hailstones are in the same league with jackalopes and other exaggerations popular in the Great Plains. While hailstones the size of softballs, 4 inches in diameter or 12.6 inches in circumference, do occur, they are rare.

Most hailstones are less than an inch in diameter. Based on possible updraft wind speeds and the air resistance of hailstones, hailstones with 5.5–6 inch diameters are considered to be the theoretical limit. This theoretical limit excludes bowling ball-sized hailstones.

The theoretical upper bound limit for the size of hailstones is derived as follows. First, the relationship of hailstone mass to its radius is determined.

$$m_{hail} = \rho_{ice}(4/3)\pi R^3 \tag{i}$$

where $m_{hail}$ = mass of hailstone, $\rho_{ice}$ =density of hail ice, and R = mean radius of the hailstone.

When a hailstone is supported and lofted upward within a strongly convective cloud system, the air friction between the updraft and the hailstone has to exceed the downward force exerted by gravity. Thus, the following relation between the force of gravity and the force created by frictional drag is derived:

$$F < (1/2)\rho_{air}C(A_{frontal})V^2 \tag{ii}$$

$$m_{hail}g < (1/2)\rho_{air}C(\pi R^2)V^2$$

$$\rho_{ice}(4/3)\pi R^3 g < (1/2)\rho_{air}C(\pi R^2)V^2$$

where F = downward force due to gravity, g = acceleration of gravity, $A_{frontal}$ = projected area of hailstone facing updraft, R = mean radius of hailstone, $\rho_{air}$ = density of air around hailstone, C = coefficient of drag, and V = velocity of air supporting hailstone.

Solving for the velocity term in Equation (ii) yields the following:

$$V > [2(\rho_{ice}/\rho_{air})(4/3)(R)g/C]^{1/2} \qquad (iii)$$

As a rule, the ratio of the density of hail ice to air at sea level is about 660 to 1. Of course, at higher elevations where hail is formed, the ratio is greater because the density of air is lower. An average ratio value of about 1376 to 1 is assumed for the higher elevations where hail is usually formed. For a spherical hailstone and assuming a Reynolds number larger than $10^5$, a typical value for "C" in Equation (iii) is about 0.2. Substituting in the other values accordingly, Equation (iii) reduces to the following:

$$V > 135[Rg]^{1/2}. \qquad (iv)$$

Given that the other factors such as the coefficient of friction or the ratio of hail ice density to air density do not change very much, Equation (iv) indicates that the updraft velocity needed to suspend a hailstone within a cloud is proportional to the square root of the radius of the hailstone. Each doubling of the radius of a hailstone requires that the associated updraft velocity increase at least 41.4%.

Using Equation (iv), when R = 1 inch, the minimum required velocity of the updraft, "V," is about 221 ft/sec or 150 mph. Likewise, when R = 3 inches, "V" is minimally about 383 ft/sec or 261 mph.

Table 21.1 shows the relative ranking of wind intensity in tornadoes and windstorms. For a 2-inch diameter hailstone, winds corresponding to an F-2-ranked storm is required. For a hailstone diameter of 6 inches, Table 21.1 shows that winds corresponding to an F-5-ranked storm are needed, which are as big as they come on earth.

**Table 21.1   Fuijita Ranking for Wind Storms**

| Rank | Wind Speed (mph) | Level of Damage |
|------|------------------|-----------------|
| F-0 | up to 72 | Light |
| F-1 | 73 to 112 | Moderate |
| F-2 | 113 to 157 | Considerable |
| F-3 | 158 to 208 | Severe |
| F-4 | 209 to 250 | Devastating |
| F-5 | 251 and more | Incredible, Atomic bomb-like effects |

Bowling balls, by the way, have a diameter of 8.5 inches, or an "R" of 4.25 inches. Using Equation (iv), this computes to an updraft velocity of at least 455 ft/sec or 310 mph. This is about Mach 0.41, or 41% of the speed of sound at sea level. A convective updraft of this magnitude has yet to be measured on earth. This is why reports of hailstones the size of bowling bowls, without actually having one to prove the point, should be considered an exaggeration. Of course, if such bowling ball-sized hailstones really did fall, the accompanying F-5 magnitude tornado or windstorm would make damage claims from the hail a moot point. Hail damage would be the least of their concerns.

## 21.3   Hail Frequency

Since hail fall occurs nearly exclusively in association with thunderstorm activity, it follows then that most hail damage to structures and property occurs in the months with the greatest number of thunderstorms. As noted in Table 21.2, the month with the greatest number of thunderstorms in the metropolitan Kansas City area is June. In fact, the sum of thunderstorms that occur in May, June, and July typically comprises nearly half of the total annual thunderstorms that occur in the region around Kansas City. Thus, it is expected that on the average, about half of the yearly hail damage occurring in the region around Kansas City happens during these three months.

In a typical year in the Kansas City area, there are 53 thunderstorms, which is slightly more than one per week on the average. Hail, of course, only falls in a fraction of these thunderstorms.

**Table 21.2   Average Annual Thunderstorm Activity in Kansas City Area**

| Month | Thunderstorms |
|-------|---------------|
| Jan. | 0.1 |
| Feb. | 0.6 |
| Mar. | 2.3 |
| Apr. | 5.3 |
| May | 8.8 |
| June | 9.9 |
| July | 7.3 |
| Aug. | 7.7 |
| Sep. | 5.8 |
| Oct. | 3.4 |
| Nov. | 1.3 |
| Dec. | 0.5 |
| Total | 53.0 |

**Table 21.3  Hail Storm Activity in Kansas City Area —
1987**

| Month | Number of Thunderstorms | Number of Hail Storms |
|-------|------------------------|----------------------|
| Jan. | 0 | 0 |
| Feb. | 0 | 0 |
| Mar. | 2 | 0 |
| Apr. | 4 | 0 |
| May | 10 | 1 |
| June | 13 | 0 |
| July | 7 | 0 |
| Aug. | 7 | 0 |
| Sep. | 4 | 0 |
| Oct. | 3 | 0 |
| Nov. | 1 | 0 |
| Dec. | 0 | 0 |
| Total | 51 | 1 |

Table 21.3 shows the reported number of thunderstorms and hail falls noted for the Kansas City area in 1987. The reader will note that while the total number of thunderstorms that year was 51, the number of associated hail falls was only 1. The year prior, 1986, had 59 thunderstorms and 3 hail falls.

In the Midwest, the probability of a hail fall being severe enough to cause damage to structures and property is relatively low for any given location. However, some locations through the years do seem to have more than their fair share of hail damage, and various reasons are usually put forward to explain this, such as, "the lake tends to draw the hail this way," or "there are no hills in this area to deflect away the hail clouds." My personal favorite is, "The iron in the soil here seems to attract the hail in this direction."

The apparent higher rate of hail fall in most cases is simply a normal statistical phenomenon, that is, a statistical cluster. By random chance alone, such groupings are expected to occur. Conversely, there are also places that never seem to receive hail. This is also a statistical cluster, but few people ever notice or complain about the absence of hail in an area. As a rule, good statistical clusters are simply accepted as good fortune by those affected. People adversely affected by bad clusters, however, usually demand a reason for its occurrence.

A similar effect occurs in the public's perception of diseases. When a higher than average incidence of a bad disease occurs in a town or county, many people are convinced there is something insidious causing the disease to occur, and will usually demand an investigation to determine the cause. On the other hand, when a community is free of the same disease, or well below the national average, they do not demand an investigation to determine the cause of their good fortune.

**Table 21.4  Probability of Hail Size**

| Hail Size (diameter) | Average Yearly Number of Hail Days | | |
|---|---|---|---|
| | 2 | 4 | 6 |
| < 0.5 in. | 61% | 44% | 25% |
| 0.5 to 1.0 in. | 28% | 43% | 58% |
| 1.0 to 1.5 in. | 4% | 4% | 5% |
| 1.5 to 2.0 in. | 3% | 4% | 5% |
| 2.0 to 2.5 in. | 2% | 2% | 3% |
| 2.5 to 3.0 in. | 0.5% | 0.6% | 0.8% |
| 3.0 to 3.5 in. | 0.2% | 0.4% | 0.8% |
| 3.5 to 4.0 in. | 0.2% | 0.4% | 0.4% |
| >4.0 in. | 0.2% | 0.4% | 0.7% |

In Kansas, the average number of days in any given year in which hail will fall is about 4 to 5. In Missouri, it is 3 to 4. Generally, the northern areas in the U.S. have fewer hail days than southern areas. Southern, warm coastal areas have the most days in which hail will likely fall.

One might suppose that if one particular year had only one hailstorm and another year had three, then claims for hail damage might be about one third less in the former than in the latter. This might be true if all claims were truthful and accurate, and all hail falls had the same intensity. However, it is not unusual for a similar number of hail damage claims to occur in both years despite one year having three times the hail falls as the other. This is because:

- there are differences in hail fall intensity.
- there are differences in hailstone size.
- some of the roof damages attributed to hail are actually due to wear, age, or preexisting hail damage.

In a 1967 study published by the Travelers Insurance Companies over 3,000 Midwestern weather reports between the years 1950 and 1966 were tabulated and assessed (see references at end of chapter). In essence, the study determined the probability of hailstone size with respect to the average annual number of days in which hail fall would occur. For example, in eastern Kansas where the average hail fall days are about 4 per year, the chances that the hailstones will be 1.5 inches in diameter or less are 91.2%. A summary of the findings of the study is given in Table 21.4.

## 21.4  Hail Damage Fundamentals

In terms of monetary losses, hail primarily damages crops, building roofs and siding, and cars. The severity of hail damage depends on two main factors:

**Plate 21.2** Hail impacts on A/C heat exchanger fins. Fins can be combed out with no significant loss of efficiency.

1. How much kinetic energy the hailstone possesses at impact, which is a function of its mass and impact velocity.
2. The fraction of energy that is actually transferred to the roof, car, or impacted object.

The amount of kinetic energy that a hailstone possesses is given by the standard equation for kinetic energy.

$$KE = (1/2)mv^2 \qquad \text{(v)}$$

where m = mass of the hailstone and v = velocity of hailstone just prior to impact.

In general, the shape of a hailstone is spherical. Since a hailstone is composed of ice, its mass can be reasonably estimated from its mean diameter and the fact that hail ice has about 89–91% the density of water, or 0.89 to 0.91 g/cm³.

From experimentation and empirical experience, it is known that when a hailstone strikes a flat surface, it contacts an area about half the diameter of the hailstone, unless of course, it has sufficient impact energy to actually penetrate the material.

With respect to impact damage, only the velocity vector component normal to the impacted surface is involved in causing damage. The velocity vector parallel to the impacted surface simply causes the hailstone to skid or briefly slide along the surface. This action can impart a significant rotation of the hailstone, but is of no consequence with respect to damage.

Combining the above factors into a single algebraic expression gives the following:

$$KE/area = (4/3)\rho Dv^2 (\cos \theta)^2 \qquad\qquad (vi)$$

where $D$ = mean diameter of hailstone, $\rho$ = density of hard hail ice, ~ 0.91 g/cm³, $v$ = velocity of hailstone, and $\theta$ = angle of approach to the impacted surface as measured from the normal.

Equation (vi) indicates that when the angle of impact with the surface deviates from the perpendicular position, the impact severity diminishes. For example, at 30 degrees from perpendicular, the impact severity is reduced by 25%. At 45 degrees, the severity is reduced 50%, and at 60 degrees, the severity is reduced by 75%. Equation (vi) also indicates that the impact energy of a hailstone increases linearly with diameter, and exponentially with velocity.

This is why there will usually be a difference in the severity of damage on a house that has a gabled roof, or structure that has several surfaces that are oriented differently. As depicted in Figure 21.1, the trajectories of hailstones A and B are such that they impact more or less perpendicular to the surface of the roof on the right side of the roof ridge. On this particular roof, impacts A and B will have the greatest intensity. On the other hand, the trajectories of hailstones C and D are such that they both impact more or less at a 45-degree angle with the surface of the roof on the left side of the roof ridge. As indicated in Equation (vi), impacts C and D will be about 50% less severe than impacts A and B.

In the example depicted in Figure 21.1, while the right side of the roof might be damaged by hail, if the impacts are severe enough, it is quite likely that the left side will be undamaged. As will be noted later in this chapter, hail damage is not a continuous phenomenon; for damage to occur, it is first necessary for the impact severity to be greater than a certain threshold.

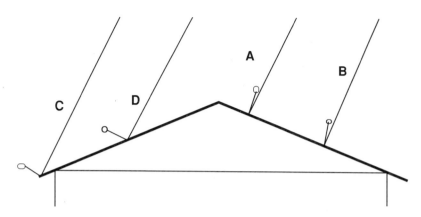

**Figure 21.1** Hail impacts on gabled roof.

These are the reasons why it is unusual to find naturally occurring evenly distributed hail damage on a gabled roof or car. The variation in impact angles afforded by the variously inclined surfaces cause variations in impact severity, and the fact that the hail impact must be at least equal to or greater than a certain severity threshold means that some areas might exhibit damage, while some areas exhibit no damage.

For hail damage to occur, the kinetic energy of the hailstone has to be transferred to the object being struck. If the hailstone strikes the surface and then rebounds with the same speed it had before, no impact energy is transferred to the impacted surface and the surface is left undamaged; the impact has been wholly elastic and the hailstone leaves with as much energy as it had when it arrived. If the hailstone strikes the surface and has no rebound velocity, all the kinetic energy of the hailstone has been either transferred to the impacted surface or has been absorbed by the hailstone itself, and damage to either the hailstone or roof is maximum.

A measure of the energy transfer upon impact is called the coefficient of restitution. The coefficient of restitution is defined as follows:

$$\varepsilon = v_2 / v_1 \tag{vii}$$

where the subscript 1 denotes the hailstone velocity prior to impact and the subscript 2 denotes the hailstone velocity after impact.

A coefficient of restitution of "1" indicates that no energy is transferred to the surface by the hailstone and the impact is elastic. A coefficient of "0" indicates that all the kinetic energy has been converted into plastic deformation of either the impacted surface or the hailstone. Given the same impact energy distributed over the same impact area, an impact with a coefficient of "1" causes no damage to the roof material or the hailstone. An impact

**Plate 21.3** Buckling of asphalt shingles around vent cap.

with a coefficient of "0" will cause the maximum damage possible to either the roof or the hailstone.

Since hailstones of appreciable size have reasonably consistent mechanical strength properties, the primary variation in the coefficient of restitution between a hailstone and an object is the material on which the hailstone impacts. A single-layer roof that has firm underlayment and decking will provide a much higher coefficient of restitution than a roof that has been applied over several layers of older roofs or which has a rotted or "soft" decking.

This is the reason why, given identical types of roof and the same hail fall, one roof has damage and the other does not. With respect to potential hail damage, the material under the roof is often as important as the roof material itself.

For example, it is reported by Greenfeld in *Hail Resistance of Roofing Products* (cited in references, p. 6) that asphalt shingles mounted on plywood were damaged along shingle edges and unsupported portions by 1.75-inch diameter hail, while the same asphalt shingles mounted on 1 in. × 6 in. tongue and groove were similarly damaged by 2-inch hailstones. Likewise, asphalt shingles with #15 felt underlayment sustained the same threshold damage by smaller hailstones, approximately 16% smaller, as the same asphalt shingles with no underlayment.

## 21.5   Size Threshold for Hail Damage to Roofs

One of the premier studies in hail damage, *Hail Resistance of Roofing Structures — Building Science Series 23*, was done by the U.S. Department of

**Plate 21.4** Hail dents in copper decoration on roof that has been in place for more than 20 years.

Commerce in 1969. Authored by Sidney Greenfeld, a research associate of the Asphalt Roofing Manufacturers' Association, the study was commissioned by the Building Research Division of the Institute for Applied Technology of the former National Bureau of Standards. The purpose of the study was to determine the susceptibility of various roofing materials to hail damage. The idea of the study was simple: shoot hailstones at a bunch of roof types, and see what happens.

In the experiments conducted by Mr. Greenfeld, ice spheres were molded and frozen to a temperature of about −12°C (−10°F). The spheres were allowed to freeze and freely expand. This prevented compressive fractures from occurring and closely modeled the way in which hailstones naturally freeze. The ice spheres were then shot at various roofing materials using a compressed air gun with a maximum muzzle velocity of 300 ft/sec or about 205 mph. The maximum diameter accommodated by the air gun was 3.25 inches. By the time the ice spheres reached their targets, they had reached within ± 10% of their free-fall terminal velocities.

A notable finding of this experiment was that while hailstones are not perfectly spherical, treating them as a smooth sphere for aerodynamic purposes well approximated the experimental results. The following table lists the hail diameter size, terminal velocity (at ground level and standard atmospheric pressure), and associated impact energies as reported by Laurie (see references).

**Table 21.5   Terminal Velocity and Impact Energy vs. Hail Diameter**

| Diameter | Terminal Velocity | Impact Energy |
|---|---|---|
| 1.00 inch | 50 mph | < 1 ft-lbf |
| 1.25 inches | 56 mph | 4 ft-lbf |
| 1.50 inches | 61 mph | 8 ft-lbf |
| 1.75 inches | 66 mph | 14 ft-lbf |
| 2.00 inches | 72 mph | 22 ft-lbf |
| 2.50 inches | 80 mph | 53 ft-lbf |
| 2.75 inches | 85 mph | 81 ft-lbf |
| 3.00 inches | 88 mph | 120 ft-lbf |

Source: *Hail and Its Effects on Buildings*, by J. Laurie, CSJR Research Report No. 176, Bull. 21, National Building Research Inst., South Africa.

Using the data listed in Table 21.5, an additional column can be added that notes the impact energy of the hailstone per area contacted. This additional column of data is listed in Table 21.5A.

An important aspect of the Greenfeld study was the definition of damage, which is quoted in full as follows:

> Damage to roofing by hail falls into two general categories: (1) Severe damage, which leads to penetration of the structure by the elements and (2) Superficial damage, which affects appearance but does not materially interfere with the performance of the roofing. While the latter is distracting and leads to insurance claims, the former is the type of damage that should be of most concern, because the possible loss can exceed the replacement cost of the roofing many fold. Thus, while the dents will be reported, only the fractures of the coating, felt or other shingle material will be called failure in this report.

The above definition is very important. Cosmetic damage should not be confused with structural damage. Cosmetic damage usually consists of

**Table 21.5A   Terminal Velocity, Impact Energy, and Impact Energy/Contact Area vs. Hail Diameter**

| Diameter | Terminal Velocity | Impact Energy | Impact Energy/Contact Area |
|---|---|---|---|
| 1.00 inch | 50 mph | < 1 ft-lbf | <2.55 ft-lbf/in$^2$ |
| 1.25 inches | 56 mph | 4 ft-lbf | 13.0 ft-lbf/in$^2$ |
| 1.50 inches | 61 mph | 8 ft-lbf | 18.1 ft-lbf/in$^2$ |
| 1.75 inches | 66 mph | 14 ft-lbf | 23.3 ft-lbf/in$^2$ |
| 2.00 inches | 72 mph | 22 ft-lbf | 28.0 ft-lbf/in$^2$ |
| 2.50 inches | 80 mph | 53 ft-lbf | 43.2 ft-lbf/in$^2$ |
| 2.75 inches | 85 mph | 81 ft-lbf | 54.5 ft-lbf/in$^2$ |
| 3.00 inches | 88 mph | 120 ft-lbf | 67.9 ft-lbf/in$^2$ |

burnish marks, skid marks, or temporary depressions that are often readily visible, but do not impair the functionality of the roof. Many types of cosmetic damage will simply disappear with time, as for example, happens with burnish and skid marks.

Structural hail damage to a roof, however, is where the roof material can no longer perform its function, which is to prevent rain and precipitation from entering the structure. Ripping off a perfectly good 30-year rated roof because it exhibits cosmetic burnish marks, only to replace it with a cheap, 10-year rated roof does not seem like a good idea, but it often happens.

The Greenfeld study examined the resistance to direct hail impacts of asphalt shingles, commercial built up roofs (BURs), asbestos cement shingles, cedar shingles, standing-seam terne metal, slate, and red clay tiles. The study reported the following findings.

- The threshold where damage from direct hail impact for most commercially prepared roofing systems begins is a hailstone diameter of 1.5–2.0 inches. Heavier shingles and roofing materials tend toward the 2.0 inch diameter threshold.
- Weathering tends to lower the resistance of roofing materials to hail damage. That is, old roofs that have lost material strength due to age, also lose their resistance to hail. (In most cases, this occurs after the roof has gone beyond its rated service life.)
- Sheet metal roofing was dented by nearly all sizes of hailstones. The threshold for unacceptable denting, however, was the same: direct impact by 1.5–2.0-inch diameter hailstones.

In considering the Greenfeld report finding that the damage threshold by hail to commercially prepared roofing systems is between 1.5 and 2.0-inch diameter hailstones, it is seen in Table 21.5A that this corresponds to an impact energy per contact area of from 18.1 to 28.0 ft-lbf/in$^2$. Bear in mind, however, that this means that damage *could* occur at this level, especially in roofing materials that are low quality or that have been poorly installed. It does not mean that damage assuredly has occurred. In general, the damage threshold for good quality materials that are properly installed is 2.0-inch diameter hail, or an impact energy per contact area of about 28.0 ft-lbf/in$^2$ or more.

## 21.6   Assessing Hail Damage

In general, damage to asphalt shingles, built up roofs (i.e., tar and gravel), asbestos cement shingles, cedar shingles, standing-seam terne metal roofs, slate, and red clay tiles that have been caused by severe hailstone impacts consist of impact craters and cracks and fissures originating at those craters.

Impact craters are where the impacts by hailstones have penetrated sufficiently into the material to breach its functional integrity and produce significant, permanent depressions or "dents." Generally, the associated cracks
and fissures are larger in width at the point of impact and diminish in width
and size with increasing distance away from the point of impact.

In asphalt shingles, hailstones sufficient in size to cause damaging impact
craters to the top surface, can also cause damage to the back surface of the
shingle, especially in the "tab" area of the shingle. The effect is similar to that
observed when a BB impacts a pane of common glass: there is an obvious
impact crater on the side on which the BB strikes, but there is also a conical
"blowout" area on the opposite side. The blowout on the opposite side is due
to momentum transfer via a compression wave through the material to the
opposite surface. Because asphalt shingles are not brittle like glass, the effect
is less pronounced but is observable in extreme cases. A similar effect has
been noted on the moon and on Mercury, where the impact of a meteor was
so severe, the shock waves caused damage at the corresponding antipode.

In the case of brittle materials, like red clay tiles, impact-associated cracks
and fissures might look like star shaped cracks, where several cracks radiate
outward from a central impact point. If impacts have been severe, the tiles
might be fractured, broken, or even have holes knocked through them. Cracks
and fissures that have been caused by hail impacts, should exhibit the same
freshly exposed surfaces as those found in the hail impact crater, that is, they
should be the same age.

**Plate 21.5** Splits and cracks and curling in wood shakes caused by weathering.
Note staple in circled area to hold down curled shake.

While often alleged to have been caused by hail or assumed to be sure indications of hail damage, the following conditions are *not* due to hail.

**Curled and cupped shingles** — This is when a shingle has curled up, especially in the portions of the shingle or wood shake that are exposed to sunlight. The effect is most pronounced in flatgrain shingles and shakes. Curling or cupping is due to drying out of the topmost layer of the shingle by ultraviolet light exposure. When the top layer dries out, it has less mass in it than before so it shrinks in size. Since the bottom layer is not exposed to ultraviolet light, it does not lose material and shrink. The tension created by the differential shrinkage between the top and bottom surfaces of the shingle or shake then causes the shingle or shake to curl or cup. Nailing a shake or shingle too high, too far from the edge, or improperly will accelerate the effect.

Cupping and curling is further exacerbated when the roof has been "shorted," that is, when the roof has been covered with fewer shakes or shingles than required by good practice. Usually this is done by increasing the amount of shingle or shake exposed to the sunlight, and decreasing the amount of shake or shingle that is covered by overlap with other shakes or shingles. Examination of the edge of a roof can often be useful in visually checking that the proper amount of overlap has been used, or by measuring the exposed portion of the shake or shingle.

For example, wood shake roofs using number 1 hand-split and resawn shakes that are 24 in. × 3/4 in. generally have an exposed portion measuring 7 1/2 inches long. Along the side of the roof, the roof thickness will appear to be four shakes thick (for a distance of about 2 inches) when counted upwards from the point of a shake along the bottom of the roof.

However, if the exposed portion of the shakes are increased just 1 inch- to 8 1/2-inches, the roof thickness along the edge will only appear to be three shakes thick when counted upward from the point of a shake along the bottom of the roof. By increasing the exposure by just 1 inch, the roofer uses about 13% fewer shakes, although the owner may have been charged a roof price based upon the proper number of shakes per "square" of roof area. By increasing the exposed area of the shakes, not only does the roof defraud the owner, but the shakes also dry out faster than normal. This means that they will curl and cup sooner, and the roof generally will wear out faster.

Shakes and shingles that have already become cupped or curled are more susceptible to cracking and splitting damage by hail impacts than shakes or shingles that are laying down in the usual way. This is because the curled portion of the shake or shingle is unsupported; it is basically a cantilever. Thus, a curled shake will absorb more of the impact energy imparted by the hailstone. However, when a wood shake or shingle roof is sufficiently curled to be more susceptible to hail damage, it is already in need of replacement regardless of the existence of any hail damage.

**Plate 21.6** Curled asphalt shingles.

**Buckled shingles** — Buckled shingles, which have the opposite appearance of curled shingles, are usually the result of improper spacing of the shingles. During the summer when roof surface temperature may be quite hot, shingles expand. If the spacing is too tight to allow expansion, the shingles have no place to go but up, so they buckle. Similarly, wood shingles that are spaced too closely may also buckle, but for a slightly different reason. After a rain, wood shakes and shingles swell. Again, if the spacing is too tight and the wood shingles swell, they buckle. Often when shingles buckle, for either reason, they do not recover and remain buckled. Often, buckling will give a roof a rippled or wavy appearance.

**Finding the stony cover of asphalt shingles in storm gutters** — While hail impacts will cause some of the stony cover over asphalt singles to loosen and wash off, the same thing also occurs after a good rain. Sometimes no rain or hail fall is needed: the stones simply loosen and fall off due to differential expansion and contraction over time. Cheap shingles tend to lose their stony covers more easily than good-quality shingles. Thus, finding stony shingle cover particles in the rain gutters is not an indication that the roof has been damaged by hail and needs to be replaced. It does, however, indicate that the asphalt shingles have lost some of their cover as they normally do in a hard rain.

The purpose of the stony cover on asphalt shingles is to protect the underlying bitumen material from ultraviolet light exposure. Often, the primary difference between 30-year rated asphalt shingles and 10-year rated asphalt shingles is simply the thickness and adhesive strength of their stony

**Plate 21.7** Asphalt-shingled roof with fungal growth on one side due to material from tree.

covers. In designing shingles, the loss of the stony cover over time, and after particularly hard rains and hail falls, is taken into consideration by the manufacturer. Finding stony material in the gutters does not automatically mean a new roof is required, or that its life has been otherwise significantly shortened.

**Cracking, splits, and checks** — Cracking of asphalt shingles and pitch membranes can be due to a number of causes. Cracks caused by hail impacts usually emanate from an obvious hailstone crater or impact point. The interior of the crack will be unweathered just like the freshly exposed surface within the hail impact. It will also have an obvious color difference from similar weathered material.

Cracks are also caused by exposure to ultraviolet light, and subsequent shrinkage. Exposure to ultraviolet light causes the lighter hydrocarbons in the bitumen mix to break down, volatilize, and outgas. The loss of this material then causes the affected material to shrink.

These cracks tend to appear in somewhat regular patterns, especially in areas where there is regular exposure to sunlight, because the shrinkage rate is regular. Generally, the cracks initiate and are widest at the top surface, where exposure to ultraviolet sunlight is maximum, and diminish in width with depth into the material, where the ultraviolet light cannot penetrate. Because the cracks open slowly, as opposed to hail-produced cracks, which open all at once, the interior of these cracks will appear weathered and oxidized. They may even be filled with various types of outdoor debris.

**Plate 21.8** Asphalt shingles that have lost their stony cover due to weathering.

In cases where cracking of the bitumen material is an issue, it is often useful to find an area of the roof that does not normally receive much sunshine, but which would have obviously received hail strikes. Such an area can then be compared to one that normally receives sunshine most of the day.

**Dented tin work** — Metal flashing, vent caps, and other light gage metal items usually exhibit hail "dings" or dents no matter what size the hailstone. Many people will point out the dings and indicate that if the metal was dinged, then their roof must be badly damaged. While this is convenient logic when a settlement check is on the line, it is not true.

Hail does not damage everything equally. Each type of material responds differently to impacts from the small spheres of ice. The thin gage sheet metal material in a vent cap, for example, is very different from the thick bitumen material used for a roof membrane. By the way, in most cases, the fact that sheet metal items are dinged does not mean they have been damaged, as long as they can continue to keep out the rain and perform their function. In some case, there may be cosmetic issues that override functional issues. In general, however, most of the sheet metal caps and devices common to a roof top cannot be seen from the ground. Thus, cosmetic damage is not an issue.

Dings in tinwork can actually be very useful in conducting a hail damage investigation. Generally, the width of the ding in the sheet metal is about 1/2 the diameter of the hailstone that made it. By measuring the dings noted in tin work, especially unobstructed vent caps that present normal surfaces to the hail velocity vectors, the relative size of the hailstones can be reasonably estimated. Further, the distribution of hail dings on various tin items on a roof indicates the vector direction of the hail fall, and therefore the wind vector that was present during the hail fall. This can be very important if there is a question about when the hail in question occurred.

For example, consider the hypothetical case where Mr. Jones discovers that his roof is hail-damaged after he buys his home. He then waits for the next hail fall, and makes a claim for hail damage at that time.

According to local weather records, the hail fall on the date of the claim was composed of hailstones no larger than 1/2-inch along with strong winds from the west. Examination of his asphalt shingles finds hail craters that are 3/4–1 inch in size, but the associated cracks and fissures appear weathered on the inside instead of freshly exposed. Examination with a hand lens finds dead bugs, leaf debris, and shingle cover stones within the cracks. Examination of the tin on his roof finds that 2-inch hailstones fell from the east at some time, but that 1/2-inch hailstones also fell at some time from the west. A little more digging into the local weather records finds that 2-inch diameter hailstones last fell in the area 2 years before Mr. Jones bought the house, and that the winds were from the east. The roof was installed 1 year before the 2-inch hail fall, but no hail had fallen in the area in the meantime.

**Gravel cover driven into a bituminous roof** — Some roofers claim that hail can drive the gravel cover of a built-up roof into the asphalt membrane and underlying felt layers, causing penetrations and subsequent leakage of the roof. Hail cannot do this. In a collision between an irregularly shaped rock in a gravel cover and an irregularly spherical hailstone, the contact is

**Plate 21.9** Hail dents in metal roof.

not sufficiently clean for all the momentum of the hailstone to be transferred to the surface rock, and then directly to the one below it, and so on. In fact, when the two do collide, the hailstone usually shatters, or glances off into another nearby rock.

The primary purpose of a gravel overlay is to protect the roof from ultraviolet light. The secondary purpose of the gravel is to absorb hail impacts. A hailstone impacting into a pile of loose gravel simply dissipates its impact energy impacting and ricocheting among the surface rocks, which in turn, have more collisions with the rocks under them. As a rule of thumb, the impact energy of the hailstone expands as a 45-degree cone through the gravel cover.

In other words, if a 1-inch diameter hailstone impacts a gravel cover that is 1-inch deep, the impact energy will be spread over an area of 0.20 square inches at the surface where contact is first made, but will then expand to an area of 4.9 square inches at the surface of the bitumen. Of course by then, all the impact energy will have been dissipated by the friction between the rocks.

This point is easily demonstrable by having a person throw a hail-sized piece of ice as hard has he can into a layer of gravel, and then examining the underlying membrane to see if any gravel has been driven into the bitumen.

The usual reason for gravel to be pushed into the roofing membrane is direct bearing loads, usually caused by people walking on the gravel or by heavy loads being supported by small "feet" or blocks resting on the gravel. A 200-pound roofer can easily mash gravel into the roof with the heel of his boot. In cases where such claims have been made, sweeping the gravel back often reveals that the only places where gravel has been mashed into the bitumen is where there has been foot traffic on the roof.

In assessing hail damage, the value of examining collateral damage should not be overlooked. Collateral damage is damage that has been caused to nearby things by the same hail fall. For example, if the roof of a house has endured a damaging hail fall, the garage in the backyard with the same kind of roof will likely exhibit the same kind of damage. Evidence of 1.25-inch diameter hailstone impacts on the roof, will likely be corroborated by finding burnish marks on the air conditioner cabinet in the side yard, also indicating a hailstone diameter of 1.25 inches.

Collateral damage can be very helpful in establishing the time frame for a particular hail fall. For example, consider the situation where Mr. Jones claims that the severe hail damage on his roof was from the most recent hail fall, yet the Smith's house across the street, which has a roof less than a year old, exhibits no sign of hail damage, and neither does the new deck recently added on to the Murphy's house next door. Can the same hail fall that damaged the Jones' house fail to cause any noticeable damage to the house across the street, or to the deck next door?

**Plate 21.10** Worn and rotted shakes with small burnish marks (circled). There is a green color, which is due to fungus eating into the shakes.

## 21.7  Cosmetic Hail Damage — Burnish Marks

The most common type of cosmetic damage that is mistaken for structural damage are burnish marks. A burnish mark is where the impact of a hailstone rubs off a small portion of the weathered exterior film of the material without damaging the material itself. The rubbing by the hailstone leaves a small, fresh surface surrounded by a larger, weathered surface area. The overall appearance of the surface is often described as "freckled," "spotted," or "pockmarked."

The use of the last descriptor is unfortunate, because while most people use it in the context of having an appearance like a person who has scattered marks on his skin from chicken pox or small pox, some people may interpret the word in its strict dictionary sense to mean that the surface is full of indentations or pits. Thus, the term "pockmarked" should not generally be used when describing the appearance of a roof exhibiting only burnish marks.

Over time and especially after exposure to the weather, many types of materials develop external films, coatings, or layers of oxidized materials. Sometimes the formation of these oxides are accompanied by noticeable discolorations whose appearance is simply described as aged or weathered. Wood, painted surfaces, aluminum siding, copper, plastics, rubber, and even roofing materials like pitch, bitumen, tar, and clay tile all do this. In fact, there are few materials, which, upon exposure to weather, do not eventually exhibit recognizable weathering characteristics of some type.

When a hailstone strikes the surface of a weathered material, especially when the approach vector is not exactly perpendicular to the surface, the hailstone momentarily skids laterally along the surface. This skidding is due to the fact that the hailstone has velocity components both perpendicular and parallel to the surface of the material.

While the hailstone is in contact with the surface, briefly compressing itself and the surface material and then rebounding, it also moves laterally across the surface of the material. Because contact with the material surface is with the bottom of the hailstone, as the hailstone moves across the surface it begins to spin. This spinning effect is easily observed after hailstones impact a sidewalk or street surface and then rebound. The same thing happens to a rubber ball when it is bounced off pavement or a wall: it develops spin when it rebounds from the surface.

While the hailstone is in contact with the surface, it abrasively cleans away a small patch of surface material, removing a portion of the weathered layer that had been present. Typically, this causes the cleaned portion of the surface to have a different appearance than the weathered portions. This is how the "freckled" or "mottled" appearance is created. The resulting marks where relatively fresh or unweathered material has been exposed are called burnish marks. Sometimes burnish marks will appear darker than the rest of the material, such as on painted metal surfaces, and sometimes burnish marks will appear lighter than the rest of the material, such as with wood.

Despite their sometimes disconcerting appearance, the simple presence of burnish marks is not a positive indicator of structure damage. In fact, when burnish marks are less than an inch in width, it is generally a good indication that no structural damage to the material has occurred. This is because a burnish mark less than 1 inch in width indicates that the hailstone that created the burnish mark was less than 2 inches in diameter. As noted earlier, the threshold for damage by perpendicular hailstone impacts is about 2 inches.

Given time and exposure to weather, most burnish marks will slowly reweather and will eventually match the predominant color of the rest of the material. Burnish marks do not impair the functionality of the material.

Of course, the separation of damage into cosmetic vs. structural does not mean that all cosmetic damage is to be ignored. Some types of cosmetic damage can cause a degradation in appearance, and the appearance or decorative value of the item may have been its primary function. If the "beauty" of a structure or item is marred by the presence of burnish marks, then perhaps some type of economic loss, albeit subjective, may have occurred. This will, of course depend on the type of structure, its decorative function, the type of material that sustained the burnish marks and its location, etc.

However, many if not most types of cosmetic damage have no tangible consequence. For example, if the material that exhibits burnish marks is a flat sheet metal roof that is surrounded by parapet walls and can only be seen by birds and airplane pilots, it is doubtful that such burnish marks are of any artistic consequence since no one sees them anyway. Likewise, burnish marks on outside air conditioner cabinets, electrical boxes, and similar such objects generally do not have any significance. These items are not considered *objets d'art*.

When burnish marks are larger than an inch in width, it is possible that material damage has occurred. To determine if the hail impacts have been severe enough to cause both burnish marks and impact damage, it is necessary to examine the burnish marks closely to look for penetrations and gouges sufficiently deep to allow water leakage or loss of material integrity. In such cases, the burnish marks provide an easy visual cue to quickly determine where to look for potential damage. Check out the largest burnish marks first. If there is no damage in the largest burnish marks, there will likely be no damage in the smaller ones.

Because burnish marks are an obvious indicator of where hail impacts have occurred, they can also be used to determine hail fall intensity. For example, a 4 ft × 4 ft sample area on the roof can be measured and divided into a square grid using chalk. The number of hail strikes and the approximate hailstone diameter for each strike in each grid can be counted, and a statistical measure of the hail intensity for this particular location can be calculated. This statistical snapshot can then be applied to the rest of the roof, which has the same basic characteristics as the sample area. In this way, damage estimates can be reasonably determined without having to inventory every square foot of the roof.

For example, it is found that in a 4 ft × 4 ft sample roof area, approximately three shingles were penetrated by hailstones larger than 2 inches in diameter. The sample area was representative of 1/2 of a 1600 ft$^2$ house roof.

**Plate 21.11** Curled wood shakes.

Thus, the side of the roof from which the sample was taken would be expected to have about 300 shingles that were damaged by hail penetrations.

## 21.8   The Haig Report

One of the most quoted, and misquoted sources of information about hail damage to wood roofs is commonly called the "Haig report." The formal title is, *Haig Engineering Red Cedar Shingle Study,* by Haig Engineering Company (2400 West Loop South, Houston, Texas 77027). The report is also summarized with explanatory commentary in the monograph, *Hail Damage to Red Cedar Shingles: A Guide for Adjusters and Roofers,* published in 1975 by the American Insurance Association.

Basically, Haig Engineering Company conducted a 10-year study of the effects of hail on red cedar shingles. The findings of that study are summarized as follows.

- Hail dents in cedar wood shingles do not cause delayed cracking in the wood. If the impacts are sufficient to cause cracking, it will occur at the time of the impact.
- Hail dents without accompanying splits or cracks in the shingle are not considered damaging to the shingle.
- Most of the cracks and splits caused by natural weathering, about 75% to 80%, occur in the first 3 years of the life of a wood roof. High-

quality wood shingles split significantly less than mediocre or poor-quality shingles.

- Hail dents in cedar wood shingles that do not cause structural damage to the material are often not detectable 5 years later. Fibers that were compressed by hail impacts often recover due to the absorption of moisture during weathering. Erosion of the shingle also helps the dent to blend in with the rest of the roof. The accompanying burnish mark also "weathers in" and is not distinguishable from the rest of the roof.
- Old cedar wood shingles that are physically sound are no more susceptible to hail damage than new ones. The key words here are "physically sound." Wood shingles that have fungal rot and have become soft are not physically sound. They are more easily damaged by hail. In fact, they often can be damaged manually with little effort. Fungal rot is often distinguishable by color. If the fungus or bacteria are eating the lignin and cellulose in the wood, the rotted area will have a whitish color. If the fungus or bacteria are eating the cellulose only, it will appear brown, dark brown, or black.
- Hail dents do not significantly cause fungal or bacterial attack to be enhanced or accelerated. Fungal growth, however, is more probable in "untreated" shingles and shakes, or when the shingles or shakes have sufficiently weathered such that any fungicide impregnated in the wood has weathered out.
- Cedar wood roofs having steep slopes generally have a longer life and are less affected by hail. In fact, the pitch of a roof is one of the more significant factors in determining the longevity of a wood roof. A roof with a 1:2 pitch will last about 3 times longer than one with a 5:24 pitch roof.
- Flat-grain cedar wood shingles are more susceptible to cracking by weathering than edge-grain shingles.
- To properly assess hail damage to a wood roof, the roof should be dry. Wood shingles and shakes tend to swell when they are wet. This may camouflage any hail damage that might exist.

Weathering effects obviously cause cracks to form in wood shakes and shingles, and are the most prevalent cause of cracks, splits, and checks. Unprotected wood that weathers will develop a rough surface, which will eventually slough off or erode. The fact that a wood roof exhibits cracks and splits is not evidence of hail damage. To verify that the cracks and splits were caused by hail, it is necessary to find that the cracks emanate from a hail crater or impact.

Cracks that develop in wood shakes and shingles by weathering are relatively easy to verify. The interior of the crack will likely be weathered also, although perhaps not to the same degree as the exposed surface. Cracks caused by hail impacts will have freshly exposed, unweathered material within the crack.

With respect to fraud, occasionally some people will try to simulate hail damage on a roof. Usually, the simulated hail damage is easily detected because it will not have a statistically normal distribution of hail sizes and ding locations, and it will not be consistent with the wind direction of the hail storm.

For example, if the ball end of a ball peen hammer is used, all the hail dings will have the same indentation diameters. In cases where the person doesn't move much while he does his work, the dings will tend to form circular patterns around where the person was sitting when he made the dings.

It is very difficult to mimic a random distribution of hail dings on a roof. Most people who try to simulate hail dings repeat certain recognizable patterns due to the fact that they are consistently right handed, hate to walk close to edges, or forgot that hail usually only affects one side of a gabled roof and make dings on all sides of a multiangled roof. This is why collateral damage assessment is so useful. If hail with a certain pattern affected one house, it should also have similarly affected the shed out back, the house next door, the car parked outside, etc.

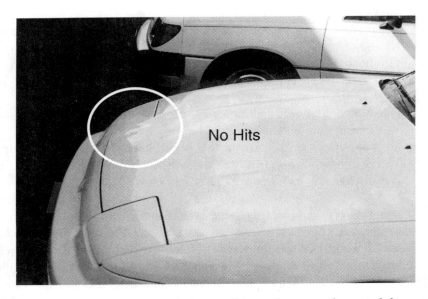

**Plate 21.12** Dents noted in circled area all have the same shape and diameter. No dents on top of hood or roof which is the most exposed and more vulnerable to denting.

## 21.9 Damage to the Sheet Metal of Automobiles and Buildings

The sheet metal covering of a car or a metal building is basically a stiff, planar membrane in the localized area where a hailstone impacts. When a hailstone impacts the sheet metal, the sheet metal is slightly depressed by the impact. Directly under the impact area, the material deflects and bends.

If the impact is light, the sheet metal will spring back in a fully elastic fashion and no permanent deformation will occur. If the impact is sufficiently severe, there will be plastic deformation in the metal and the sheet metal will exhibit a permanent concave "ding" or impact depression. If the impact is very severe, the hailstone will tear the material in the center of the contact area leaving a hole at the center of the "ding."

If it is assumed that the localized area where a hailstone strikes the sheet metal covering of a car is approximated by a circular plate supported at the edges, and that the force exerted by the hailstone is more or less concentrated within a relatively small circular area in the middle of that circular plate, then the maximum bending stress in the plate will occur at the center of the plate directly under the point of impact. The maximum bending stress in the plate is given by the following formula.

$$\sigma = 3P/\pi t^2 \tag{viii}$$

where $\sigma$ = maximum stress in the middle of circular impact area, $P$ = force exerted by hailstone, and $t$ = thickness of plate.

Hot-rolled steel sheet metal has a yield point of about 42,000 lbf/in². A common sheet metal thickness used for many things, including cars, is about 0.060 inches (#16 gage). Thus, using the above formula finds that the minimum concentrated force needed to initiate plastic deformation in 0.030-inch thick hot-rolled steel sheet metal is:

$$P = (42,000 \text{ lbf/in}^2)(3.14159)(0.060 \text{ in})(0.060 \text{ in})/3 = 158 \text{ lbf}$$

The distortion energy per unit volume of a material that is subject to two principal tensile stresses that are equal is given by the following formula.

$$U = (1 - \mu)\sigma^2/E \tag{ix}$$

where $U$ = distortion energy per unit volume, $\mu$ = Poisson's ratio for the material, $E$ = Young's modulus, and $\sigma$ = principle stress.

Again, if the material is hot-rolled steel sheet metal, and the yield point of the material is 42,000 lbf/in², then the amount of distortion energy required to initiate yielding is estimated as follows.

$$U = (1 - 0.22)(42,000 \text{ lbf/in}^2)^2/30,000,000 \text{ lbf/in}^2$$

$$U = 45.86 \text{ lbf-in/in}^3 = 3.82 \text{ lbf-ft/in}^3$$

For sheet metal that is 0.060-inches thick, the amount of distortion energy per area needed to initiate yield is then 0.23 lbf-ft/in².

In impacting sheet steel, hailstones do not cause the sheet steel to significantly compress in the area of contact. However, the hailstones themselves certainly do compress in the area of contact. Often, the impact is sufficient to cause the hailstone to fracture. After a hail fall, it is common to find many broken or fractured hailstones around sidewalks and other hard surfaces. Concrete is particularly unyielding with respect to hail impacts. (In a collision contest between concrete and ice, who would you bet on to survive unscathed?)

Rebound tests of spherical ice pellets indicate that 80–90% of the impact energy is absorbed by the ice itself, depending upon how rigid or "mushy" the ice is. Ice spheres approximately 1 inch in diameter often break apart at impact speeds of just 20 ft/sec, which contains a total impact energy of 0.12 lbf-ft or a contact impact energy of 0.0367 ft-lbf/in². Presuming that 80–90% of the impact energy is absorbed by the ice itself, this leaves only 10–20% of the impact energy available for rebound velocity, spin energy, and absorption by the sheet metal.

If it is very conservatively assumed that half of all the energy not absorbed by the hailstone is absorbed by the sheet metal, then approximately 5–10% of the impact energy carried by the hailstone will go into the sheet steel. Returning to the hot-rolled steel sheet metal that is 0.060-inches thick, the amount of kinetic energy needed by a hailstone to initiate yielding in the sheet metal is 2.3 to 4.6 ft-lbf/in². Checking Table 21.5A, this corresponds to hailstones approximately 1 inch in diameter or larger.

If it is assumed less conservatively, but more realistically, that most of the energy not absorbed by the hailstone is consumed by the rebound and spin of the hailstone, then approximately 1–2% of the impact energy carried by the hailstone goes into the sheet metal. Using the same steel sheet metal as before, this corresponds to a hailstone kinetic energy of 11.5 to 23 ft-lbf/in² to initiate yielding. Checking Table 21.5A again, this corresponds to a hailstone size of 1.25–1.75 inches in diameter.

Thus, impacts from hailstones of about 1.25–1.75 inches in diameter or larger are sufficient to initiate plastic deformation in 0.060-inch thick steel sheet metal. Of course at this level, the amount of plastic deformation is not readily visible. At the point where plastic deformation initiates, no plastic deformation has actually taken place. For plastic deformation to be readily visible, the hailstone has to be somewhat larger than 1.25–1.75 inches in diameter.

As a comparison, tests by the National Building Research Institute in South Africa (CSIR Report #176) report that corrugated 24-gage galvanized steel sheet metal that is used for roofing, visibly dents at impact energies of 77.8 ft-lbf, but does not puncture until energies of about 311 ft-lbs are achieved. In checking Table 21.5A, the impact energy level corresponds to a hail size of between 2.5 and 2.75 inches in diameter. It should be noted, however, that because the type of sheet metal used in this test was corrugated, the hail would be impacting at an angle. Correcting for the corrugation angle to "normalize" it for a straight-on impact with flat sheet metal, the impact energy would be 38.9 ft-lbf/in². In comparing this to Table 21.5A, it corresponds to a hail diameter of between 2.0 and 2.5 inches.

Similarly, the same report indicates that vitreous enameled #16 gage sheet metal (0.060 inches thick) exhibited barely visible dents from hailstones impacting at 45 degrees with an energy of 77.8 ft-lbf/in². Correcting for angle, this corresponds to a direct hit energy level of 38.9 ft-lbf/in². Table 21.5A indicates that this corresponds to a hailstone diameter between 2.0 and 2.5 inches in diameter.

In the Greenfeld study, it was reported that standard standing-seam terne metal roofing exhibited objectionable denting when hailstones were 2.0 inches in diameter. Cracking of the plywood deck under the metal occurred when 2.5-inch diameter or larger hailstones were used.

In short, analysis indicates that the typical hail size threshold for cars and sheet metal buildings to exhibit hail "dings" and dents is about the same as that required to damage most roofing systems. However, depending upon the thickness of the sheet metal, and its location with respect to support (firm vs. soft support), the overall range of hail diameter size that can cause objectionable denting is greater.

Because a car has many surfaces at many angles to the hail trajectory vector, it is expected that when a car is damaged by hail, there will be a wide range of hail dings. Those surfaces normal to the hail vector will be dented the deepest. Those surfaces that angle away from the hail vector will be less severely impacted, and portions of the car that are in the "hail shadow" of the car will not be damaged at all.

When a car is inspected for hail damage, it is recommended that a good loupe be used to examine the center of the hail "ding." People who attempt to simulate hail damage will often use the ball end of a ball peen hammer. Of course, this means that all the hail indentations will have the same diameter, instead of having various randomly distributed ding diameters. Further,

impact with a steel ball peen hammer is much different than that by a
hailstone. The ball of the hammer will usually scratch and cause striations
in the paint due to the rotation of the hammer during the impact. Balls of
ice are not hard enough to do this.

Some individuals are astute enough to use several different sized ball
peen hammers to simulate the various sizes of hail, and will clean and wax
the car prior to any inspection because the wax will fill in the striations caused
by the metal to metal grinding. For this reason, it is often useful to clean hail
dents with an appropriate solvent prior to examination, especially if the car
is unusually clean and gleams.

In one rather amusing case, a person tried to simulate hail damage to a
car by using a golf ball. He placed the golf ball against the surface of the car,
and then used a hammer to make the dent. Unfortunately for him, however,
dents produced by hammered golf balls not only leave dimpled dents, but
occasionally leave an imprint of the brand name of the golf ball — spelled
backwards of course.

Similarly, if the car was parked during the hail fall, then the dings on the
car should more or less match up to the distribution and intensity of dings
on other nearby metal items, such as mailboxes, metal sheds, and other cars.
In other words, hail dings should be deepest on surfaces normal to the
direction in which the hail originated, and the directions of hail origin should
coincide (if the car has not been moved).

Because hail damage and hail damage fraud have been such a bane to
both the car owners and insurers (and such a boom for bootleg car part
manufacturers), some car manufacturers have started using various types of
plastic panels on those portions of the car that typically receive the most
"dings" from hail or other items. Hailstone impacts, or dings caused by
adjacent car doors being opened, don't normally cause dings in these panels.

Probably the most publicized car company that has done this is Saturn.
The plastic panels on various Saturn models are sufficiently resilient to absorb
the impact without plastic deformation. Resilience is the work required to
deform an elastic material to the yield point divided by the volume of the
material in which the work was done. By allowing more of the panel to deflect,
the impact energy is distributed through a much larger volume of material.
The per unit volume distortion energy is therefore kept low.

## 21.10 Foam Roofing Systems

Sprayed polyethylene foam (SPF) roofing systems, commonly called foam
roofing systems, are relatively uncommon now in the U.S. except in the South.
Historically, they have not fared well with respect to hail. Part of this is due
to poor installation techniques, poor quality control of the properties of the

foam when it has been mixed in the field, or simply to unexpected problems encountered in the inherent properties of the materials used in either the foam or the coating. Because of these bad experiences, some local building codes have actually "outlawed" the application of foam roofing systems, and some insurance companies do not insure foam roofs.

Some of the foam roofing systems were found to be particularly vulnerable to ultraviolet light. Within an unexpectedly short time, UV light caused the top surface of the material to become brittle and lose strength. Consequently, the membrane was easily breached by hail or footsteps. Once breached, water infiltrated the foam and the roof deteriorated quickly.

Foam roofs were popular in the late 1970s and early 1980s due to their low cost and their promise to help thermally insulate the roof of otherwise uninsulated buildings. Most of those foam roofs have worn out or been removed by now. Some new types of high density foam roofs have been marketed recently with an improved coating to protect it from UV damage and hail impacts. Some types of contemporary foam roofing systems are also using traditional gravel covers to protect the foam from UV light and hail impacts. These new types of foam roofs do appear to be performing better than the old types when installed correctly.

## Further Information and References

*Advanced Mechanics of Materials*, by Seely and Smith, 2nd ed., John Wiley & Sons, Inc., New York, 1952, pp. 223–228. For more detailed information please see Further Information and References in the back of the book.

*Atmosphere*, by Schaefer and Day, Peterson Field Guides, Houghton Mifflin Co., New York, 1981. For more detailed information please see Further Information and References in the back of the book.

"Fronts and Storms," in *Van Nostrand's Scientific Encyclopedia*, 5th ed., Van Nostrand Reinhold Co., pp. 1105–1110, and also "Weather Observation and Forecasting," pp. 2319–2327, and "Winds and Air Movement," pp. 2339–2343. For more detailed information please see Further Information and References in the back of the book.

*Hail Damage to Red Cedar Shingles*, American Insurance Association, 1975. For more detailed information please see Further Information and References in the back of the book.

*Hail and Its Effects on Buildings*, CSIR Research Report No. 176, Bulletin 21, 1-12, National Building Research Institute, Pretoria, South Africa, 1960. For more detailed information please see Further Information and References in the back of the book.

*Hail Resistance Tests of Aluminum Skin Honeycomb Panels for the Relocatable Lewis Building*, Phase II, by Mathey, U.S. Department of Commerce, National Bureau of Standards, NBS Report 10-193, April 10, 1970. For more detailed information please see Further Information and References in the back of the book.

*Hail Resistance of Roofing Products — Building Science Series 23*, by Sidney Greenfeld, U.S. Department of Commerce, National Bureau of Standards, August 1969. For more detailed information please see Further Information and References in the back of the book.

*Prospective Weather Hazard Rating in the Midwest with Special Reference to Kansas and Missouri*, by Friedman and Shortell, Research Department, Travelers Insurance Companies, August 28, 1967. For more detailed information please see Further Information and References in the back of the book.

*Reader's Digest New Complete Do-It-Yourself Manual, Reader's Digest*, 1991. For more detailed information please see Further Information and References in the back of the book.

"SPF Roof Systems: Field Survey and Performance Review," by Dr. Dupuis, P.E., *Professional Roofing*, March 1996, pp. 32–36.

*Weather*, by Ralph Hardy, TY Books, 1996. For more detailed information please see Further Information and References in the back of the book.

"Wood Roofing: Basics for Craftsmanship," by Jim Carlson, *Professional Roofing*, June 1996, pp. R4–R9. For more detailed information please see Further Information and References in the back of the book.

# Blaming Brick Freeze-Thaw Deterioration on Hail

# 22

I don't mind it because when they (the press) throw bricks at me, I'm a pretty good shot myself, and I usually throw 'em back at 'em.

— **Harry Truman,** *1958*

## 22.1   Some General Information about Bricks

With apologies to Gertrude Stein, it is not true that a brick is a brick is a brick. While masonry bricks come in all sizes and shapes, modern baked clay bricks used on the exterior of buildings, i.e., face bricks, come in three distinct grades: SW (severe weather resistant), MW (moderate weather resistant), and NW (no weather resistance).

Face bricks are called this because they were once reserved for facing walls that would be exposed to the weather. Building bricks or common bricks were originally the all-purpose bricks. However, these days the term, "building brick" usually refers to an off-grade face brick, which is perhaps visually imperfect for facing work, but still suitable for use in unexposed areas.

Other types of bricks that occasionally become confused with face bricks are firebricks and paving bricks. Firebricks are yellowish in color, and are valued for their resistance to heat. They are generally used to line fireplaces, brick ovens, hearths, and brick-lined combustion chambers. Firebricks may have the same shape and size as facing bricks.

Paving bricks, as their name implies, were originally used to pave roadways and sidewalks. These days however, they are often used for driveways, patios, and decorative walkways. Paving bricks are generally more durable and harder than face bricks. They are usually slightly different in shape. Whereas a standard facing brick is 2 1/4 in. × 3 3/4 in. × 8 in., a pavement brick is usually 2 1/4 in. × 3 4/4 in. × 7 3/4 in. The shorter length makes it stockier and a little less prone to breakage by bearing loads.

Sometimes the three grades of facing bricks, SW, MW, and NW, are respectively referred to as hard, medium, and soft. Unfortunately, the grade of brick is usually not marked or otherwise noted on the brick itself. Consumers purchasing bricks from local hardware stores or lumberyards do not usually know, nor are they usually told, what grade of bricks they are buying. To most consumers, all bricks seem to be more or less the same, except perhaps for texture, size, color, and most importantly, price.

Even more unfortunate is the fact that some contractors do not know that bricks have three distinct grades, and that each grade of brick has an appropriate application. These unknowledgeable contractors purchase bricks at the lowest possible price, and use whatever grade is bought for any and all applications where brick and mortar are required. Thus, brick that is used indoors for decorative trim around a hearth or planter by the contractor may be the same brick that is used outdoors in a wall or facade, or perhaps to construct the exterior of a fireplace chimney.

The use of inferior brick, or brick of an inappropriate grade for a particular application is usually not readily determinable by casual inspection. In lieu of having some documentation to indicate a brick's pedigree, close visual inspection of the porosity with a hand lens and field measurements of the brick's density or specific weight are good indicators of its type. For a definitive determination, a simple water absorption test and appropriate compression testing should be done.

## 22.2  Brick Grades

Table 22.1 compares the physical properties of the three grades of facing brick as required by ASTM C-62.

Grade NW is the most inferior grade of facing brick with respect to compressive strength and moisture absorption. It is also generally the cheapest to purchase. This type of brick is intended for use as backup masonry or for interior masonry.

When used in exposed locations, grade NW is supposed to be used only where there is no frost and there is less than 20 inches of rain per year, or where the rainfall is less than 15 inches per year if frost occasionally occurs. This is because this grade of brick has the highest porosity and can retain

**Table 22.1  Brick Grades**

| Grade | Average[a] Compressive Strength | Average[a] Percent Water Absorption by Weight | Specific Gravity |
|---|---|---|---|
| SW | 3000 lb/in$^2$ | 13–14 | 2.0–2.2 |
| MW | 2500 lb/in$^2$ | 19–20 | 1.8–2.0 |
| NW | 1500 lb/in$^2$ | no limit | 1.6–1.8 |

[a] "Average" in this case usually means the average properties of five bricks taken at random from a lot. Individual bricks within a sample group of five may vary from these averages.

significant amounts of water. When saturated, 25–35% of the weight of a type NW brick may be due to absorbed water.

Type NW brick has the lowest compressive strength of the three types. Despite the fact that the bearing loads on bricks typically do not exceed 100 psi, when used in load bearing walls where SW brick should have been used, NW bricks may crack and break out, especially after a period of heavy rains where the wall has become significantly heavier due to water absorption in the face brick. Such cracking and break out occurs at the base of the wall, where the accumulated bearing loads are highest, or just below locations where significant point bearing loads are transferred to the wall from floor joists, roof joists, or similar.

Type MW is to be used where the brick is exposed to temperatures generally below freezing, but the brick is unlikely to be permeated with water, as would occur below grade in damp earth, for example. When used in moderate climates as recommended, the loss of material due to 50 cycles of freezing and thawing is less than 3%.

Type SW is used where a uniform and high degree of resistance to weathering is required. The brick will be exposed often to freezing, and may be permeated with water when freezing occurs. When subjected to 50 freezing and thawing cycles, the loss of material is not greater than 1.0%.

## 22.3   Basic Problem

There is a fundamental problem with contractors using the wrong types of brick. Let's say a wholesale lot of type NW bricks is bought by a contractor, usually a residential builder. The contractor then uses these type NW bricks for chimneys, exterior decorative walls, decorative retaining walls, or other exterior applications.

The house is located where there are long periods of hard freezing temperatures in the winter, for example, Illinois, Missouri, Kansas, Iowa, or Maine. During a rainy part of the fall or winter, the exposed portions of the bricks absorb water from either direct rain impingement or by rainwater running down the brickwork from other locations.

With the exposed portions of the bricks laden with up to 25–35% water by weight, a hard freeze occurs. As every school child knows, when water freezes, it expands. Specifically, it expands almost 10% in volume. A cube of water exactly 1.0000 in.$^3$ at 4°C will, upon freezing, expand to a cube containing 1.0989 in.$^3$ of water.

Water trapped in the pores of the brick along the exposed portions will cause internal cracking in the brickwork, usually along a plane roughly parallel to the exterior surface through which the water entered the brick. After several such freeze-thaw cycles, flat plate-like pieces of the exterior face of the brick are ready to fall out.

Late spring and late summer are when most of the hail falls in the U.S. Because the pieces of brick broken out by freeze-thaw cycling are ready to fall off at the slightest provocation, they often fall out during a severe thunderstorm with accompanying hail. While pieces do fall out in storms having high winds, often the pieces of brick are flung away from the wall. However, during a hail fall, impacts by hailstones knock the pieces loose and they usually fall down at the base of the wall.

Thus, after the hail fall has subsided and the homeowner is checking his home for damages, he notices pieces of brickwork missing from the walls and the corresponding pieces of missing brick are on the ground at the base of the wall. The assumption is then made that since these pieces had not been visibly missing from the wall prior to the hail fall, then hail impacts must have caused it to happen. This is another example of correlation being mistaken for causation.

This also occurs when MW type bricks are used in cold climate areas, but it takes longer for the cracking to occur because the brick absorbs less water and the cracking, therefore, occurs at a slower rate.

## 22.4  Experiment

To demonstrate that hailstone impacts do not damage bricks that are in good condition, select some bricks and get some ice cubes. Place the ice cubes on the bricks and hammer the ice cubes into the bricks without letting the hammer itself strike the bricks. Let the ice be the go-between. The ice, of course, will fracture and break apart but the brick will not be damaged.

Thus, even if hailstones slam into a brick wall at speeds sufficiently hard to break them into smithereens, they will not damage a brick in good condition. This is because, of course, bricks are inherently stronger than ice. Ice is a relatively weak material. The only way that pieces of brick can be knocked out of a brick wall by hailstones, is when the bricks are already fractured and are hanging together by a thread.

## Further Information and References

*ASTM C62, Physical Requirements of Clay Brick*, and *ASTM C73, Physical Requirements for Sand Lime Brick*, American Society of Testing Materials. For more detailed information please see Further Information and References in the back of the book.

*Handbook of Design*, "Principles of Brick Engineering," by Plummer and Reardon, Structural Clay Products Institute. For more detailed information please see Further Information and References in the back of the book.

*New Complete Do-It-Yourself Manual*, published by *Reader's Digest*, 1991, pp. 162–179. For more detailed information please see Further Information and References in the back of the book.

# Management's Role in Accidents and Catastrophic Events

# 23

It is not the critic who counts; not the man who points out how the strong man stumbles, or where the doer of deeds could have done them better. The credit belongs to the man who is actually in the arena, whose face is marred by dust and sweat and blood; who strives valiantly; who errs, and comes short again and again, because there is no effort without error and shortcoming; but who does actually strive to do the deeds; who knows the great enthusiasm, the great devotions; who spends himself in a worthy cause; who at the best knows in the end the triumph of high achievement, and who at the worst, if he fails, at least fails while daring greatly, so that his place shall never be with those cold and timid souls who know neither victory nor defect.

— **Theodore Roosevelt,** *Address at the Sorbonne in Paris, April 23, 1910*

## 23.1 General

Some companies rarely have accidents or catastrophic events. Others seem to have accidents and misfortunes on a regular basis. Why? Is one plant luckier than another? Are the people who work in one plant inherently more "accident prone" than those who work in another? Or, do the people in one plant just make more mistakes, perhaps more serious mistakes, than the people in another plant?

Everyone makes mistakes. It's an important part of the learning process. No one plays a Beethoven concerto the first time he is introduced to the piano. Even child prodigies require practice. Practice, by the way, is the period of time set aside wherein it is okay to make mistakes. Practice allows a person the freedom to experiment and make mistakes while developing proficiency skills.

Most processes, procedures, designs, and even theater plays are not perfect the first time they are attempted. They are usually revised several times until the obvious mistakes and problems are solved. Often, pilot plants, mock-ups, model studies, computer simulations, prototypes, test runs, and out-of-town openings are required to work out the big problems. Later on, reports from the field, customer complaints, and crowd reaction is used to make further adjustments to improve the product. It's an evolutionary process that never seems to be finished. Each succeeding improvement depends on detecting and acting on the mistakes noted in the previous attempt.

To deny that mistakes occur, or to promote a company policy of "zero tolerance for mistakes" is unrealistic. Mistakes are inevitably made even by the most experienced, highly-trained, and well-motivated individuals. In companies that have a "no tolerance" attitude about mistakes and a system of punishment for the person or department that makes them, three things commonly occur that promote the eventual occurrence of serious accidents and catastrophic events.

1.  Mistakes are actively hidden, covered over, and denied.
2.  Blame finding and scapegoat hunting becomes the dominate company culture characteristic.
3.  Employees engage in passive-aggressive sabotage and malicious compliance actions to get back at management.

Mistakes, per se, are not accidents or catastrophic events. Mistakes are simply a type of accident precursor. This means that they can lead to accidents if not handled appropriately. This is what occurs after a mistake is made that determines whether it becomes an accident or a catastrophe.

Except for willful sabotage, a human mistake is, by definition, simply an incorrect judgment or decision. If a work process or task is represented by a decision tree, a person makes a mistake when he reaches a decision-making point in the process and takes the wrong branch. The mistake turns into an accident if the subsequent actions in a decision pathway lead to an accident. If somewhere along the decision tree between where the mistake is made and an accident occurs, the mistake is corrected and the work process is diverted back into the correct decision pathway, the accident can be averted and the mistake corrected.

This is the basis for many of the modern quality control concepts in manufacturing management. Previously, quality control tended to be an "after the fact" part of the process. Usually at the end of the production line, a person would take samples of the finished product and look for flaws or deficiencies. If the detected deficiencies or flaws were statistically less than some maximum acceptance criteria, the process was allowed to continue. If the detected flaws were higher than the allowable amount, then the line was shut down and someone would "troubleshoot" the problem until once again the occurrence of defective products was below the predetermined statistical set point.

This was the basis for many quality control systems, including those employed by the U.S. Department of Defense. Military Standard 414 dated June 11, 1957, for example, states the following in its introduction:

This Standard was prepared to meet a growing need for the use of standard sampling plans for inspection by variables in Government procurement, supply and storage, and maintenance inspection operations. The variables sampling plans apply to a single quality characteristic which can be measured on a continuous scale, and for which quality is expressed in terms of percent defective. The theory underlying the development of the variables sampling plans, including the operating characteristic curves, assumes that measurements of the quality characteristic are independent, identically distributed normal random variables.

More modern management concepts employ localized self-correcting techniques. In these techniques, the mistakes are not detected at the end of the production process, after the product has already been built. That is too late. By then, materials and labor have already been invested in a defective product that can't be sold.

Modern management concepts are concerned about correcting mistakes at the point where they are made. This is done by adding a localized feedback loop at each important process step. After a decision has been made in the process, the results of that decision are compared to a standardized outcome. If it meets the requirements of the standard, the work process is allowed to continue to the next step. If it does not meet the requirements of the standard at that step, the process decision is reviewed and revised until it does.

In control theory parlance, the above can be described as follows. In older production systems, there was only one signal comparator and one feedback loop for the whole system. Corrections could only be made after the signal had already passed through the entire system. Systematically, this made the entire system slow to correct for out-of-tolerance signals that developed within the system, and could lead to undesirable instabilities.

In newer systems, each important process step has a comparator and a feedback loop. In this way, corrections are made at each important step of the process. Out-of-tolerance behavior is detected and corrected on the spot where and when it occurs, and the overall system is more stable and dependable. Figure 23.1 schematically depicts how this works.

As it is with production quality, so it is with safety. Traditional management systems tend to look at the number of accidents that occur in a year at their plant and compare their performance to the published industry average. They then go back and troubleshoot their work processes until the safety deficiencies causing the higher than normal number of accidents are found and corrected. As with older quality control systems, the feedback loop is at the end of the process, well after the accidents have occurred. The problem is considered solved when the number of accidents in the next year drops below some acceptance criterion, usually the published average for that process or industry.

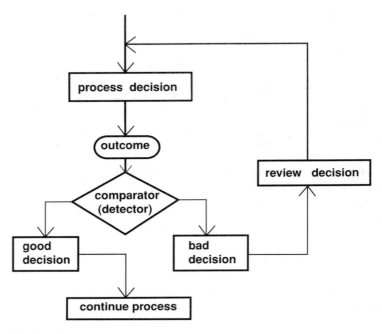

**Figure 23.1** Self-correcting process step.

Modern safety management strategies endeavor to detect and correct mistakes where and as they occur. In this way, the mistake does not propagate through the system until an accident occurs. Consistent with this strategy, most modern approaches to safety management assume the following to be true.

- Humans make mistakes; even the most experienced and well-trained individuals make mistakes.
- Decision-making junctions or judgment points in a work process where mistakes are likely to occur are predictable, especially from past experience, and can be managed to prevent a mistake from propagating through the system and turning into an accident.
- Management and employees are partners who share the responsibility for safety in the workplace.
- Being social creatures, humans react and respond to their work environment. This means that their behavior responds to both peer encouragement and positive reinforcement, and peer discouragement and negative reinforcement. Long-term results are best achieved with positive reinforcement techniques. Short-term results are easier to achieve with negative reinforcement techniques, but the improvements are short lived and eventually degenerate over the long term.

**Table 23.1   Fault List — Human Error vs. Management Error**

| Human Errors | Management Errors |
|---|---|
| Inexperience | Heavy workload |
| Lack of knowledge | Emphasis on deadlines |
| Lack of proficiency | Distractions, poor environment |
| Incorrect assumptions | Unclear instructions/standards |
| Fatigue/overwork | Poor training content/practices |
| Complacency | Insufficient resources provided |
| "Make it work" attitude | "Just get it done" attitude |
| "Take a chance" attitude | "I don't want to know" attitude |

## 23.2   Human Error vs. Working Conditions

Some safety researchers have categorized the causation of accidents and catastrophic events into two categories: those caused by unsafe acts or human error, and those caused by unsafe conditions, i.e., faulty equipment, poor working conditions, etc. Based on this generalization, individuals are held responsible for human error, and management is held accountable for working conditions, tools, resources, and so on.

Generally the errors caused by humans and errors caused by management are divided as shown in Table 23.1.

While this is an easy way to classify the cause of an accident, because it is likely that at least one of the items on the list will have occurred in any given accident, it is oversimplified. It presumes that management actions and human error are separate and distinct. They are not. Each affects the other, although sometimes in remarkably subtle ways.

## 23.3   Job Abilities vs. Job Demands

One theory of accident-related human behavior asserts that accidents precipitated by human error occur when the momentary demands of the job or task exceed the momentary ability of the person to accomplish it.

This can be restated by the following simple inequality.

$$J_D < J_A \quad \text{no accident occurs} \tag{i}$$

$$J_D > J_A \quad \text{accident occurs}$$

where $J_D$ = momentary job demands and $J_A$ = momentary job ability.

The momentary demands of a job can include:

- job-specific manual skills.
- job-specific knowledge, instructions, and procedures.
- the exercise of job-specific, knowledge-based judgment.
- job-specific tools, parts, and resources.
- the ability to respond to changing job conditions as the job progresses.
- the ability to respond to changing instructions and expectations while the job progresses.
- coping with momentary environmental factors such as lighting levels, sound levels, temperature, bodily discomfort, perceived risk of danger, and various distractions.
- job time constraints and deadlines.
- job outcome expectations by supervision.

Human performance, of course, can vary from day to day. Thus, if there is only a small margin between the average ability of a person to perform a task and the average demands of the task (i.e., when $J_D \cong J_A$), when that person has an "off" day or momentary decrease in ability, his chances of causing an accident are relatively high.

On the other hand, if there is a significant margin between the average ability of a person to perform a task and the average demands of the task, even if that person has an "off" day, his chances of causing an accident are relatively low. There is enough margin to absorb the variation in ability.

Of course, the above relationship also works in reverse. Again, if there is only a small margin between the average ability of a person to perform a task and the average demands of the task, when the demands of the job momentarily increase, the chances of an accident occurring increase. But, as before, if there is a significant margin between the average ability of a person to perform a task and the average demands of the task, there is enough margin to absorb the variation in job demands.

One of the primary functions of the management of a company is to select personnel to do the work. This includes setting the requisite minimum standards for experience, training, and skill level for a particular position. Most companies call this the minimum job qualifications, and it is usually a part of the overall job description for a particular position.

Likewise, one of the primary functions of the management of a company is to provide a place to work. This includes providing the necessary tools, resources, conditions, and general work environment that helps to minimize the momentary job demands. Furthermore, another basic function of management is to provide supervision to monitor variations in work demands and personnel abilities and make appropriate adjustments.

In short, it is the management of a company that sets up the basic human factors accident risk equation as noted in Equation (i). It determines the average job ability level by its personnel selection and management policies, it determines the average job demand level by its work rules, working conditions, and general work management practices, and it monitors the margin between the two factors through supervision.

## 23.4  Management's Role in the Causation of Accidents and Catastrophic Events

In general, because management is responsible for the overall conditions of the work place, management actions and policies can be an important factor in the causation of accidents and catastrophic events. This is because management:

- allocates the resources for the work tasks, including the proper tools and safety equipment.
- is required to provide equipment and machinery that meets current safety standards and practices (grandfathering is not allowed when it comes to OSHA requirements).
- sets standards for on-the-job adherence to safety standards and practices with either positive reinforcement or negative reinforcement practices by supervision.
- selects and qualifies personnel for particular work tasks and assignments.
- selects and provides the appropriate training for specialized work tasks when needed.
- maintains records necessary to determine if federal, state, and local safety standards and codes are being met, and actively adhered to, and to spot trends and problems.
- implements procedures and standards to correct unsafe actions or practices, or to promote general safety and safe practices.

Management sometimes can obviate, reduce, stop, or mitigate accidents and catastrophic events by developing appropriate procedures and practices. In that regard, management has or could exert significant influence over the following four factors relevant to accidents and catastrophic events occurring.

- Identification and reduction of accident precursors. This means that the conditions that favor mistakes being made are identified and reduced as much as possible. An example of a condition that is an accident precursor is allowing fatigued employees to work where a high degree of accuracy is demanded, or assigning untrained employees a task requiring practiced skills and judgment.

- Identification and reduction of accident initiators. An accident initiator is an action that can cause an accident. Removing a belt guard on an operating machine is an example of an accident initiator. Regular inspections of belt guards by supervision to look for "missing" guards and replace them is an example of identifying accident initiators and reducing them. Sometimes skillful design can remove an accident initiator entirely by making it impossible or highly unlikely for the event to occur. For example, if the inadvertent operation of a switch is an accident initiator, the redesign of the switch so that it cannot be inadvertently operated obviates the potential of this event occurring.

- Development of and promotion of accident arrestors. An accident arrestor is a device or procedure that attempts to come between the accident initiator and the accident itself. It stops an accident initiator from becoming an accident. An example of an accident arrestor is a contact switch on a machine guard. Even if the guard is removed while the machine is operating (an accident initiator), the switch turns off the machine to prevent a person from getting caught by the moving belt or rotating pulleys. Another example is a GFCI, or ground fault current interrupter. With a GFCI installed in a bathroom outlet, even if a person inadvertently drops an electric shaver into a basin full of water, the danger of electrical shock is removed or reduced because the GFCI automatically opens the circuit at the outlet.

- Development and promotion of accident mitigators. This is a device or procedure that mitigates the consequences of an accident once it has occurred. A fire sprinkler system is an example of an accident mitigator. Another example is an emergency eye wash device. It doesn't stop the accident from occurring, but it does reduce the damage that is done once the accident has occurred. An example of a procedure being an accident mitigator, is a fire escape plan.

## 23.5  Example to Consider

The following is a brief account of a catastrophic event where six men were killed. It is presented to stimulate discussion about management's role in the prevention of accidents. As you read it, consider how you might answer the following questions with regard to the management of the fire department, construction company, security guard company, and local unit of government.

- What were the accident precursors in this case? Were they anticipatable? What could have been done to reduce or eliminate them?
- What were the accident initiators in this event?
- Were there any accident arrestors? Could there have been any, either by procedure or with the use of equipment?
- What were the accident mitigators? Do you have any recommendations concerning procedures or equipment that could have prevented the six firemen from being killed?
- Develop a flow chart or decision tree that represents this accident. Discuss the decisions made at each point and how a self-correction feedback loop could have prevented the mistakes from becoming a catastrophic event.

A fire and explosion event occurred in Kansas City, Missouri, on November 29, 1988, that killed six firemen. The explosion occurred at a highway construction site, located about 650 feet northeast of the present day intersection of 87th Street and U.S. 71. The immediate area surrounding the blast was undeveloped at the time and was populated with trees and scrub brush.

The sequence of events leading up to the explosions began with the discovery of prowlers by construction site security guards shortly after 3:00 A.M. on November 29, 1988. The two guards went to investigate the prowlers. While doing so, the guards observed that a vandalism-type fire had been set in the pickup truck of one of the guards on the other side of the construction site. Both of the guards rushed to the pickup truck. The pickup truck fire was reported to the Kansas City Fire Department at 3:40 A.M.

A second fire was then noted by the guards in the area of two construction company semitrailers parked at the site. They were both located a short distance from the pickup truck. Both of the semitrailers contained ammonium nitrate fuel oil (ANFO) construction explosives. The second fire occurring at the two semitrailers was also reported to the fire department by the guards.

A pumper truck was dispatched to the site and put out the fire in the pickup truck by about 4:00 A.M. When this was done, the pumper truck went over to assist a second pump truck, which had already been dispatched to put out the fire at the location of the two semitrailers.

The first explosion occurred about 4:06 A.M. in the area of the semitrailers. Reportedly, the blast overpressure near the scene was sufficient to move parked vehicles several feet. A bright, white fire continued to burn at the scene after the explosion. At 4:48 A.M., a second explosion occurred. A similar bright fire also continued to burn at the site after the second explosion.

Investigative authorities entered the explosion site area sometime after the second explosion, about 6:00 A.M., when it was deemed safe to do so. All the firemen who had manned the two pumpers were dead, and the two pumpers were wholly destroyed. Because of their closeness to the explosions, there were few remains of the firemen to be found.

The first semitrailer parked at the construction site was known to contain 17,000 pounds of aluminized ANFO in 5- and 6-inch diameter blasting socks. The first semitrailer also contained 3500 pounds of nonaluminized ANFO in 50-pound bags.

The second semitrailer, which was parked near and roughly parallel to the first semitrailer, contained 30,000 pounds of aluminized ANFO in 30-pound socks. Thus, there had been 50,500 pounds, or just over 25 tons, of ANFO stored at the site divided between the two semitrailers. The construction of both of the semitrailers was such that their structures offered no significant containment of the resulting explosions.

When the explosion site was examined after the blasts, three distinctive craters were found. Two fully separate craters, one 38 feet across and 6 feet deep, and another 20 feet across and 4.5 feet deep, located where the first semitrailer had been. A third crater, which was measured to be 49 feet across and 6.5 to 7.0 feet deep, was located where the second semitrailer had been. This third crater was not regular in shape, but had a smaller overlapping crater, or a crater extension of 20 feet at one end.

Thus, in the two semitrailer locations, there were patterns of a large crater next to a smaller one corresponding to each semitrailer. In the first semitrailer location, the large and small craters were fully separate from one another. In the second semitrailer location, they overlapped. The overlapping was due to the significantly larger amount of explosive material in the second semitrailer, which was large enough to cause the resulting craters to overlap.

At the time, ANFO was considered "safe around fires." In checking the references on which the "safe around fires" conclusion had been made, it was found that the experiments had involved relatively small amounts of ANFO, much less than the amounts involved in this case. In those experiments, various amounts of ANFO were burned and in no cases did the ANFO detonate. From that data, it was concluded that ANFO was safe in a fire. However, this accident indicated that a size effect required consideration. In amounts above a certain threshold, burning ANFO can obviously detonate when involved in a fire.

Some years after the event, the people who allegedly set the fires resulting in the explosions were caught, tried, and convicted. The craters and physical evidence that the event happened has long ago been obliterated by real estate development in the area. A memorial to the memory of the firemen was eventually erected.

## Further Information and References

*Handbook of Human Reliability Analysis with Emphasis on Nuclear Power Plant Applications*, Final Report, NUREG/CR-1278, 1983.

*Human Factors in Engineering and Design*, by Sanders and McCormick, 6th ed., McGraw Hill, New York, 1987. For more detailed information please see Further Information and References in the back of the book.

*Military Standard: Sampling Procedures and Tables for Inspection by Variables for Percent Defective*, MIL-STD-414, June 11, 1957. For more detailed information please see Further Information and References in the back of the book.

"New Methods of Consumer Product Accident Investigation," by Drury and Brill, *Proceedings of the Symposium — Human Factors and Industrial Design in Consumer Products*, Tufts University, 1980, pp. 196–211. For more detailed information please see Further Information and References in the back of the book.

*Procedures for Performing a Failure Mode, Effects and Criticality Analysis (FMECA)*, MIL-STD-1629A, November 24, 1980. For more detailed information please see Further Information and References in the back of the book.

*To Engineer is Human*, by Henry Petroski, Vintage Books, 1992. For more detailed information please see Further Information and References in the back of the book.

# Further Information
# and References

## Chapter 1

"Chemist in the Courtroom," by Robert Athey, Jr., *American Scientist*, 87(5), September-October 1999, pp. 390–391, Sigma Xi. This is a short article describing a forensic chemist's experiences as an expert witness.

*The Columbia History of the World*, Garraty and Gay, Eds., Harper and Row, New York, 1981. Pages 681–686 discuss Aristotle's inductive method.

"Daubert and Kumho," by Henry Petroski, *American Scientist*, 87(5), September-October 1999, pp. 402–406, Sigma Xi. Mr. Petroski, who has an established reputation concerning how advances in engineering are gained through thoughtful assessment of failures (see *To Engineer is Human*) has written a concise article concerning the legal foundations of what constitutes expert witness testimony. "Daubert and Kumho" is the name of the most recent legal case that defines the standards for the admissibility of expert testimony.

*The Engineering Handbook*, Richard Dorf, Ed., CRC Press, Boca Raton, FL, 1995. Pages 1897–1906 briefly discuss several systematic approaches to failure and hazard analysis.

*Forensic Engineering*, Kenneth Carper, Ed., Elsevier, New York, 1989. This is an introductory text that explains and defines forensic engineering. There are fourteen chapters, each written by a different author, that describe some facet of forensic engineering. Chapter 3, authored by Paul E. Pritzer, P.E., is about forensic engineering with respect to fires, and discusses briefly the "investigative team" concept. I also recommend chapter 10 (photodocumentation), chapter 11 (reports), and chapter 12 (expert witness).

*Galileo's Revenge*, by Peter Huber, Basic Books, New York, 1991. This book discusses the apparent proliferation of "junk science" testimony in the court room in many high profile cases.

*General Chemistry*, by Linus Pauling, Dover Publications, New York, 1970. Pages 13–15 discuss the modern scientific method as espoused by Dr. Pauling, a Nobel Prize Laureate.

*Introduction to Mathematical Statistics*, by Paul Hoel, John Wiley & Sons, New York, 1971. Pages 190–224 discuss general principles for statistical inference, one of the methodologies employed in modern inductive reasoning.

*On Man in the Universe*, Introduction by Louside Loomis, Walter Black, Inc., Roslyn, NY, 1943. Discussion of Aristotle's method of inductive logic on page xiv.

*Procedures for Performing a Failure Mode, Effects and Criticality Analysis (FMECA)*, MIL-STD-1629A, November 24, 1980. This is the formal procedure used by all departments and agencies of the U.S. Department of Defense. Like many military publications, this one can be hard to follow at first if the reader has not previously been exposed to military jargon. The document lays out the military method of determining potential failure modes in a system and the resulting consequences. If used in reverse, the method can be useful in diagnosing failures in systems.

*Reporting Technical Information*, by Houp and Pearsall, Glencoe Press, Beverly Hills, California, 1968. With respect to the topics discussed in this chapter, pages 224–225 provide a similar format used for the reporting of physical research findings. Pages 47–56 provide an excellent discussion of audience analysis, and pages 57–68 provide useful information for the organization of findings and observations. Overall, this text provides an excellent foundation for technical writing.

*Reason and Responsibility*, Joel Feinburg, Ed., Dickenson Publishing, Encino, CA, 1971. The a priori method of reasoning is discussed at length in the essay authored by A. J. Ayer, "The A Priori," beginning on page 181.

*To Engineer is Human*, by Henry Petroski, Vintage Books, 1992. This book is about failure with respect to engineering design and has an excellent chapter about forensic engineering (chapter 14). I highly recommend this book for its insights into the workings of engineering failure. It is also very readable.

"Trial and Error," by Saunders and Genser, *The Sciences*, September/October 1999, 39(5), 18–23, the New York Academy of Sciences. This is an interesting short article about defining "reasonable doubt" in terms of decision theory and uncertainty. The authors discuss how the vagueness of the legal term "reasonable doubt" can lead juries to different verdicts despite the fact that they hold similar beliefs concerning the defendant's guilt.

"When is Seeing Believing?" by William Mitchell, *Scientific American*, Feb. 1994, 270(2), pp. 68–75. This is recommended reading for all persons involved in the presentation and examination of evidence. The article describes some of the ways in which computer graphics can be used to provide fake photographic evidence. Unfortunately, one of the darker sides of the computer revolution is that photographic evidence, once considered quite reliable, is now becoming increasingly easier to fake. It will not be long until unscrupulous forensic experts use computerized faking techniques to persuade juries as to the "scientific" evidence in a case, especially when the monetary stakes are sufficiently high.

## Chapter 2

ANSI A58.1, *American National Standard Minimum Design Loads for Buildings and Other Structures*, American National Standards Institute. This is likely the most thorough and up-to-date criteria for wind design with respect to buildings. Some cities have adopted the ANSI A58.1 standard in lieu of the UBC or BOCA wind design standards.

*BOCA National Building Code*, 1993 edition. The section on designing for wind is similar to that found in the UBC.

*Kansas Wind Energy Handbook*, Kansas Energy Office, 1981. A handbook prepared by the now defunct Kansas Energy Office that discusses wind power potentials in the state. The book contains excellent wind data specific to Kansas.

*Peterson Field Guide to the Atmosphere*, by Schaefer and Day, Houghton-Mifflin, 1981. Appendices 9 and 10 are about the Beaufort wind scale, and the relationship between wind speed and pressure.

*UBC*, International Conference of Building Officials, 1991 ed., Appendix D, Figure 23-1. Map of the U.S. showing the basic wind speed for minimum design.

*Wind Power Climatology of the United States — Supplement*, by Jack Reed, Sandia Laboratories, April 1979. Summary of wind power statistics in the U.S. Page 17 gives the comparison of the "windiest" states.

Winds and Air Movement, in *Van Nostrand's Scientific Encyclopedia*, Fifth Edition, Van Nostrand Reinhold Company. Pages 2339–2343 have a good discussion of wind, especially the mathematics that govern relationships between wind speed and other factors.

## Chapter 3

"The Essentials of Lightning Protection," *Record*, May–June 1981, pp. 12–18. A short article that discusses the damaging effects of lightning surges and shows some nice photographs of actual damage. Protection measures and arrestors are also discussed.

"In Search of Electrical Surges," by Ivars Peterson, *Science News*, December 12, 1987, 132, pp. 378–379. A short article describing how many surges in the home are self-generated, and how multiple grounding in networks can actually exacerbate surge damage.

*Lightning*, by Martin Uman, Dover Publications, New York, 1984. An excellent summary of research and knowledge about lightning and its properties.

*Mechanics of Structures with High Electric Currents*, by Francis Moon, National Science Foundation, Princeton University, Princeton, NJ, July 1973. Pages 35–41 discusses the effects of a lightning magnitude surge on materials used in power switches and electrical contacts.

*Note on Vulnerability of Computers to Lightning*, by R. D. Hill, General Research Corporation, Santa Babara, CA, April 1971. Discussion of lightning damage to unprotected computers.

## Chapter 4

"Air Blast Phenomena, Characteristics of the Blast Wave in Air," reprint of chapter II from, *The Effects of Nuclear Weapons*, by Samuel Glasstone, Ed., April 1962,

U.S. Atomic Energy Commission, pp. 102–148. Don't let the title of the book fool you into thinking this is only about nuclear weapons. The article is excellent in its explanation of how air blast waves from detonations propagate.

*An Introduction to Mechanical Vibrations*, by Robert Steidel, Jr., John Wiley & Sons, New York, 1971, pp. 65–70, 83. This is a very good college textbook on vibrations. The pages cited explain in detail "Rayleigh's energy method," which was referred to in this chapter.

*Blasting Damage — A Guide for Adjusters and Engineers*, American Insurance Association, 1972. This is a very good introduction to diagnosing blasting effects in structures. It is basically an abridged version of Bulletin 656, whose citation follows.

*Blasting Vibrations and Their Effects on Structures*, by Nicholls, Johnson, and Duvall, United States Department of the Interior, Bureau of Mines, Bulletin 656, 1971. This is generally considered the primary source for blasting effects on nearby buildings, and is the basis for most state and local blasting regulations.

*Damage from Air Blast*, by Windes, Bureau of Mines Report of Investigation 3708, 1943. This was a follow-up to Bulletin 442 and established the first overpressure thresholds for glass damage.

*Engineering Analysis of Fires and Explosions*, by R. Noon, CRC Press, Boca Raton, FL, 1995. Chapters 6, 7, and 10 discuss explosions, and chapter 6 has a brief discussion of the thermodynamics of explosions and pressure fronts.

*Explosive Shocks in Air*, by Kinney and Graham, Springer-Verlag, 2nd ed., 1985. This is probably the best text on air blasts written for engineers and physicists. For readers already familiar with the thermodynamics, Chapters 6–8, 10, and 11 are recommended. These chapters deal specifically with topics of concern to engineers involved in diagnosing air blast damage.

*Investigation of Fire and Explosion Accidents in the Chemical, Mining, and Fuel Related Industries — A Manual*, by Joseph Kuchta, United States Department of the Interior, Bureau of Mines, Bulletin 680, 1985. This is an excellent resource for deflagration type explosions.

*NFPA 495: Explosive Materials Code*, National Fire Protection Association, 1990 Edition. These are the basic regulations that deal with the handling of explosive materials. It is the "building code" of explosive storage facilities.

*Seismic Effects of Quarry Blasting*, by Thoenen and Windes, Bureau of Mines, Bulletin 442, 1942. One of the first major studies to establish threshold acceleration criteria with respect to blast damage to residential structures.

# Chapter 5

*Formulas for Stress and Strain*, by Raymond Roark, 2nd ed., McGraw-Hill, New York, p. 300. This is the premier reference work for stress and strain formulas. The first edition was printed in 1938, and the second edition was printed in 1943. I am fortunate enough to have found a second edition at a flea market for a quarter. More modern editions, of course, are authored by Roark and Young,

and some versions are now available coupled with standard mathematics software, e.g., *MathCad 8.0®*.

*Mechanical Engineer's Pocket Book*, by William Kent, 9th ed., John Wiley & Sons,
New York, 1915, p. 372, "Strength of Lime and Cement Mortar," and p. 370,
"The Strength of Brick, Stone, Etc." This, and similar old engineering texts and
manuals, are a valuable resource when attempting to understand construction
and mechanical design practices prior to World War I. Because engineering
practices have changed much in the intervening years, modern handbooks may
only have limited value in the diagnosis of problems that predate their publication by many decades.

*U.S. Bureau of Standards, Technical Paper 123*, published in 1919. This paper lists the
estimated life of stonework in buildings. For limestone, the life is estimated at
perhaps 40 years. For high grade silceous Ohio sandstone, the life is estimated
at one to several centuries. Despite the fact that some may perhaps consider this
reference too old to be of any real use, I find the paper very interesting because
most of the estimates appear accurate.

## Chapter 6

*All About OSHA*, U.S. Department of Labor Occupational Safety and Health Administration, OSHA 2056, 1995 (revised). This is an excellent booklet that explains
succinctly OSHA's mandate, limitations, and scope. I recommend that it be read
by employees, employers, supervisors, and workplace managers.

*Code of Federal Regulations*, Parts 1900 to 1910, and Part 1926. This is the primary
OSHA code for construction and industry. Every major employer should have
ready access to a full copy of the code when safety or workplace responsibility
questions arise.

*Human Factors in Engineering and Design*, by Sanders and McCormick, 6th ed.,
McGraw-Hill, New York, 1987. An excellent introductory text in ergonomics for
upper level undergraduate students in engineering, the text covers the general
topics most needed by engineers involved in safety and the design of equipment.

*The One Best Way*, by Robert Kanigel, Viking Press, New York, 1997. A good book
that discusses Frederick Taylor's life and work, and the effects of "Taylorism" on
the modern notions of efficacy.

## Chapter 7

*ATF Arson Investigation Guide*, published by the Department of the Treasury. Mostly
this is a compendium of investigative check lists and federal resources used in
the investigation of arsons by the ATF. The book has a nice glossary of terms in
the back, and there are a number of excerpts from the federal laws concerning
arson and explosions.

*Code of Federal Regulations 29*, Parts 1900 to 1910, and Part 1926 (two volumes).
These are the basic OHSA regulations that govern LPG, gas fuels, welding,
storage of explosives, fire protection, and prevention, etc. This is usually the first

set of regulations checked when it is suspected that a compliance problem led
to a fire or explosion.

*The Condensed Chemical Dictionary*, revised by Gessner Hawley, 9th ed., Van Nostrand Reinhold Company, 1977. Newer editions of this standard reference text
are also available. This is a very useful reference concerning chemical terms,
chemicals, and trade name items. Thumbnail descriptions of the chemical's
properties and hazards are listed.

*Fire and Explosion Investigations*-1992 ed., NFPA 921. Like its sister publication NFPA
907M, NFPA 921 is sometimes cited in courtroom proceedings as "the standard"
for fire investigation work. As its scope indicates, the purpose of the publication
"... is to establish guidelines and a recommended practice for the systematic
investigation or analysis of fire and explosion incidents." It is not an engineering
text, but it does contain very useful qualitative information, explanations, and
investigative procedures concerning fire C&O investigations. Like NFPA 907M,
it is recommended that people involved in fire investigation work be familiar
with NFPA 921.

*Fire and Explosion Protection Systems*, by Michael Lindeburg, P.E., Professional Publications, Belmont, CA, 1995. While brief, this is a good overview of fire protection systems used in buildings and how they are designed. The book considers
requirements of the major building codes.

*Fire Investigation Handbook*, U.S. Department of Commerce, National Bureau of
Standards, NBS Handbook 134, Francis Brannigan, Ed., August 1980. This is a
very readable text on fire analysis. The text is a practical amalgam of theory and
experience and contains a number of very instructive photographs. It is often
used as a basic text in fire academies and training schools. Persons "in the
business" should have a copy on hand since it is often referenced as an authoritative source in court proceedings.

*General Chemistry*, by Linus Pauling, Dover Publications, New York, 1988. I believe
this is the best reference text on general chemistry available, and at a very modest
price. Dr. Pauling, of course, received the Nobel Prize in chemistry in 1954. With
respect to fires and explosions, I recommend the following: chapter 10, chemical
thermodynamics; chapter 11, chemical equilibrium; chapter 15, oxidation-
reduction reactions; and chapter 16, the rate of chemical reactions. I also recommend the various discussions about entropy in the text.

*Handbook of Chemistry and Physics*, Robert Weast, Ph.D., Ed., 71st Edition. CRC
Press, Boca Raton, FL. This is the motherlode of information concerning chemical compounds, heats of reaction, solubility, etc. It is hard to imagine a practical
reference shelf without this volume. I have listed a specific edition, but new ones
come out annually.

*Investigation of Fire and Explosion Accidents in the Chemical, Mining, and Fuel Related
Industries — A Manual*, by Joseph Kuchta, United States Department of the
Interior, Bureau of Mines, Bulletin 680, Washington DC, 1985. Despite the
formidable title, this is an excellent engineering reference source for fires and

deflagration-type explosions. The text is not a primer. To properly understand the book, the reader should have previous knowledge in chemistry and basic thermodynamics. The text is to the point, well written, and contains a cornucopia of reference information.

*NFPA Life Safety Code (NFPA 101)*, National Fire Protection Association, February 8, 1991. Also listed as ANSI 101. This is the general fire code adopted by many units of government around the U.S. and other countries. It covers fire exit requirements, safety equipment, fire detection systems, etc.

*A Pocket Guide to Arson and Fire Investigation*, Factory Mutual Engineering Corporation, 3rd ed., 1992. A small booklet that can provide handy information while in the field, it fits in a shirt pocket or briefcase.

*Principles of Combustion*, by Kenneth Kuo, John Wiley & Sons, New York, 1986. This is a first-rate upper-level undergraduate or graduate level engineering text about combustion. While the notation is sometimes overdone, there is a rich range of topics and the technical depth is excellent. I recommend this to engineers who have already had basic thermodynamics and heat transfer, and are ready for the next level.

*Rego LP-Gas Serviceman's Manual*, Rego Company, 4201 West Peterson Avenue, Chicago, IL, 60646, 1962. This is a handy reference that will fit in a shirt pocket or briefcase. It contains most of the basic information about LPG installations for residences. The section on leak testing an LPG installation is the same technique used after a fire or explosion to check out the remaining pipework.

*Standard Handbook for Mechanical Engineers*, by Baumeister and Marks, 7th ed., 1967. A standard reference book for mechanical engineers, this handbook contains information on fuels, explosives, machinery, gases, materials, circuitry, switch gear, and many types of equipment that may be involved in fires and explosions. This is very useful when a person is trying to understand what used to be there before the fire or explosion and how it worked. Newer editions than the one listed are also available.

*Thermodynamics*, by Ernst Schmidt, Dover Publications, New York, 1966. This is a classic engineering text on thermodynamics, which also contains excellent chapters on thermochemistry. Chapter XII, Combustion, is particularly relevant to the topic of the chemistry of fire. For the money, this is a first-rate text.

## Chapter 8

*Fire Investigation Handbook*, U.S. Department of Commerce, National Bureau of Standards, NBS Handbook 134, Francis Brannigan, Ed., August 1980. This is a very readable text on fire analysis. The text is a practical amalgam of theory and experience and contains a number of very instructive photographs. It is often used as a basic text in fire academies and training schools. Persons "in the business" should have a copy on hand since it is often referenced as an authoritative source in court proceedings. Sections 4.4 and 4.5 discuss electrical shorting, and there are some excellent demonstrative photographs on pp. 113–118.

*Manual for the Determination of Electrical Fire Causes*-1988 ed., NFPA 907M. This
   is a useful check-list type document for the investigation of electrically caused
   fires. It is not an engineering type technical text, but it does contain very useful
   qualitative information and investigative procedures concerning electrically
   caused fires. The document is sometimes cited in courtroom proceedings as
   being "the standard" for electrical fire investigations. This is an exaggeration of
   the document's technical importance, and certainly short shrifts other authoritative texts on the subject. However, because of this attitude, it is recommended
   that people involved in fire investigation work be familiar with it.

*National Electric Code*, published by the National Fire Protection Association, Quincy,
   MA., various editions, published approximately every 3 years. When trying to
   diagnose electrical problems in residential and commercial installations, this is
   the "bible" of how it should have been done in the first place. The *National
   Electric Code* is actually a portion of a much larger body of fire-related standards
   and guidelines published by the NFPA, and is alternately listed as NFPA 70-1987.
   The last four numbers in the designation indicate the specific edition year. It is
   worth noting that some localities adopt specific editions of the NEC and may
   not automatically change to the newest edition.

*SPD — Electrical Protection Handbook*, published by Bussmann-Cooper Industries.
   While this is a manufacturer's product publication to promote Bussman Fuses,
   it offers a very good explanation of how fuses operate and how they are used to
   protect various types of electrical systems. Section 1 contains topics on overcorrects, overloads, and short circuits.

## Chapter 9

"Air Blast Phenomena, Characteristics of the Blast Wave in Air," reprint of Chapter
   II from *The Effects of Nuclear Weapons,* Samuel Glasstone, Ed., April 1962, U.S.
   Atomic Energy Commission, pp. 102–148. Don't let the title fool you into
   thinking this is only about nuclear weapons. The article is excellent in its explanation of how blast waves from detonations propagate.

*ATF Arson Investigation Guide*, published by the U.S. Department of the Treasury.
   Mostly this is a compendium of investigative check lists and federal resources
   used in the investigation of arsons by the ATF. The book has a nice glossary of
   terms in the back, and there are a number of excerpts from the federal laws
   concerning explosions.

*Chemical Thermodynamics*, by Frederick Wall, W. H. Freeman and Co., San Francisco,
   1965. Pages 69–70 have an excellent basic explanation of how to determine the
   temperature within an explosion at constant volume.

*Code of Federal Regulations 29, Parts 1900 to 1910, and Part 1926* (two volumes).
   These are the basic OHSA regulations that govern LPG, gas fuels, welding,
   storage of explosives, fire protection and prevention, etc. This is usually the first
   set of regulations checked when it is suspected that a compliance problem led
   to a fire or explosion.

*The Condensed Chemical Dictionary*, revised by Gessner Hawley, 9th ed., Van Nostrand Reinhold Co., 1977. Newer editions of this standard reference text are also available. This is a very useful reference on chemical terms, chemicals, and trade name items. Thumbnail descriptions of the chemical's properties and hazards are listed.

*Explosive Shocks in Air*, Kinney and Graham, Second Edition, Springer-Verlag, 1985. This is probably the best text on air blasts written for engineers and physicists. For those already familiar with basic thermodynamics, I recommend chapters 6–8, and 10–11. These chapters deal specifically with topics of concern to engineers involved in diagnosing the hows and whys of blast damage. However, I think the whole book deserves a reading for those serious students of fires and explosions.

*Fire and Explosion Investigations*-1992 ed., NFPA 921. Like its sister publication NFPA 907M, NFPA 921 is sometimes cited in courtroom proceedings as "the standard" for fire investigation work. As its scope indicates, the purpose of the publication "… is to establish guidelines and a recommended practice for the systematic investigation or analysis of fire and explosion incidents." It is not an engineering text, but it does contain very useful qualitative information, explanations, and investigative procedures concerning fire C&O investigations. Like NFPA 907M, it is recommended that people involved in fire investigation work be familiar with NFPA 921.

*Fire Investigation Handbook*, U.S. Department of Commerce, National Bureau of Standards, NBS Handbook 134, Francis Brannigan, Ed., August 1980. Section 4.6 is a short, but very useful description of explosion types and how to quickly identify an explosion by the type of damage produced.

*General Chemistry*, by Linus Pauling, Dover Publications, New York, 1988. I believe this is the best reference text on general chemistry available, and at a very modest price. Dr. Pauling, of course, received the Nobel Prize in chemistry in 1954. With respect to explosions, I recommend the following: chapter 10, chemical thermodynamics; chapter 11, chemical equilibrium; chapter 15, oxidation-reduction reactions; and chapter 16, the rate of chemical reactions. I also recommend the various discussions about entropy in the text.

*Handbook of Chemistry and Physics*, Robert Weast, Ph.D., Ed., 71st ed., CRC Press, Boca Raton, FL. This is the mother lode of information concerning chemical compounds, heats of reaction, solubility, etc. It is hard to imagine a practical reference shelf without this volume. I have listed a specific edition, but new ones come out annually.

*Hydrogen-Oxygen Explosions in Exhaust Ducting*, Technical Note 3935, by Paul Ordin, National Advisory Committee for Aeronautics, Lewis Flight Propulsion Laboratory, Cleveland, Ohio, April 1957. The NACA was the immediate predecessor to NASA. This paper describes how hydrogen can be made to detonate instead of deflagrate. While the paper was originally intended for use in rocket engines, it has application when considering tunnels or long ducts that have hydrogen-oxygen atmospheres.

*Investigation of Fire and Explosion Accidents in the Chemical, Mining, and Fuel Related Industries — A Manual*, by Joseph Kuchta, U.S. Department of the Interior, Bureau of Mines, Bulletin 680, 1985. Despite the formidable title, this is an excellent engineering reference source for fires and deflagration type explosions. The text is not a primer. To properly understand the text, the reader should have previous knowledge in chemistry and basic thermodynamics. The text is to the point, well written, and contains a cornucopia of reference information.

*NFPA 495: Explosive Materials Code*, National Fire Protection Association, 1990 ed. These are the basic regulations that deal with the handling of explosive materials. It contains things like minimum distances from storage facilities to inhabited areas, etc. It is the "building code" of storage facilities for explosive materials.

*Pressure Integrity Design Basis for New Off-Gas Systems*, by C.S. Parker and L.B. Nesbitt, General Electric, Atomic Power Equipment Department, May 1972. This paper was originally intended for use in atomic power plants, where there is a hazard of hydrogen explosions in the off-gas systems of boiler water reactors. An off-gas system is one in which the gases formed by disassociation in the steam flow from the reactor to the turbine are extracted from the steam flow and then dispersed to the atmosphere. Like the previously cited paper, this paper also discusses hydrogen-oxygen explosions in long ducts, and how they can transform from deflagrations to detonations.

*Principles of Combustion*, by Kenneth Kuo, John Wiley & Sons, New York, 1986. This is a first-rate upper-level undergraduate or graduate level engineering text about combustion. While the notation is sometimes overdone, there is a rich range of topics and the technical depth is excellent. I recommend this to engineers who have already had basic thermodynamics and heat transfer, and are ready for the next level.

"Sensitivity of Explosives," by Andrej Macek, U.S. Naval Ordnance Laboratory, *Chemical Reviews*, 1962, Vol. 41, pp. 41–62. Excellent discussion of how explosives degrade and how their corresponding characteristics change.

*Thermodynamics*, by Joachim Lay, Merrill Publishing, 1963. Chapter 6 (pp. 125–140) has an excellent discussion of gas flow that is applicable to pressurized gas explosions, and the thermodynamic fundamentals relating to explosive over-pressure are succinctly explained in chapter 17 (pp. 490–493).

## Chapter 10

*The Engineering Handbook*, Richard Dorf, Ed., CRC Press, Boca Raton, FL, 1995. A short but useful discussion of the effects of explosions on people is contained on p. 1899. The same article also briefly discusses simple gas expansion type explosions, and the role of the ASME Boiler Code.

*General Chemistry*, by Linus Pauling, Dover Publications, New York, 1988. A good discussion of the kinetic theory of gases begins on p. 323, which includes an explanation of diffusion. A discussion of Dalton's law of partial pressures begins on p. 309.

*Thermodynamics,* by Ernest Schmidt, Dover Publications, New York, 1966. In chapter XVIII, section 117, pp. 493–495 there is an excellent discussion of Ficke's law.

## Chapter 11

ASTM E1387, *Standard Test Method for Flammable or Combustible Liquid Residue in Extracts from Samples of Fire Debris by Gas Chromatography.* As mentioned, this is the standard often cited, which describes the "proper" way to collect and test samples for the possible presence of accelerants.

*General Chemistry,* by Linus Pauling, Dover, New York, 1970. Beginning on page 100 is a good explanation of how a mass spectrograph works, and how it is used to identify unknown materials.

*Van Nostrand's Scientific Encyclopedia,* 5th ed., Van Nostrand Reinhold Co., New York, 1976. Pages 539–544 contain a meticulous explanation of chromatograph, especially column type gas chromatography.

## Chapter 12

"The Amateur Scientist" by Jearl Walker, *Scientific American,* February 1989, pp. 104–107. This is an excellent article that discusses modulated braking and braking strategies to minimize braking distance.

"The Amateur Scientist" by Jearl Walker, *Scientific American,* August 1989, pp. 98–101. This is an interesting article that discusses the wave motion of highway traffic flow.

"The Physics of Traffic Accidents" by Peter Knight, *Physics Education,* 10(1), January 1975, pp. 30–35.

*Bicycle Accident Reconstruction: A Guide for the Attorney and Forensic Engineer,* by James Green, P.E., 2nd ed., Lawyers and Judges Publishing Co., Tucson, AZ, 1992. This is one of just a handful of publications that deal with vehicular/bicycle accidents. The book seems to be directed more to lawyers than engineers. It also repeatedly extolls the virtues of qualified bicycle mechanics to the point where the reader might come to believe that no one but a bicycle mechanic "really" understands how a bicycle works. Despite this bias, however, the book has useful information about the principle of conspicuity as applied to bicyclists. Some of this information is applicable to motorcyclists.

*Consumer Reports.* Various issues of this magazine discuss performance factors of popular cars and vehicles. Information regarding acceleration, braking distances, curb weight, and so on are listed.

*Engineering in History,* by Kirby, Withington, Darling, and Kilgour, published by Dover Publications, 1990. This book is an excellent read for students of engineering, who generally are not provided any historical context for their major subjects in undergraduate studies. With respect to the development of automobile transportation, I recommend Chapter 12, pp. 405–414, wherein a brief but useful history is given.

*Insurance Institute for Highway Safety/Highway Loss Data Institute.* These two organizations are essentially the same, and are supported by the insurance industry.
They collect data on automotive safety topics, e.g., crash rates, relative crashworthiness, effects of regulations, antilock brake devices, and so on. They publish
several papers and brochures each year on their findings and related trends.

*Manual on Uniform Traffic Control Devices for Streets and Highways*, National Joint
Committee on Uniform Traffic Control Devices, U.S. Department of Commerce,
Bureau of Public Roads, June 1961. This is the "bible" of traffic signs, traffic lights,
sign and light placement, road markings, and general signage requirements.

*MathCad*, Version 8.0, Addison-Wesley, 1999. While not strictly a "book," this software is extremely useful in working out the equations concerning accident
reconstruction. An interesting sample problem concerning braking is included
in some sample files under BRAKE.MCD. An iterative energy method is
employed to obtain a plot of remaining kinetic energy versus time during stopping. The sample problem is well worth noting.

*Physics*, by Arnold Reimann, Barnes and Noble, New York, 1971. While there are
many excellent texts that discuss friction and energy with respect to braking,
the section on friction in this text, pp. 266–275, is unusually clear and straightforward.

*Research Dynamics of Vehicle Tires*, Vol. 4, by Andrew White, Research Center of
Motor Vehicle Research of New Hampshire, Lee, New Hampshire, 1965. While
this book is out of print, copies are still available through libraries. This particular volume of the series is about improper mounting of tires, and the hazards
associated with it. It may be the best all around discussion of this subject in the
literature. I was especially impressed with the photographs showing what a tire
can do when it explodes during mounting on a wheel rim.

*Statistical Abstract of the United States*, U.S. Printing Office. This collection of facts
and figures is published annually by the U.S. Census Bureau. Within it are hoards
of information relating to traffic accidents, registered vehicles, road miles driven,
and so on.

*The Traffic Accident Investigation Manual*, by Baker and Fricke, 9th ed., 1986, Northwestern University Traffic Institute. This is an often referenced text used to train
police officers. The emphasis is on collecting and preserving accident scene
information. The accident analysis sections are rudimentary but sound. The
qualitative information contained in the text is very good. It is a good idea to
be acquainted with this text since it is so commonly referred to in court proceedings, police reports, and so on. There are several good chapters on tire mark
identification, skid marks, and determination of speeds from skid marks.

*The Way Things Work*, by David Macaulay, 1988, Houghton Mifflin, New York. The
sections about friction, clutches, brakes, and syncromesh transmissions have
very instructive illustrations.

*The Way Things Work: An Illustrated Encyclopedia of Technology*, 1967, Simon and
Schuster, New York. This is a handy compendium of short articles and explanatory diagrams of many appliances and machines. With respect to accident recon-

struction, there are many fine articles and illustrations in the several volumes that deal with automotive brakes, steering systems, transmissions, and so on.

*Work Zone Traffic Control, Standards and Guildlines*, U.S.D.O.T., Federal Highway Administration, 1985, also A.N.S.I. D6.1-1978. This booklet describes the standards for traffic control in roadway construction areas. Such information is a must in assessing vehicular accidents occurring in construction areas.

## Chapter 13

*Mathematics in Action*, by O.G. Sutton, Dover Publications, 1984. Chapter 3, "Ballistics or Newtonian Dynamics in War" adds further analytical detail to the subject of falling objects.

## Chapter 14

*Accident Reconstruction: Automobiles, Tractor-Semitrailers, Motorcycles and Pedestrians*, Society of Automotive Engineers, Warrendale, PA, February 1987. This is a collection of papers representative of the direction of research in this field in 1987. I recommend two papers on braking, 870501 and 870502.

"The Amateur Scientist" by Jearl Walker, *Scientific American*, February 1989, pp. 104–107. Excellent article discussing modulated braking and braking strategies to minimize braking distance.

*Consumer Reports.* Various issues discuss performance factors of popular cars and vehicles. Information regarding acceleration, braking distances, curb weight, and so on are listed.

*Detroit News*, Five part series of articles on SAIs, December 13–17, 1987.

*An Examination of Sudden Acceleration*, U.S.D.O.T., National Highway Traffic Safety Administration, Report Number DOT-HS-807-367, Washington DC, January 1989. This study exhaustively examines sudden acceleration incidents (SAIs), which include stuck accelerator cases, runaway car engines, and so on. The report contains information relating to the famous Audi accelerator case and other useful test data.

The Insurance Institute for Highway Safety and the Highway Loss Data Institute are essentially the same, and are supported by the insurance industry. They collect data on automotive safety topics, e.g., crash rates, relative crashworthiness, effects of regulations, antilock brake devices, and so on. They publish several papers and brochures each year on their findings and related trends.

*Manual on Uniform Traffic Control Devices for Streets and Highways*, National Joint Committee on Uniform Traffic Control Devices, U.S. Department of Commerce, Bureau of Public Roads, Washington DC, June 1961. This is the "bible" of traffic signs, traffic lights, sign and light placement, road markings, and general signage requirements.

The Occurrence of Accelerator and Brake Pedal Actuation Errors During Simulated Driving, by S. B. Rogers and W. Wierwille, in *Human Factors*, 1988, 30(1), pp. 71–81.

"The Physics of Traffic Accidents" by Peter Knight, *Physics Education,* 10(1), January 1975, pp. 30–35.

*Research Dynamics of Vehicle Tires,* Vol. 4, by Andrew White, Research Center of Motor Vehicle Research of New Hampshire, Lee, New Hampshire, 1965. While this book is out of print, copies are still available through libraries. This particular volume of the series is about improper mounting of tires, and the hazards associated with it. It may be the best all around discussion of this subject in the literature. I was especially impressed with the photographs showing what a tire can do when it explodes during mounting on a wheel rim.

*SAE Handbook, Volume 4: On-Highway Vehicles and Off-Highway Machinery,* Society of Automotive Engineers, Warrendale, PA. This handbook is regularly updated from year to year and contains a wealth of information with respect to accident reconstruction. Standard test procedures, performance standards, and design specifications relating to vehicles and equipment are covered. It is fundamentally important to be familiar with this document.

"Unintended Acceleration" by Tom Lankard, *Autoweek,* January 19, 1987.

*The Way Things Work,* by David Macaulay, 1988, Houghton Mifflin, New York. The sections about friction, clutches, brakes, and synchromesh transmissions have very instructive illustrations.

*The Way Things Work: An Illustrated Encyclopedia of Technology,* 1967, Simon and Schuster, New York. This is a handy compendium of short articles and explanatory diagrams of many appliances and machines. With respect to accident reconstruction, there are many fine articles and illustrations in the several volumes that deal with automotive brakes, steering systems, transmissions, and so on.

*Work Zone Traffic Control, Standards and Guidelines,* U.S.D.O.T., Federal Highway Administration, Washington DC, 1985, also A.N.S.I. D6.1-1978. This booklet describes the standards for traffic control in roadway construction areas. Such information is a must in assessing vehicular accidents occurring in construction areas.

## Chapter 15

*Automobile Side Impact Collisions – Series II* by Severy, Mathewson and Siegel, University of California at Los Angeles, SAE SP-232. This is an excellent monograph on side collisions, especially their effects on vehicle occupants.

*Mechanical Design and Systems Handbook,* Harold Rothbart, Ed., Chapter 16, McGraw-Hill, New York, 1964, Impact. This chapter, especially Sections 16.1 through 16.3, has a first rate analysis of impacts. It discusses in detail the use of the momentum method and the application of the coefficient of restitution.

*Motor Vehicle Accident Reconstruction and Cause Analysis,* by Rudolf Limpert, 2nd ed., The Michie Company, Charlottesville, Virginia, 1984. Despite the fact that this text is published as one of a series of litigation-related texts for lawyers, it is a first-class textbook for accident reconstruction. The approach is technically sound and is relatively mathematically sophisticated when appropriate.

*Symposium on Vehicle Crashworthiness Including Impact Biomechanics*, Tong, Ni, and Lantz, Eds., American Society of Mechanical Engineers, AMD-Vol. 79, BED-Vol. 1, 1986. This is a collection of papers presented at the ASME Winter Annual Meeting. With respect to automotive crashes, the paper, which begins on p. 91, is particularly useful in its discussion of barrier collisions.

*The Traffic Accident Investigation Manual*, by Baker and Fricke, 9th ed., Northwestern University Traffic Institute, 1986. This is an often referenced text used to train police officers. The emphasis is on collecting and preserving accident scene information. The accident analysis sections are rudimentary, but sound. The qualitative information contained in the text is very good. It is a good idea to be acquainted with this text since it is so commonly referred to in court proceedings, police reports, and so on.

# Chapter 16

*Atlas of Stress Strain Curves*, Howard Boxer, Ed., ASM International, Metals Park, Ohio, 1987. This is a useful source of information about metals when a person is calculating expended crush energy. Stress-strain pull test charts for most of the alloys used in automobiles and vehicles are listed.

*Automobile Side Impact Collisions – Series II,* by Severy, Mathewson and Siegel, University of California at Los Angeles, SAE SP-232. This is an excellent monograph on side collisions, especially their effects on vehicle occupants.

*Basic Principles and Laws of Mechanics*, by Alfred Zajac, D.C. Heath and Co., Chicago, 1966. This is Volume I of a two-volume textbook set on physics for the first-year undergraduate or advanced undergraduate. Chapters 8, 9, and 10 provide an excellent introduction into the use of Lagrangian equations with generalized coordinates.

*A Crash Test Facility to Determine Automobile Crush Coefficients*, by Miyasaki, Navin, and MacNabb, University of British Columbia, SAE Paper 880224. The paper describes an inexpensive barrier-type crash facility developed at the University of British Columbia to determine crush coefficients.

"Differences between EDCRASH and CRASH3," by Day and Hargens, Engineering Dynamics Corp., SAE 850253. This paper compares two of the popular computer programs used to reconstruct vehicular accidents. Many of the "black box" programs offered on the market for the diagnosis of vehicular accidents are versions of these.

*Dynamics*, by Pestel and Thomson, McGraw-Hill, New York, 1968, pp. 349–353. This citation provides an excellent derivation of Euler's equations of motion.

"Energy Basis for Collision Severity," by K. L. Campbell, GM Safety Research and Development Laboratory, July 1974, SAE 740565. This is the paper upon which a number of current computer simulation programs are based. Mr. Campbell assumed that the force required to cause crush is a linear function of the crush depth, that is, "$F = mc + b$," where "m" and "b" are constants associated with the particular vehicle, and "c" is the crush depth.

*Energy Methods in Applied Mechanics*, by Henry Langhaar, John Wiley & Sons, Inc., New York, 1962. Langhaar's text has been a mainstay in mechanicals for some time. Section 6.2 provides a good derivation of Euler's columnar buckling formula, and further discusses postbuckling behavior.

*Field Accidents: Data Collection, Analysis, Methodologies, and Crash Injury Reconstructions*, Society of Automotive Engineers, February 1985. One of several in a series published by SAE. This is an excellent collection of papers dealing with accident reconstruction. Article 850437 is especially useful with respect to crush stiffness coefficients.

*Formulas for Stress and Strain*, by Raymond Roarck, McGraw-Hill, New York, 1943. While I realize that much newer editions are available, I personally prefer the layout of the older versions of this venerable reference. A software version is currently available to go with the T-K Solver program, and there is also a version on MathCad. The text has excellent sections on columnar buckling, and plates and shells buckling. Table XVI, item M shows a thin-walled cylinder under uniform longitudinal compression, and the shell shown in item K in the same table is very similar to a car door panel undergoing compression from a frontal impact.

The Insurance Institute for Highway Safety and the Highway Loss Data Institute are organizations supported by the insurance industry. They collect data on automotive safety topics, e.g., crash rates, relative crashworthiness, effects of regulations, anti-lock brake devices, and so on. They publish several papers and brochures each year on their findings and related trends.

*Measuring Protocol for Quantifying Vehicle Damage from and Energy Basis Point of View*, by Tumbas and Smith, SAE Paper 880072. Excellent paper that describes quantifying the crush damage to a vehicle to determine its speed prior to impact. The methodology is largely based upon K. L. Campbell's paper, which was previously cited.

*Motor Vehicle Accident Reconstruction and Cause Analysis*, by Rudolf Limpert, 2nd ed., 1984, The Michie Company, Charlottesville, Virginia. This text provides an abbreviated version of the energy method beginning in Section 28-5, "Impact Analysis." The graph of velocity vs. crush deformation on p. 417 is well worth noting, especially the general shape of the curve.

NHSTA. Each year the NHSTA crashes each type of car that is sold in the U.S. into barriers and publishes the results. The data is sometimes hard to find, but it is very valuable for determining the specific crush factors for each make and model. The NHSTA also has a web page that is worth checking out.

"An Overview of the Way EDSMAC Computes Delta-V," by Day and Hargens; Engineering Dynamics Corp., SAE 880069. As its title suggests, this paper discusses the internal workings of one of the computer programs that is used to reconstruct vehicular accidents.

*An Introduction to the Use of Generalized Coordinates in Mechanics and Physics*, by William Byerly, Dover Co., New York, 1965. This is an excellent concise text on the use of the Langrangian and Hamiltonian methods, and doesn't cost very much.

*Symposium on Vehicle Crashworthiness Including Impact Biomechanics*, Tong, Ni, and Lantz, Eds., American Society of Mechanical Engineers, AMD-Vol. 79, BED-Vol. 1, 1986. This is a collection of papers presented at the ASME Winter Annual Meeting. With respect to automotive crashes, the paper beginning on p. 91 is particularly useful in its discussion of barrier collisions.

## Chapter 17

*Motor Vehicle Accident Reconstruction and Cause Analysis*, by Rudolf Limpert, 2nd ed., 1984, The Michie Co., Charlottesville, Virginia. Chapters 18 and 22 discuss vehicular turning dynamics. Special attention should be paid to Section 22-2.7, which is about turning dynamics of car-trailer combinations. As noted before, this is a sound text about vehicular accident reconstruction. People directly involved in vehicular accident reconstruction should be familiar with it.

*The Traffic Accident Investigation Manual*, by Baker and Fricke, 9th ed., 1986, Northwestern University Traffic Institute, LCCC No. 86-606-16. Topics 817 and 828 are germane with respect to this chapter. However, as noted before, since this text is often referred to by police officers involved in vehicular accident reconstruction, it is best to have read the book.

## Chapter 18

*Bicycle Accident Reconstruction: A Guide for the Attorney and Forensic Engineer*, by James Green, P.E., 2nd ed., Lawyers and Judges Publishing Co., Tucson, AZ, 1992. This is one of just a handful of publications that deal with vehicular/bicycle accidents. The book seems to be directed more to lawyers than engineers. It also repeatedly extols the virtues of qualified bicycle mechanics to the point where the reader might come to believe that no one but a bicycle mechanic "really" understands how a bicycle works. Despite this bias, however, the book has useful information about the principle of conspicuity as applied to bicyclists. Some of this information is applicable to motorcyclists.

*Statistical Abstract of the United States*, U.S. Printing Office. This collection of facts and figures is published annually by the U.S. Census Bureau. Within it is a mother lode of information relating to types of traffic accidents, numbers of registered vehicles, road miles driven, and so on.

## Chapter 19

*Atlas of Stress Strain Curves*, Howard Boxer, Ed., ASM International, Metals Park, Ohio, 1987. Stress-strain pull test charts for most of the alloys used in automobiles and vehicles are listed, including tungsten.

*The Condensed Chemical Dictionary*, 9th ed., Gessner Hawley, Ed., Van Nostrand Reinhold Co., New York, 1977. The basic chemical properties of tungsten are listed on p. 896. For basic data on most chemicals, materials, and chemical processes, this is a very useful reference text.

*Handbook of Chemistry and Physics*, 51st ed., 1970–1971, CRC Press, Boca Raton, FL. The various chemical and physical properties of tungsten are listed in various sections in this standard library reference.

*Handbook of Physics*, edited by Condon and Odishaw, McGraw-Hill, New York, 1967. This is an excellent reference text and its range of topics is extensive. Beginning on p. 6–37 and continuing to p. 6–39 is a good article about tungsten filament illumination.

*Materials Handbook*, by Brady and Clauser, 11th ed., McGraw-Hill, New York, 1977. A good article about the general properties of tungsten begins on p. 810.

*Response of Brake Light Filaments to Impact*, by Dydo, Bixel, Wiechel, and Stansifer of SEA Inc., and Guenther of Ohio State University, SAE Paper 880234. In this paper, vehicular impact on hot and cold filaments was recreated in the laboratory. There are excellent photographs that accompany the paper.

*The Traffic-Accident Investigation Manual*, by Baker and Fricke, 9th ed., Northwestern University, 1986. Topic 823, pp. 23–3 through 23–45, "Lamp Examination for On or Off in Traffic Accidents," is the methodology employed by most police departments in assessing lamp filament damage after an accident. There are some excellent photographs that accompany the article.

Van Nostrand's Scientific Encyclopedia, 5th ed., Van Nostrand Reinhold Co., New York, 1976. A more extended discussion of the chemical properties of tungsten is provided beginning on p. 2243.

## Chapter 20

*ATF Arson Investigation Guide*, published by the Department of the Treasury. Mostly this is a compendium of investigative check lists and federal resources used in the investigation of arsons by the ATF. The book has a nice glossary of terms in the back, and there are a number of excerpts from the federal laws concerning arson and explosions.

*Fire Investigation Handbook*, U.S. Department of Commerce, National Bureau of Standards, NBS Handbook 134, Francis Brannigan, Ed., August 1980. This is a very readable text on fire analysis. The text is a practical amalgam of theory and experience and contains a number of very instructive photographs. It is often used as a basic text in fire academies and training schools. Persons "in the business" should have a copy on hand since it is often referenced as an authoritative source in court proceedings.

*Investigation of Motor Vehicle Fires*, by Lee Cole, Lee Books, 1992. The book is useful in that it has a lot of qualitative information useful to the novice in this field. The text is a collection of anecdotal experiences of the author in the investigation of vehicle fires. Technical particulars are not fully explained, and there seems to be a lot of information included in the text that appears to have dubious association with car fires (e.g., trends in the use of various metals used in car production). Despite some shortcomings, however, the book is worth a reconnoiter with regards to on-site investigative tips and is similar to the manual published by the National Automobile Theft Bureau.

*Manual for the Investigation of Automobile Fires*, National Automobile Theft Bureau. This booklet is not easy to locate. It was apparently self-published by the NATB and only a handful of copies were made. My copy had no publication or copyright date. I found that the booklet has some useful tips concerning the examination of a fire-damaged vehicle and the collection of evidence, especially with respect to what to look for in an arson case. It was presumably prepared for insurance adjusters and private investigators.

*A Pocket Guide to Arson and Fire Investigation*, Factory Mutual Engineering Corp., 3rd ed., 1992. A small booklet that can provide handy information while in the field, it fits in a person's shirt pocket or briefcase.

## Chapter 21

*Advanced Mechanics of Materials*, by Seely and Smith, 2nd ed., John Wiley & Sons, Inc., New York, 1952, pp. 223–228. These pages show how the circular plate equations relating to hail impacts on sheet metal are derived. The circular plate formulas aside, the book is a classic text on mechanics of materials.

*Atmosphere*, by Schaefer and Day, Peterson Field Guides, Houghton Mifflin Co., New York, 1981. Pages 270–271 contain four black-and-white plates that are excellently prepared showing the onion-skin construction of a hailstone, and its crystalline patterns. Page 204 shows a black-and-white plate of a hail fall and a hail shaft, as seen from cloud height. There are also several other excellent, interesting photographs relating to hail.

"Fronts and Storms," in *Van Nostrand's Scientific Encyclopedia*, 5th ed., Van Nostrand Reinhold Co., pp. 1105–1110, and also "Weather Observation and Forecasting," pp. 2319–2327, and "Winds and Air Movement," pp. 2339–2343. While brief, the explanations and mathematical descriptions of wind and weather phenomenon are excellent.

*Hail Damage to Red Cedar Shingles*, American Insurance Association, 1975. This is often called the "Haig study," after the company that did the research work, Haig Engineering in Dallas, Texas. This report documents a 10-year study on the effects of hail on cedar shingles and shakes and is the industry benchmark document for assessing hail damage to wood shake roofs.

*Hail and Its Effects on Buildings*, CSIR Research Report No. 176, Bulletin 21, 1-12, National Building Research Institute, Pretoria, South Africa, 1960. While this paper can be difficult to obtain, it is considered the first paper that examined scientifically rather than anecdotally the effects of hail on roofing materials.

*Hail Resistance Tests of Aluminum Skin Honeycomb Panels for the Relocatable Lewis Building*, Phase II, by Mathey, U.S. Department of Commerce, National Bureau of Standards, NBS Report 10-193, April 10, 1970. This is an excellent, unbiased report concerning the quantification of hail damage to aluminum metal panels.

*Hail Resistance of Roofing Products — Building Science Series 23*, by Sidney Greenfeld, U.S. Department of Commerce, National Bureau of Standards, August 1969. This is the fundamental paper that established hail damage thresholds for roofing materials in the U.S., and it is also one of the few papers about hail damage to roofs that is considered reliable and unbiased.

*Prospective Weather Hazard Rating in the Midwest with Special Reference to Kansas and Missouri*, by Friedman and Shortell, Research Department, Travelers Insurance Companies, August 28, 1967. This is an excellent paper describing the chances of damaging hail occurring when a hail day occurs.

*Reader's Digest New Complete Do-It-Yourself Manual, Reader's Digest*, 1991. For evaluating contractor construction techniques of wood and asphalt shingle roofs, this is an excellent text. Pages 394 through 395 show the proper way of installing asphalt shingles over a wood deck, especially the proper overlap.

"SPF Roof Systems: Field Survey and Performance Review," by Dr. Dupuis, P.E., *Professional Roofing*, March 1996, pp. 32–36.

*Weather*, by Ralph Hardy, TY Books, 1996. General purpose introductory text to weather phenomenon, including hail and thunderstorms, this book concentrates on nomenclature, definitions, and basic concepts.

"Wood Roofing: Basics for Craftsmanship," by Jim Carlson, *Professional Roofing*, June 1996, pp. R4–R9. This is a good, brief article explaining how wood shakes should be installed. Figure 1 is especially useful in examining wood shake roofs along the edge of a roof, to determine if the roofer "shorted" the owner by using fewer shakes which good practice dictates. When "shorting" occurs, not only is the owner defrauded in terms of paying for shakes not actually installed, but the roof becomes more susceptible to hail damage and ages faster.

## Chapter 22

*ASTM C62, Physical Requirements of Clay Brick*, and *ASTM C73, Physical Requirements for Sand Lime Brick*, American Society of Testing Materials. These are the specifications for brick, and the standardized way to determine their classification.

*Handbook of Design*, "Principles of Brick Engineering," by Plummer and Reardon, Structural Clay Products Institute. This is everything a brick aficionado would ever want to know.

*New Complete Do-It-Yourself Manual*, published by *Reader's Digest*, 1991, pp. 162–179. As its name implies, this is a well illustrated do-it-yourself manual. The pages cited refer to a section on bricks and mortar from the builder's point of view.

## Chapter 23

*Handbook of Human Reliability Analysis with Emphasis on Nuclear Power Plant Applications*, Final Report, NUREG/CR-1278, 1983.

*Human Factors in Engineering and Design*, by Sanders and McCormick, 6th ed., McGraw Hill, New York, 1987. Chapter 21, beginning on p. 606, deals with human error and how to prevent it.

*Military Standard: Sampling Procedures and Tables for Inspection by Variables for Percent Defective*, MIL-STD-414, June 11, 1957. Methodology for determining the number of defectives in a lot. This is a typical quality control document that deals with the occurrence of defects after they have been built into the product.

"New Methods of Consumer Product Accident Investigation," by Drury and Brill, *Proceedings of the Symposium — Human Factors and Industrial Design in Consumer Products*, Tufts University, 1980, pp. 196–211. This paper proposes that accidents occur in industrial settings when the momentary demands of the job exceed the momentary abilities of the individual assigned the job.

*Procedures for Performing a Failure Mode, Effects and Criticality Analysis (FMECA)*, MIL-STD-1629A, November 24, 1980. This is the formal procedure used by all departments and agencies of the U.S. Department of Defense. Like many military publications, this one can be hard to follow at first if the reader has not previously been exposed to military jargon. The document lays out the military method of determining potential failure modes in a system and the resulting consequences. If used in reverse, the method can be useful in diagnosing failures in systems.

*To Engineer is Human*, by Henry Petroski, Vintage Books, 1992. This book is about failure with respect to engineering design and has an excellent chapter about forensic engineering (Chapter 14). I highly recommend this book for its insights into the workings of engineering failure. It is also very readable.

# Index

## A

ABS, see Antilock brake systems
Abnormal psychology, 204
Absorptivity, 118
Academic qualifications, one-upmanship
    concerning, 9
Accelerant(s)
    burn through bedroom floor, 202
    chemical analysis for, 221
    detection of after fire, 218
    hydrocarbon, 219
    liquid, 212
    pour pattern of, 207, 213
Acceleration
    angular, 313
    retardation, by air resistance, 246
    tests, 256
Accelerators, stuck, 254
Accident(s), see also Motorcycle accidents,
    visual perception and
    involving motorcycles and four-wheel
        vehicles, 343
    job-related, 92
    management's role in causation of, 419
    marine, occurring at night, 351
    prone people, 413
    truck, 267
Accidents and catastrophic events,
        management's role in, 413–423
    example to consider, 420–422
    human error vs. working conditions, 417
    job abilities vs. job demands, 417–419
    management's role in causation of
        accidents and catastrophic events,
        419–420
Acetone, explosive limit of, 184
Acoustical energy, 197
Activation energy, 103, 109
ADA, see Americans with Disabilities Act

Aerosol products, flammability of, 114
Air
    concussion, 54, 62
    condition compressor oil, 368
    flow, constriction of over house, 29
    resistance, 244, 246, 248, 304
    shock wave damage, 57
Alligatoring patterns, as fire spread
        indicator, 130
Alternating current, single-phase, 155
Aluminum
    conductors, 171
    electrical cable, 166
    foil, absorptivity of, 119
    lower explosive limits for, 184
    melting point of, 153
    siding, 396
Ambient temperature, 141, 142
Americans with Disabilities Act (ADA), 101
Ammonia, explosive limit of, 184
Ammonium nitrate fuel oil (ANFO), 52, 53,
    179, 421, 422
ANFO, see Ammonium nitrate fuel oil
Angular acceleration, 313
Angular momentum, 287, 288
Angular velocity rotations, 314
Antilock brake systems (ABS), 236, 237
Applied voltage, 142
A prioi biases, 13–14
Aquifer ground, 42
Army Air Corps, 91
Arson(s), 361
    first- and second-degree, 202
    precursors, 209
    primary motive for, 210
    for profit, 204
    reporting immunity laws, 211
    vehicle, 362
Arson and incendiary fires, 201–222
    arsonist profile, 203–204

arson reporting immunity laws, 211–212
basic problems of committing arson for
    profit, 204–206
daisy chains and other arson precursors,
    209–211
detecting accelerants after fire, 218–221
liquid accelerant pour patterns, 212–214
prisoner's dilemma, 206–207
spalling, 214–218
typical characteristics of arson or
    incendiary fire, 207–209
Asphalt shingles, 30
    buckling of, 384
    cracking of, 391
    curled, 390
    stony cover lost due to weathering, 392
ASTM, 111, 112
Atmospheric inversion layer, 57
Attorneys, 17
Audi 5000, 95
Audi Victims Network (AVN), 96
Automatic transmission
    efficiencies, 252
    problems, 259
Automobiles, damage to sheet metal of, 401
Automotive ergonomics, 95
Automotive fires, 361–371
    electrical fires, 368–369
    fuel-related fires, 364–367
    mechanical and other causes, 370–371
    other fire loads under hood, 368
    vehicle arson and incendiary fires,
        362–364
AVN, see Audi Victims Network

**B**

Back of envelope calculation, 5
Banking, negative, 332
Basement cracks, due to drainage water from
    roof gutters, 71
Battery explosions, 361
Bayes' theorem, 38
Beading, 152, 168
Beaufort scale, 33, 34
Before and after method, 48
Bernoulli equation, 26, 29, 192
Biases, a priori, 13–14
Billiard ball model, 316

Bimetallic element, 158
Bituminous roof, gravel cover driven into,
    393
Black bodies, 118
Blast
    energy, 58
    epicenter, 56
    frequency, relationship of displacement of
        structure to, 68
    hole
        area, 51
        P wave front of, 58
    monitoring, with seismographs, 59
    survey companies, 49
Blasting
    criteria, safe, 63
    damages typical of, 66
    discriminate cracks caused by, 49
    formula, safe, 62
    location, 60
    noise, human perception of, 64
    quarry, 61
    safe, 61
    study, by U.S. Bureau of Mines, 60, 61
Blasting damage, evaluating, 47–74
    air concussion damage, 54–57
    air shock wave damage, 57–58
    blasting study by U.S. Bureau of Mines,
        Bulletin 442, 60–61
    blasting study by U.S. Bureau of Mines,
        Bulletin 656, 61–62
    blast monitoring with seismographs,
        59–60
    continuity, 72–73
    damages typical of blasting, 66–69
    effective surveys, 49–50
    flyrock damage, 51–53
    ground vibrations, 58–59
    human perception of blasting noise and
        vibrations, 64–65
    OSM modifications of safe blasting
        formula in Bulletin 656, 63–64
    pre-blast and post-blast surveys, 47–49
    safe blasting formula from Bulletin 656,
        62–63
    surface blast craters, 53–54
    types of damage caused by blasting, 50–51
    types of damage often mistakenly
        attributed to blasting, 69–72
Blizzard, 31

BOCA, see Building Officials and Code
        Administrators
Boltzman's constant, 193
Bootlegger's turn, 266
Box method, 127
Brake(s)
    emergency, 256, 267
    engine vs., 255
    failure, 231, 261
    fluid, 368
    power, 257
Breakers, 46, 156
Breeze box fan, 169
Brick freeze-thaw deterioration, blaming on
        hail, 407–411
    basic problem, 409–410
    brick grades, 408–409
    experiment, 410
    information about bricks, 407–408
Brick grades, 408
Brown lung, 89
Buckling
    elastic, 318
    load, Euler, 319, 326
    point, 318
Bug killers, electric, 182
Buick Century, 285
Building(s)
    code(s), 2
        contractor compliance with, 24
        enforcement, 81
    damage to sheet metal of, 401
    decimation of by gas explosion, 176
    wiring system, 135
Building collapse, due to roof leakage, 75–88
    deferred maintenance business strategy,
        80–82
    lime mortar, 77–80
    restoration efforts, 87
    roof leaks, 80
    structural considerations, 84–87
    structural damage due to roof leaks, 82–84
    typical commercial buildings 1877–1917,
        75–77
Building Officials and Code Administrators
        (BOCA), 23
Built up roofs (BURs), 77, 387
Bumper imprint, in crush damage, 278
Burglaries, 204
Burning

downward, 106
rate, lateral, 106
velocity, 104, 107, 108, 109
Burnish mark
    hailstone, 385
    size of, 397
    on steel A/C cabinet, 374
BURs, see Built up roofs
Business strategy, deferred maintenance, 80,
        81
Butane, minimum ignition energy of, 181

## C

CADD, see Computer-assisted drafting and
        design
Calcium hydroxide, 79, 83
Calcium oxide, reaction of with water, 79
Car(s)
    ashtrays, 363
    crush damage to side of, 316
    driver
        older, 247
        20/50 vision of, 348
    fires in, 370
    hail ding on, 403
    lemon, 362
    postimpact momentum of, 278
    pre-impact momentum of, 278
    rebellion, fear of, 255
    referred to as old tin cans, 320
    sheet metal covering of, 401
    stiffness coefficients
        of full sized, 325
        of medium, 325
        of small, 324
    tires, frictional coefficients for, 2226
    upside down value of, 362
Carjackers, 362
Cartesian coordinate system, 297, 300
Cash flow problems, 362
Catalytic converter, 370
Catastrophic events, see Accidents and
        catastrophic events, management's
        role in
Cause and effect, underlying reasons for
        presuming, 39
CB radio coaxial cables, 172
Cedar wood shingles, 398, 399

Ceilings, cracking pattern in, 70
Cellar, 78
Center of gravity, 281, 282
    during rollover, 307
    elevation of, 312
    position of, 305, 311
Center of mass, rotation of body around,
        286
Centrifugal force, 330
Centroid method, 124, 128
Charge pattern, 60
Chemical attack, 169
Chemical vapors, 170
Chevrolet Citation, 322
Chromium, 351
Cigarette fires, 361, 363
Circuit
    breakers, 157, 158
        high voltage, 160
        underload type, 159
    completion of, 156
    divider, 41
Civil War Draft Riots, 204
Claims adjuster, 16
Closed drum test, 113
Cloud formation, 374
Coal mines, 89
Coefficient of friction, 235, 237, 254, 310
Coefficient of restitution, 275, 277, 288
Coffeemakers, 144, 161
Coincidence argument, converse of, 38
Collaborative International Pesticide
        Analytical Council Handbook, 114
Collateral damage, 394
Collision
    elastic, 273, 274
    fully elastic, 289
    half-elastic, 281
    head-on, 292
    plastic, 276
Color
    recognition deficiencies, red-green, 92
    -related recognition mistakes, 93
Column(s)
    buckling, Euler's formula for, 85
    post-buckling behavior of, 318
Combustible(s)
    fuel, 103
    gases, minimum ignition energies for, 181
    ignition of, 126

Combustion, activation energy needed to
        initiate, 109
Commercial buildings, masonry, 75
Computer(s), 144
    -assisted drafting and design (CADD),
        129
    simulations, 413
Concrete
    absorptivity of, 119
    blocks, green, 72
    floor, flame spread rating of, 111
    oxidation of, 69
    slab, flammable liquid on, 215
    substandard, in foundation walls, 71
Conduction, 114
    law, 194
    melting of, 148
Conductor(s)
    aluminum, 171
    copper, 171
    cross-sectional area of, 166
    current-carrying, 154
    deformation of, 167
    drippings, 135
    loop, resistance of, 154
    Romax type, 165
Conservation of momentum, 290
Conservative system, 297
Constant deceleration, 278
Construction explosives, detonation of, 52
Consulting companies, 20
Consumer Reports, 264
Contingency fee arrangement, 15
Contractors, using wrong types of brick, 409
Convection, 114
    effects, 108
    forced, 117
    free, 117
Convective flow, turbulent, 141
Copper conductors, 171
Copy machines, 144
Cork sheets for walls, flame spread rating of,
        111
Cornering, 336
Corrosion
    galvanic, 170
    of lugs and terminals, 170
Corruption, 15
Court
    cases, 17

examination, 13
Crash data, evaluation of actual, 322
Crime, concealment of, 204
Cruise control, 258, 259
Crush
    constant, 317
    damage, to side of car, 316
    depth measurements, 326
    uneven frontal, 326
Current
    -carrying conductors, 154
    solenoid, 158
    -time plots, 159
Curved skids, 230
Curves and turns, 329–341
    cornering and side slip, 336–337
    load shifting, 334–335
    measuring roadway curvature, 339–340
    motorcycle turns, 340
    side vs. longitudinal friction, 335–336
    transverse sliding on curve, 329–332
    turning radius, 338–339
    turning resistance, 337–338
    turnovers, 333–334

D

Daisy chains, 209
Damage, see also specific types
    air concussion, 54
    air shock wave, 57
    collateral, 394
    flyrock, 51
    freeze-thaw, 217
    lung, 195
    markers, 56, 185
    pattern, outside-to-inside, 137
    rollover, 308
    typical of blasting, 66
Dashboards, plastic materials used in
        modern automobile, 369
Data error, considerations of, 229
Daylight perception, 347
Dead short, 152
Deceleration, constant, 278
Deferred maintenance business strategy, 80,
        81
Deflagration type explosions, 109, 175, 183
Detonating explosion, 175, 178

Devils' Night, 204
Dielectric insulation, 145
Diffusion, 192
Disease, public's perception of, 379
Doctor Admirabilis, 9
Doctor Faustus legend, 10
Doorbells, old-fashioned, 182
Downward burning, 106
Downward displacement, 305
Drive axle, 262
Driver(s)
    licenses, state requirements for, 349
    older car, 347
    professional, 330
    reflexes of, 349
Dry pot situations, 164
Dynamic coefficient, 254
Dynamite, 52, 179, 187

E

Ear drums, ruptured, 195
Earplugs, 94, 98
Eccentric loading, 83
Economics, 14
Efficiency expert, 91
Elastic analysis, 318
Elastic buckling, 318
Elastic collision
    definition of, 274
    properties of, 273
Elastic energy, 321, 327
Electrical cable, aluminum, 166
Electrical codes, 2
Electrical fires, 361, 368
Electrical resistance, of fusible alloy metal
        strip, 157
Electrical shorting, 135–173
    beading, 152–156
    common places where shorting occurs,
        165–173
    example situation involving overcurrent
        protection, 161–162
    fuses, breakers, and overcurrent
        protection, 156–161
    granfathering of GFCIs, 163
    ground fault circuit interrupters, 162–163
    lightning type surges, 165
    other devices, 163–165

parallel short circuits, 146–149
series short circuits, 149–152
thermodynamics of simple resistive
    circuit, 138–146
Electric bug killers, 182
Electric cords, flexible, 168
Electric igniters, furnace, 182
Electricity consumption, 40
Electric motors, 181
Electronic components, computerized
    control, 46
Embezzlement, 203
Emergency brake, 256, 267
Emissivity, 118
Emulsion photographs, 49
Energy
    acoustical, 197
    activation, 109
    dissipated, 230
    elastic, 321, 327
    expansion, 197
    kinetic, 26, 197, 234, 239, 297, 302, 381
    losses, 119, 279, 280
    potential, 26, 233, 234, 299, 302
    radiant, absorbers of, 121
    sloshing of, 26
    source, 103
    transfer, 118
Energy methods, 295–328
    evaluation of actual crash data, 322–323
    flips, 310–316
    going from soda cans to old "can you
        drive?", 320–322
    low velocity impacts, 323–324
    modeling vehicular crush, 316–318
    post-buckling behavior of columns,
        318–319
    representative stiffness coefficient,
        324–326
    rollovers, 304–310
    theoretical underpinnings, 297–303
    types of irreversible work, 303–304
Engine
    block, 367
    compartment
        fire spread in, 367
        loose wiring in, 369
    coolants, 368
    drag, 304
    horsepower rating, 249, 250

limitations, 247
    oil, 368
    thermal efficiency, 265
Engineer, as expert witness, 14–16
Engineering
    handbook equations, 66
    investigation, 20
    mechanics, basic equation of, 223
English units system, 27, 247
Equivalence calculation, 187
Equivalent resistive load, 150
Ethane, explosive limit of, 184
Euler
    buckling load, 319, 326
    equations, 313, 314, 315
    formula, for column buckling, 85
Expansion energy, 197
Expert witness, 8, 14–16
Explosion(s), 175–190
    basic parameters, 182–184
    battery, 361
    deflagrations and detonations, 178–182
    detonating, 175
    fire-affected area of, 195
    gas, 176
    high pressure gas expansion, 177–178
    overpressure front, 185–188
    site, examination of, 422
Explosion determining point of ignition of,
        191–199
    diffusion and Fick's law, 192–194
    energy considerations, 197–198
    epicenter, 196–197
    flame fronts and fire vectors, 194–195
    pressure vectors, 195–196
Explosive(s)
    detonation of construction, 52
    first cut estimate of amount of, 186
    pressure, point of greatest, 3
    yield, 186
Eyewitness
    information, 6–8
    reports, impartial, 7


F

Face bricks, 407
Face shields, 98
Failure analysis, 1, 2

Fall down debris, 123
False low points, 123
Fatigue fracture, 12
Fault list, human error vs., management
    error, 417
Fiberboard, flame spread rating of, 111
Fiber insulation wrap, 161
Fick's law, 192, 193, 194
Filament
    discoloration vs. temperature, 354
    stretching effect, 357
Fire(s)
    brick, absorptivity of, 119
    chemistry, 132
    cigarette, 361, 363
    code flammability spread rating, 111
    conviction of people setting, 422
    -damaged areas, division of into severity
        zones, 128
    detection of accelerants after, 218
    determining position of switch during,
        110
    electrical, 361, 368
    endurance standard time-temperature
        curve, 112
    escape plan, 420
    -fighting activities, 133
    fuel, 361, 364, 365
    garage, 130, 361
    incendiary, see Arson and incendiary fires
    involving flammable liquids, 153
    load, 214
    method of detecting shorting after, 152
    pattern, unnatural, 208
    pickup truck front seat, 364
    -retardant chemicals, 113
    safety codes, 2
    scene, initial reconnoiter of, 122
    severity, weighting factors for, 129
    spread
        in engine compartment, 367
        indicators, 130
    suspicious, 203
    vectors, 194
    velocity numbers, 106
    warehouse, 127
Fire, determining point of origin of, 103–134
    burning velocities and flame velocities,
        107–110

    burning velocities and V patterns,
        104–107
    centroid method, 124–125
    combination of methods, 133
    fire spread indicators, 130–133
    flame spread ratings of materials, 110–114
    ignition sources, 125–127
    initial reconnoiter of fire scene, 122–123
    little heat transfer theory, 114–118
    radiation, 118–122
    warehouse or box method, 127–128
    weighted centroid method, 128–130
Firebricks, 407
Fireproof hotels, 104
First cut verified information, 11
First-degree arson, 202
Fixed barrier, vehicle impacting of, 291
Flame
    fronts, 194
    speed, 109
    spread rating, 110, 111
    velocities, 107
Flammable liquid(s)
    on concrete slab, 215
    fires involving, 153
    ignition of atomized, 180
Flexible elecrtic cords, 168
Flips, 310, 315
Floor
    decking, 78
    usage, 76
Flue gases
    carbonic acid vapor in, 78
    water vapor and unburned particles
        contained in, 121
Fluid mechanics, 116
Flyrock
    damage, 51
    debris, trajectory of, 52
Foam roofing systems, 404
Forced convection, 117
Forcing function, 67
Ford Mustang, 345
Forensic engineering
    applying scientific method to, 10–12
    definition of, 1–3
    investigation, reporting results of, 16–21
Forward flip, 315
Foundation walls, 71, 78

Four-wheel vehicles, accidents involving
    motorcycles and, 343
Free convection, 117
Freeze-thaw damages, 217
Freezing rain, 81
French Revolution, 204
Friction
  circle, 335
  coefficient of, 235, 237, 254, 310
  forces, in free-rolling vehicle, 245
  resistance, 248, 262
  side vs. longitudinal, 335
  static, 227
  tire, 226
Frontal crush, uneven, 326
Front-wheel
  drive, 262
  slip angle, 337
Fuel
  -to-air ratio, 108
  fires, 361, 364
  leaks, 365
  lines, leak in main, 367
Fujita ranking, for wind storms, 377
Full sized cars stiffness coefficients, 325
Furnace(s)
  electric igniters, 182
  paint finish on, 131
  pilot lights in, 182
  testing of, 5
Fuses, 46, 156
Fusible alloy, 157
Fuzzy central limit theorem, 29


# G

Gables roof, hail impact on, 383
Galvanic corrosion, 170
Galvanization, 131
Garage fires, 130, 361
Gas
  accumulation of, 6
  dryers, pilot lights in, 182
  explosion, 6, 7
    building decimated by, 176
    high pressure, 177
  Kinetic theory of, 192
  leak, 196
  range company, 12

Gaseous fuel, explosive limits of, 191
Gasoline
  explosive limit of, 184
  fumes, ignition of, 361
  stations, self-service, 371
  use of as accelerant, 207
GAWR, see Gross axle weight ratings
GFCI, see Ground fault circuit interrupters
Glass, melted, 357, 358
Goggles, 98
Grashof number, 117
Graupel, 375
Great Plains, exaggerations popular in,
    376
Greenfield study, 403
Gross axle weight ratings (GAWR), 232
Ground fault circuit interrupters (GFCI),
    162, 163, 420
Ground vibrations, 58
  damage to residential structures from,
    61
  human response to, 62, 65
  intervening deep pond shielding house
    from, 73
Gunpowder, 179
Gypsum plaster in ceiling, flame spread
    rating of, 111


# H

Haig report, 398
Hail
  dents, in copper roof decoration, 385
  diameter, impact energy versus, 386
  ding, on car, 403
  formation, 375
  frequency, 378
  size, probability of, 380
  soft, 375
  storm activity, in Kansas City area, 379
Hail damage, 373–406
  assessing hail damage, 387–395
  cosmetic hail damage, 395–398
  damage to sheet metal of automobiles and
    buildings, 401–404
  foam roofing systems, 404–405
  fundamentals, 380–384
  Haig report, 398–400
  hail frequency, 378–380

hail size, 375–378
    size threshold for hail damage to roofs,
        384–387
Hailstone
    stories of, 376
    velocity, 383
Half-elastic collision, 281
Halogen headlight bulb, 352
Hard rubber, absorptivity of, 119
Hard science, 14
Headlight(s)
    brands of, 346
    bulb, halogen, 352
    perception, 345
Head-on collision, 292
Head space analysis, 220
Heat
    choke, 143
    transfer coefficient, 116, 142
    transfer theory, 114
Heater leak, 198
Heavy rains, 50
Helmets, 98
Hertzian-type compressions, 217
High pressure gas expansion explosions,
    177
Highways, two-lane, 347
Home(s)
    foundations of modern, 68
    owners' association, 49
Honda Civic, 322, 323
Hooke's law, 273, 276, 317
Horse sense arguments, 48
Hotel, fire proof, 104
Hot shock, 356, 358, 359
Hot water tank
    pilot lights in, 182
    V pattern on, 105
House
    construction of air flow over, 29
    geometry, speed-up–slow-down–speed-
        up effect due to, 31
    shield of from ground vibrations, 73
    side view of wind going over, 27
Human
    error, working conditions vs., 417
    factors engineering, 91
    performance, 418
Hurricane, 34
    Andrew, 24

brackets, 30
Hydrocarbon
    accelerant, 219
    outgassed, 220
Hydrogen gas, 193
Hydrometer, 373

## I

Ice pellets, rebound tests of spherical,
    402
IEEE, see Institute of Electrical and
        Electronic Engineers
Ignition
    of combustibles, 126
    sources, 125
Incendiary devices, finding of, 208
Incendiary fires, see Arson and incendiary
        fires
Industrial management, 89
Inertia, moment of, 283, 284
Initial resistance, 142
Inquiry, a priori method of, 10
Installation methodologies, 2
Institute of Electrical and Electronic
        Engineers (IEEE), 45
Insulation
    dielectric, 145
    fiber, 161
    weak spot, 167
Insurance money, collection of, 201
Investigation pyramid, 3–6
Iron, lower explosive limits for, 184
Irreversible work
    total energy dissipated as, 296
    types of, 303

## J

Jackknifing, truck, 267
Job
    abilities, job demands vs., 417
    -related accidents, 92
    -specific knowledge, 418
Joist
    splices, 78
    tie-ins, effect of, 86

# K

Kansas, windy reputation of, 26
Kansas City area, hail storm activity in, 379
Kerosene, use of as accelerant, 207
Kinetic energy, 26, 197, 234, 239, 297, 302
　　hailstone, 381
　　vehicle, 223
Kinetic theory of gases, 192

# L

Labor organizations, 90
Lagrangian term, 299
Lampblack, absorptivity of, 119
Lamp filament damages, interpreting,
　　351–359
　　brittleness in tungsten, 355
　　ductility in tungsten, 355–357
　　filaments, 351–353
　　melted glass, 357–358
　　other applications, 357
　　oxidation of tungsten, 353–355
　　sources of error, 358–359
　　turn signals, 357
Lateral burn rate, 106
Lateral tire friction, 266
Lawnmower, storage of next to furnace, 132
Legal system, 8–9
　　adversarial, 12
　　scientific method and, 12–13
Light filaments, commercial incandescent,
　　351
Lightning, 182
　　access, to well pump, 40
　　strike
　　　damage to well pump by, 42
　　　essential circuit arrangement of, 41
　　surge, bone fide, 45
　　type surges, 165
　　well pump failure due to, 44
Lightning damage, to well pumps, 37–46
　　converse of coincidence argument, 38–39
　　correlation not causation, 37–38
　　failure due to lightning, 44–46
　　lightning access to well pump, 40–43
　　underlying reasons for presuming cause
　　　and effect, 39–40
　　well pump failures, 43–44

well pumps, 40
Lime mortar, 77
Linoleum, flame spread rating of, 111
Liquid(s)
　　accelerant pour patterns, 212
　　flammable, 153, 180, 215
Load
　　-bearing wall, 80
　　shift, 235, 334
Longitudinal friction, side vs., 335
Lorentz transformations, 271
Love waves, 59
Low velocity impacts, 323
Lug(s), 169
　　corrosion of, 170
　　loose, 171
Lumber, shrinkage and expansion of, 70
Lung damage, 195

# M

Machines and people, putting together,
　　89–101
　　Audi 5000 example, 95–96
　　background, 89–92
　　employer's responsibilities, 99–100
　　guarding, 97–99
　　manufacturer's responsibilities, 100–101
　　new ergonomic challenges, 101
　　sequencing, 95
　　sound, 93–95
　　vision, 92–93
Magnesium, lower explosive limits for, 184
Malfeasance, 15
Manganese, lower explosive limits for, 184
Manifold temperatures, 367
Marine accidents, occurring at night, 351
Masonry side walls, 76
Material storage specifications, 2
Maximum climb, 261
Maxwell's equations, 154, 167
Mechanical efficiency, 252
Mechanical equipment codes, 2
Mechanics of materials, 318
Medium cars stiffness coefficients, 325
Melt curve, 158
Melted glass, 357, 358
Mercury fulminate, 179
Metal roof, hail dents in, 393

Metal ventilation ductwork, 131
Methane, 184, 196
Methanol
    explosive limit of, 184
    minimum ignition energy of, 181
Mickey Mouse ears, 94
Midvale Steel Company, 90
Mine shafts, horizontal, 89
Mirandizing, of suspect, 203
Mirror symmetry, 125
Model Penal Code, 201
Modern homes, foundations of, 68
Molly Maguires, 90
Molten steel, absorptivity of, 119
Molybdenum, 351
Moment of inertia, 283, 284, 313
Momentum methods, 271–294
    analysis of forces during fixed barrier
        impact, 278–279
    angular momentum equations, 287–288
    basic equations, 272–273
    center of gravity, 281–283
    coefficient of restitution, 275–276
    discussion of coefficient of restitution
        methods, 293
    energy losses, 279–281
    estimation of collision coefficient of
        restitution from fixed barrier data,
        291–292
    moment of inertia, 283–285
    properties of elastic collision, 273–275
    properties of plastic collision, 276–277
    solution of velocities using coefficient of
        restitution, 288–291
    torque, 285–287
Motor
    burn out, 172
    -generator sets, 164
    grounded out, 43
    windings, 165
Motorcycle
    accidents involving four-wheel vehicles
        and, 343
    number of fatalities per road mile of, 344
    turns, 340
Motorcycle accidents, visual perception and,
    343–350
    background information, 344–345
    daylight perception, 347–348
    difficulty finding solution, 349

    headlight perception, 345–347
    review of factors in common, 348–349
Movie directors, 19

N

Narrative report, 16
National Building Research Institute, 403
National Highway Traffic Safety
    Administration (NHTSA), 260, 261
National Oceanic and Atmospheric
    Administration (NOAA), 23
Natural gas
    leaks, 180
    minimum ignition energy of, 181
    space heater, 191
Negative banking, 332
Neutral steering, 339
Newton's laws, 223, 272
NHTSA, see National Highway Traffic Safety
    Administration
Nitroglycerine, 179
NOAA, see National Oceanic and
    Atmospheric Administration
Nusselt number, 117

O

Occupational Safety and Health
    Administration (OSHA), 91, 97
    requirements, 419
    safety standards, 99, 100
Office space, internal heat load of, 144
Office of Surface Mining (OSM), 63
Open cup flash point, 113
Operator skill, 91
OSHA, see Occupational Safety and Health
    Administration
OSM, see Office of Surface Mining
Outboard motor shaft, 40
Out-of-tolerance behavior, 415
Ovens, pilot lights in, 182
Overcurrent protection, 156
    equipment, 160
    example situation involving, 161
Overpressure
    front, 55, 185
    negative, 55

Oversteering, 339
Oxidation reaction, self-sustaining, 103
Oxidizing agent, 103

## P

Parallel resistance equations, 149
Parallel short circuits, 146, 147
Paving bricks, 407
Peeling, 251
Peel-out, 257, 265
Peer discouragement, 416
Pensky-Martens closed test, 114
People, see Machines and people, putting
    together
Personal attacks, one-upmanship
    concerning, 9
Personal gain, 204
Photographs, emulsion, 49
Picket Night, 204
Pickup truck(s), 283, 309
    front seat, fire in, 364
    vandalism-type fire set in, 421
Pill test, 111
Pilot lights, in hot water tanks, 182
Pilot plants, 413
Planimeter, 124
Plastic
    collision, properties of, 276
    deformation, of hailstone, 383
    oxidation of, 69
Plywood
    deck, cracking of, 403
    paneling, flame spread rating of, 111
Poisson's ratio, 84, 401
Police reports, 234, 293
Polytropic gas constant, 177
Post-blast surveys, 47, 49
Post-Civil War Reconstruction, 75
Postimpact momentum, of car, 278
Potential energy, 26, 233, 234, 299, 302
Potholes, repeated hard road shocks from,
    358
Pour pattern, of accelerant, 207, 212, 213
Powder, flammable, 214
Power brakes, 257
Power transmission equipment, 97
Prandtl number, 117
Pre-blast surveys, 47, 49

Pre-impact
    momentum, of car, 278
    velocities, 296
Pressure
    vectors, 195
    versus time plot, 183
    wave, 188, 198
Primary shorting, 136
Prisoner's dilemma, 206
Product codes and specifications, 2
Propane, 196
    explosive limit of, 184
    minimum ignition energy of, 181
    piping, 5
    systems, as cause of explosion, 4
Property owners' complaints, 48
Protective clothing, 98
Psychology, 14, 89, 204
Public perception, of diseases, 379
Puffers, 108
Pullback devices, 98
Punch presses, 101
P-V work, 186
Pyramid method of investigation, 6

## Q

Quality control concepts, in manufacturing
    management, 414
Quarry blasting, 61

## R

Radiant energy intensity, 121
Radiation, 118
Radiator
    fan, 367
    steam, 115
Rain(s)
    freezing, 81
    heavy, 50
Ramp effects, 241
Rayleigh
    energy method, 67
    waves, 59
Real estate transaction activity, tracking of,
    210
Rear-wheel drive, 262

Reconstruction
  experts, 1
  hypothesis, 12
  post-Civil War, 75
Red-green color recognition deficiencies,
    92
Report identifiers, 18
Residential structures, see Wind damage, to
    residential structures
Resistive circuit, thermodynamics of simple,
    138
Resistive load, 139, 140, 147
Respirators, 98
Restitution, coefficient of, 275, 277, 288
Restoration efforts, 87
Revenge, 204
Reverse engineer, 1
Reynolds number, 117, 178
Rioting, 204
Road shocks, repeated hard, 358
Roadway
  bank of, 331
  curvature, measuring, 339
  topographical drawings, 234
Rolling resistance, 248, 304
Rollover
  center of gravity during, 307
  damage, 308
  mechanics in, 303
  side, 304
Roman Senate, 18
Romax type conductors, 165
Roof(s)
  asphalt-shingled, 391
  bituminous, 393
  built up, 77
  cedar wood, 399
  hail damage to, 384, 386
  hail impact on gabled, 383
  joist, 78
  leakage, see Building collapse, due to roof
      leakage
  metal, 393
  simulated hail damage on, 400
  slope, 76
  water leakage damage to, 83
Roofing
  paper, absorptivity of, 119
  systems, foam, 404
Root cause analysis, 1

**S**

Sabotage
  passive-aggressive, 414
  willful, 414
Safe blasting, 61, 63
Safety rules, 2
SAIs, see Sudden Acceleration Incidents
Sandia Laboratories, 25
Scaling factor, 53
Scapegoat hunting, 414
Scientific management, father of, 90
Scientific method, 9–10
  applying of to forensic engineering, 10–12
  legal system and, 12–13
Secondary shorting, 136, 137
Second-degree arson, 202
Seismographs, blast monitoring with, 59
Self-correcting techniques, of modern
      management concepts, 415, 416
Self-service gas stations, 371
Septic tank, deficient soil percolation in, 69
Sequential analysis, of fire spread indicators,
    130
Series short circuits, 149, 150
Setaflash closed test, 113
Severity zones, fire-damaged areas divided
      into, 128
Sewer lines, problems in, 69
Shear waves, 72
Sheet metal shearers, 101
Shellac finish on paneling, flame spread
      rating of, 111
Shingles
  buckling of asphalt, 384
  cedar wood, 398, 399
  cheap, 390
  curled and cupped, 389, 390
Shock waves, 197
Short, dead, 152
Short circuits
  parallel, 146
  series, 149, 150
Shortcut electrical pathway, 146
Shorting
  locations of occurrence, 165
  primary, 136
  secondary, 136, 137
Side impact, 278
Side rollover, 304

Side slip, 336
Sideways frictional force, 330
Skid(s), see also Skids, simple
    curved, 230
    deceleration, calculation of, 229
    equation, construction of, 230
    marks, 332
    soil, 228
    trajectory, 228
Skidding
    over multiple surfaces, 227
    speed reduction by, 229
    work done by, 225
Skids, simple, 223–238
    antilock brake systems, 236–237
    basic equations, 223–224
    brake failures, 231–233
    calculation of skid deceleration, 229
    changes in elevation, 233–234
    considerations of data error, 229–230
    curved skids, 230–231
    load shift, 235–236
    multiple surfaces, 227–229
    simple skids, 224–225
    speed reduction by skidding, 229
    tire friction, 226–227
Slaked lime, 79
Sliding
    frictional forces, 225
    transverse, on curve, 329
Slip
    definition of, 237
    factor, 252
Slippage, general expression for, 331
Small cars stiffness coefficients, 324
Smoke inhalation, 202
Smolders, oxygen-starved, 126
Sniffers, 218
Snowfall, 81
Snow pellets, 375
Sociology, 14
Soda cans
    compression of, 319
    crush load threshold of, 320
Soft decking, 384
Soft hail, 375
Soft science, 14
Soil skid, 228
Solenoid current, 158
Space heater, natural gas, 191

Spalling, 214, 217
Speed reduction, by skidding, 229
SPF, see Sprayed polyethylene foam
Sprayed polyethylene foam (SPF), 404
Spring models, 317
Statementizing, 7
Static coefficient, 254
Static friction, 227
Steam
    absorptivity of, 119
    radiator, 115
Steel-toed shoes, 98
Steering
    control, loss of, 267
    neutral, 339
    wheel angle, 338
Stefan-Boltzmann
    constant, 118
    equation, 119, 120
Stiffness coefficient, 319, 324
Storm gutters, 390
Stove(s)
    electric igniters, 182
    pilot lights in, 182
Structural damage, due to roof leaks,
    82
Structural hail damage, 387
Structural support system, 78
Stuck accelerators, 254
Sudden Acceleration Incidents (SAIs), 255,
    257
Sulfur, lower explosive limits for, 184
Surface blast craters, 53
Surges, lightning type, 165
Suspicious fire, 203
Switch, determining position of during fire,
    110

**T**

Tag closed test, 113
Tailer bulb, with filament distention, 356
Taillight bulb, 352
Tangential velocity, 286
Tank, collapse of caused by wind, 25
Taurus, maximum speed for, 253
Technical experts, 17
Telephones, 144
Temperature difference, 115

Terminal(s), 169
  corrosion of, 170
  loose, 171
Test pavement, 254
Thermal conductors, 115
Thermal expansion, 216
Thermal protection switches, 145
Thermostats, 182
Thickness of material, 115
Thunderstorm(s), 37
  classic towering, 373
  hail requiring of, 375
  months with greatest number of, 378
Timbers, deformed, 83
Time delay devices, 159
Tin work, dented, 392
Tire
  blow out, 37
  friction, 226, 266
TNT, see Trinitrotoluene
Toaster
  electrical cord, 147
  heating element of, 150
  resistance, 148
  total resistance of, 151
Torque, 285
Trailers, 207
Transmission efficiency
  curve, 264
  estimating, 263
Transverse sliding, on curve, 329
Trees, uprooted, 34
Trichloroethylene, minimum ignition
    energy of, 181
Trinitrotoluene (TNT), 179
Trip wires, 99
Tropopause, 374
Truck
  accidents, 267, 344
  tire, coefficient of friction for soft, 334
Tucker, Albert, 206
Tungsten, 351
  brittleness in, 355
  ductility in, 355
  oxidation of, 353
Tungsten trioxide
  coloring pigment, 353
  common uses for, 354
  smoke residue, 353
Turbulent convective flow, 141

Turn
  forces during, 33
  signals, 357
Turning
  radius, 338
  resistance, 337
Turnovers, conditions for, 333
Turpentine, use of as accelerant, 207
Two-lane highways, 347


U

UBC, see Unified Building Code
Understeering, 339
Uneven frontal crush, 326
Unified Building Code (UBC), 23
Uninterruptible power service (UPS), 164
UPS, see Uninterruptible power service
Uranium, lower explosive limits for, 184
U.S. Bureau of Mines, 56, 60, 61
U.S. Department of Defense, 414
U.S. Department of Transportation
    (USDOT), 334, 335
USDOT, see U.S. Department of
    Transportation
User-friendly software, 95
Utility systems, tampering with, 209


V

Vacuum chambers, 104
Vandalism, 204, 421
Vapors
  ignition of, 180
  rapid expansion of, 5
Vehicle
  accidents
    involving motorcycles and four-wheel,
      343
    occurring at night, 351
  analysis, 252
  arson, 362
  crush, modeling of, 316
  driven off dock, 304
  with electric fuel pumps, 366
  falls, simple, 239–246
    air resistance, 244–246
    basic equations, 239–241

ramp effects, 241–244
fire, common element in incendiary, 363
forward speed of, 329
free fall, 239
frictional forces in free-rolling, 245
horizontal speed of, 242
impact
    with embankment, 311
    with fixed barrier, 291
kinetic energy, 223
mass of, 249
maximum velocity, 250
slickness of, 244
work done in crushing of, 295
Vehicle performance, 247–269
bootlegger's turn, 266–268
brakes vs. engine, 255–257
braking, 253–254
cruise control, 258–259
deviations from theoretical model,
    251–252
engine limitations, 247–251
estimating engine thermal efficiency,
    265
estimating transmission efficiency,
    263–265
example vehicle analysis, 252–253
lateral tire friction, 266
linkage problems, 258
maximum climb, 261–263
miscellaneous problems, 260
NHTSA study, 260–261
peel-out, 265–266
power brakes, 257
stuck accelerators, 254–255
transmission problems, 259–260
Vertical velocity, 240
Visual perception, see Motorcycle accidents,
    visual perception and
Volkswagen bug, 232
Voltage source, 149
V pattern, 104, 106, 122


W

Wall(s)
cork sheet for, 111
cracking pattern in, 70
foundation, 78
load-bearing, 80
masonry side, 76
outlet, disengaging plug from, 168
pressure calculations, multiple factor in,
    28
salvage of, 87
sockets, loose electrical plugs in, 181
V patterns on, 122
Warehouse
fire, 127
fire proof, 104
method, 127
Warning buzzers, 94
Water leakage, 82
due to drainage water from roof gutters,
    71
from roof, 86
Water table changes, 69
Watts Riots, 204
Weather statistics, 38
Weighted centroid method, 128
Well pump(s), see also Lightning damage, to
    well pumps
failure, 39
    due to age and wear, 43
    due to lightning, 44
lightning access to, 40
motors, temperature sensor within, 43
replacement, 39
schematic of metal water pipe to, 41
Wheel slip, 251, 337
Wind
blowing, 32
-obstructing barriers, 33
side view of going over house, 27
speed
    estimating of from localized damages,
        33
    influence of local geography on, 33
    measurement, 32
    perpendicular, 28
    variation of with height, 32
storms, Fujita ranking for, 377
tank collapse caused by, 25
velocity, 27
Wind damage, to residential structures,
    23–35
basics about wind, 26–32
code requirements for wind resistance,
    23–26

estimating wind speed from localized
  damages, 33–34
variation of wind speed with height,
  32–33
Windows, good quality, 30
Windwagon Smith, 25
Witching Night, 204
Witness, engineer as expert, 14–16
Wood
  char, as fire spread indicator, 130
  hygroscopic, 82
  rot, 83
  shakes, curled, 388, 398
Work
  definition of, 230

rules, 100
Working conditions, human error vs.,
  417

**Y**

Young's modulus, 66, 85, 86, 215, 216

**Z**

Zinc
  lower explosive limits for, 184
  melting temperature of, 213